Birkhäuser

Trends in Mathematics

Research Perspectives Ghent Analysis and PDE Center

Volume 3

Research Perspectives Ghent Analysis and PDE Center is a book series devoted to the publication of extended abstracts of seminars, conferences, workshops, and other scientific events related to the Ghent Analysis and PDE Center. The extended abstracts are published in the subseries Research Perspectives Ghent Analysis and PDE Center within the book series Trends in Mathematics. All contributions undergo a peer-review process to meet the highest standard of scientific literature.

Volumes in the subseries will include a collection of revised written versions of the communications or short research announcements or summaries, grouped by events or by topics. Contributing authors to the extended abstracts volumes remain free to use their own material as in these publications for other purposes (for example a revised and enlarged paper) without prior consent from the publisher, provided it is not identical in form and content with the original publication and provided the original source is appropriately credited.

Duván Cardona • Joel Restrepo •
Michael Ruzhansky
Editors

Extended Abstracts 2021/2022

Methusalem Lectures

Birkhäuser

Editors
Duván Cardona
Department of Mathematics: Analysis,
Logic and Discrete Mathematics
Ghent University
Ghent, Belgium

Joel Restrepo
Department of Mathematics: Analysis,
Logic and Discrete Mathematics
Ghent University
Ghent, Belgium

Michael Ruzhansky
Department of Mathematics: Analysis,
Logic and Discrete Mathematics
Ghent University
Ghent, Belgium

ISSN 2297-0215 ISSN 2297-024X (electronic)
Trends in Mathematics
ISSN 2948-1724 ISSN 2948-1732 (electronic)
Research Perspectives Ghent Analysis and PDE Center
ISBN 978-3-031-48578-7 ISBN 978-3-031-48579-4 (eBook)
https://doi.org/10.1007/978-3-031-48579-4

This book is published under the imprint Birkhäuser, www.birkhauser-science.com by the registered company Springer Nature Switzerland AG
The registered company address is: Gewerbestrasse 11, 6330 Cham, Switzerland

Paper in this product is recyclable.

Preface

This volume belongs to the new Birkhäuser series Research Perspectives Ghent Analysis and PDE Center, which is devoted to the publication of extended abstracts of seminars, conferences, workshops, and other scientific events related to the Ghent Analysis and PDE Center in Belgium. Volumes in this subseries include a collection of revised written versions of the communications or short research announcements or summaries, grouped by events or by topics.

The Ghent Analysis and PDE Center provides the atmosphere to do research in mathematics that focuses on different areas of analysis as well as the study of partial differential equations (PDEs) and their applications. The details and activities of the center can be found in the following link:

https://analysis-pde.org

The activities of the center have been supported by grants from the Research Foundation Flanders (FWO) as well as by the Methusalem programme of Ghent University.

This book provides the most recent results in analysis, PDEs, and geometric analysis by some of the leading worldwide experts, prominent junior and senior researchers in the latter topics which were invited to be part of the Ghent Analysis and PDE Center' Seminars in the last two years 2021–2022. The contributions are from the speakers of the Methusalem Colloquium, the Methusalem Junior seminar, and the Geometric Analysis seminar. All of them took place online or in person at the Department of Mathematics: Analysis, Logic and Discrete Mathematics at Ghent University. Also, we include colloquia and mini-courses from visiting and invited guests. In general, those seminars are held in a hybrid form and the venue for the in-person form is at Campus Sterre S8 (Ghent University, Belgium). On the one hand, we gather the modern developments in these areas by some already established and well-known global experts in the fields. On the other hand, we offer the opportunity for outstanding young researchers in various areas of analysis and PDEs to share their ideas as well as broader mathematical subjects.

The volume has two main directions, complemented by a few related applied aspects:

1. Geometric analysis. The book includes studies and investigations of modern techniques for elliptic and subelliptic PDEs that have been used to establish new results in differential geometry and differential topology. These topics involve the research in microlocal analysis, geometric analysis, harmonic analysis, and related topics. We consider different problems which have relevant geometric information for different applications in mathematical physics and other problems of classification. We also include contributions of several junior and senior authors specialised in geometric and topological properties of spaces, such as submanifolds of the Euclidean space, Riemannian manifolds, symplectic manifolds, and vector bundles. The aforementioned works as well as the fundamental works on the index theory, K-theory, and their applications to non-commutative geometry, and K-theory, in view of the Atiyah and Singer solution of the Gelfand conjecture (their celebrated Atiyah-Singer index theorem), have shown to be relevant in different problems of the geometric analysis and for problems of classification of manifolds in differential geometry as well as in many other contexts of the mathematical physics.
2. Analysis and PDEs. This part presents recent results in fundamental problems for solving partial integro-differential equations in different settings, e.g. Euclidean spaces, manifolds, Banach spaces, and many other settings, discussions about the global and local solvability by using micro-local and harmonic analysis methods. The study of new techniques and approaches, which can either come from the physical perspective or the mathematical point of view, is also included. Several connected branches arise in this regard. In particular, we focus on approaches of spectral theory combined with elliptic operators, the study of higher-order PDEs of different types and natures involving classical and well-known operators such as Schrödinger-type, Dirac type, Laplace-type, etc., as well as inverse scattering problems, nonlinear equations, hyperbolic PDEs, and stability results. Functional estimates (global and local), which provide information about the existence, uniqueness, large-time behaviour, blow-up, etc., of a solution to a particular equation are also of importance here. We also cover some topics in boundary problems, dispersive equations, and the differential geometry.

3. Additionally to the above categories, there will be a section dedicated to applied mathematics, where we have some contributions related to other modern aspects of applied mathematics with respect to Parkinson's disease diagnostics as well as models of viral infection spreads by using partial differential equations, and a statistical problem.

We are grateful to all the authors who have contributed to this volume.

Ghent, Belgium

Duván Cardona
Joel Restrepo
Michael Ruzhansky

Contents

Part I
Geometric Analysis

Analysis on Noncompact Manifolds and Index Theory: Fredholm Conditions and Pseudodifferential Operators

Ivan Beschastnyi, Catarina Carvalho, Victor Nistor, and Yu Qiao

Abstract We provide Fredholm conditions for compatible differential operators on certain Lie manifolds (that is, on certain possibly non-compact manifolds with nice ends). We discuss in more detail the case of manifolds with cylindrical, hyperbolic, and Euclidean ends, which are all covered by particular instances of our results. We also discuss applications to Schrödinger operators with singularities of the form $r^{-2\gamma}$, $\gamma \in \mathbb{R}_+$.

Keywords Differential operator · Pseudodifferential calculus · Fredholm operator · Riemannian manifold · Lie manifold · Manifolds with cylindrical ends · Conformally compact manifold · Schrödinger operator

2020 Mathematics Subject Classification 35J57, 45E10, 47B35

1 Introduction

We review in this note an approach to *Analysis on non-compact manifolds* that is based on *Lie algebras of vector fields* and *compactifications of manifolds.* This note is a revised and enhanced version of the talk delivered by V.N. at the "Methusalem

I. Beschastnyi
CIDMA, University of Aveiro, Aveiro, Portugal

C. Carvalho
CEAFEL, Department of Matemática, Instituto Superior Técnico, University of Lisbon, Lisbon, Portugal
e-mail: catarinaccarvalho@tecnico.ulisboa.pt

V. Nistor (✉)
Université de Lorraine, CNRS, IECL, Metz, France
e-mail: victor.nistor@univ-lorraine.fr

Y. Qiao
School of Mathematics and Statistics, Shaanxi Normal University, Xi'an, China
e-mail: yqiao@snnu.edu.cn

D. Cardona et al. (eds.), *Extended Abstracts 2021/2022*, Research Perspectives Ghent Analysis and PDE Center 3, https://doi.org/10.1007/978-3-031-48579-4_1

Colloquium and Geometric Analysis Seminar," at Ghent University, in June 2022. The mathematical results included here are based, to a large extent, on the paper [4], but they are also enhanced by recent results from [3]. Among the questions that we will touch upon are:

- the invertibility of differential operators (Hadamard well-posedness),
- the regularity of solutions of elliptic PDEs,
- the construction of the algebras of pseudodifferential operators containing the inverses of the elliptic operators in question, and, most notably,
- the *Fredholm alternative,* the topic around which we organize our paper.

Informally stated, our main result on the Fredholm alternative is:

Theorem 1 *On a "nice" manifold, an adapted (pseudo)differential operator is* Fredholm *if, and only if,*

(i) *it is* elliptic *and*
(ii) all its "limit operators" are invertible.

A more formal statement will be included below (Theorem 4). In the above result, by "elliptic" we mean a pseudodifferential operator whose principal symbol is invertible outside of the zero section. On a compact manifold, there are no "limit operators," and the Fredholm property reduces to ellipticity. In the non-compact case, the *limit operators do appear and play an important role.* The appearance of limit operators is thus one of the main differences between the analysis on compact and on non-compact manifolds. A large part of this note is devoted to presenting a more formal version of the above theorem (Theorem 1), as well as to explaining what the statement of that theorem becomes for some standard classes and then for some new classes of manifolds. We consider in this regard manifolds that have ends that are asymptotically of one of the following forms:

(i) cylindrical, (ii) hyperbolic, or (iii) Euclidean (more general: conical).

In particular, we pay special attention to the description and properties of the limit operators for these manifolds. The new type of examples is related to Schrödinger operators with potentials with non-integer order singularities at the origin. These examples rely on a generalization of the setting of [4] that leads to more general pseudodifferential calculi [3].

Let us stress that, although we are advertising mostly smooth (non-compact) manifolds, the results presented here can be useful also for (pseudo)differential operators on singular spaces, since one can treat singular spaces by doing analysis on their smooth part, which is regarded as a non-compact manifold, possibly with a different metric, usually conformally equivalent to the original one. This is the case with *the passage from a manifold with conical points to a manifold with cylindrical ends,* which was classically done via the so called "Kondratiev transform" $r = e^t$ and is explained in the text.

Compared to the original talk, we have removed some repetitions and we have included some additional details, examples, and results; we have tried, however, to maintain as much as possible the style (and idiosyncrasies) of the original talk.

In particular, the plan of the presentation is to successively discuss the Fredholm alternative and related topics for:

(i) *Smooth, compact manifolds* (the classical case).
(ii) *Manifolds with cylindrical ends* (an almost classical case). The main examples here are the Laplacian in polar and generalized spherical coordinates.
(iii) *Other classes of manifolds*, for which we stress the similarities and differences to manifolds with cylindrical ends. The main examples here are the Laplacian in cylindrical coordinates and in flat, Euclidean coordinates.

As in the original talk (in any talk, for that matter), it was not realistic to include a complete list of references. This is unfortunate, because very many people have worked on analysis on non-compact manifolds. Since we are not including enough references, let us stress that, *unless explicitly stated otherwise,* none of the results below belong to us. We thank Cipriana Anghel, Sergiu Moroianu, Elmar Schrohe, and Jörg Seiler for useful discussions.

2 Motivation and Some Classical Results

Let us see what Theorem 1 becomes in some classically well-understood cases, namely those of compact manifolds and of manifolds with cylindrical ends.

2.1 A Classical Case: Smooth, Compact Manifolds

For pedagogical reasons, it will be useful to recall first the Fredholmness result on smooth, compact manifolds without boundary. For simplicity, we will assume from now on that all our manifolds are *smooth and complete Riemannian* (except the manifolds with conical points). Thus we can define the Sobolev spaces on these manifolds using the powers of the Laplacian. Let P be an order m, classical, pseudodifferential operator on a compact manifold M. Then $P : H^s(M) \to H^{s-m}(M)$ is bounded. In general, we will consider operators acting between sections of vector bundles E and F. The following result is classical:

Theorem 2 *Assume that* M *is* smooth, compact, without boundary, *and that* P *is an order* m*, classical, pseudodifferential operator. Then* $P : H^s(M; E) \to H^{s-m}(M; F)$ *is* Fredholm *if, and only if, it is* elliptic.

We stress again that, in the compact case, *there are no limit operators.* As we will see in the next subsection, this is *not* the case in the non-compact case.

2.2 *Motivating Example: Cylindrical Ends*

The simplest, motivating example of a non-compact manifold is that of a manifold with *cylindrical ends.* We begin with three examples before giving the formal definition.

2.2.1 Polar Coordinates

The Laplacian on $\mathbb{R}^2$ in *polar coordinates* (r, θ) is

$$\Delta_{\mathbb{R}^2} u = r^{-2}\big((r\partial_r)^2 u + \partial_\theta^2 u\big).$$

Ignoring r^{-2}, we obtain the differential operator $(r\partial_r)^2 + \partial_\theta^2$ acting on $[0, \infty) \times S^1 \ni (r, \theta)$, which is a *degenerate elliptic* differential operator acting on a manifold with boundary. We follow Melrose's observation [10] that this operator is generated by the vector fields

$$r\partial_r \text{ and } \partial_\theta ,$$

which are tangent to the boundary $\{0\} \times S^1$, an observation that will be useful in generalizations.

This example can be treated with the help of the *Kondratiev transform* $t = \log r$, which transforms $r\partial_r$ into ∂_t. The operator $(r\partial_r)^2 + \partial_\theta^2$ then becomes $\partial_t^2 + \partial_\theta^2$. Let S^{n-1} be the *unit sphere* in $\mathbb{R}^n$. The Kondratiev transform maps the domain $(0, \infty)\times S^1$ to $\mathbb{R}\times S^1$. The advantage of the Kondratiev transform is that the resulting operator $\partial_t^2 + \partial_\theta^2$ is nothing but $\Delta_{\mathbb{R}\times S^1}$, the *Laplacian* for the metric $(dt)^2 + (d\theta)^2$ on $\mathbb{R} \times S^1$. The Fourier transform in t maps ∂_t to $i\tau$ and hence it maps $\partial_t^2 + \partial_\theta^2$ to $-\tau^2 + \partial_\theta^2$. In turn, this operator can be understood via the spectral theory of ∂_θ^2 on S^1, which is well known. Anticipating, $\mathbb{R} \times S^1$ is the simplest non-trivial example of a "manifold with cylindrical ends." For general manifolds with cylindrical ends, the Fourier transform will be applied not to the operator itself, but rather to its limit operators, which are also $\mathbb{R}$-invariant operators.

2.2.2 The Black-Scholes Operator

An example that can be treated similarly is that of

$$\partial_t u + \frac{\sigma^2}{2}(x\partial_x)^2 u + \Big(r - \frac{\sigma^2}{2}\Big)x\partial_x u - ru ,$$

the *Black-Scholes operator*. This example is, in a certain sense, even simpler than that of the Laplace operator in polar coordinates, since it is an example in one dimension.

2.2.3 Generalized Spherical Coordinates

A related example is the *Schrödinger operator* $\Delta_{\mathbb{R}^n} + Z/\rho$ on $\mathbb{R}^n$, $Z \in \mathbb{C}$, in *(generalized) spherical coordinates* $(\rho, x') \in (0, \infty) \times S^{n-1}$:

$$-\Big(\Delta_{\mathbb{R}^n} + Z\rho^{-1}\Big)u = -\rho^{-2}\big((\rho\partial_\rho)^2 u + (n-2)\rho\partial_\rho u + \Delta_{S^{n-1}} u + Z\rho u\big).$$

In this example, the operator is considered on $[0, \infty) \times S^{n-1}$ and is generated by $\rho\partial_\rho$ and vector fields independent of ρ (which yield the Laplacian on the sphere S^{n-1}). The relevant geometry for this operator is again that of manifolds with cylindrical ends, since it can be generated by vector fields that are tangent to the boundary $\{0\} \times S^{n-1}$ of $[0, \infty) \times S^{n-1}$. Note that, after ignoring the factor ρ^{-2}, the resulting operator no longer has a singular potential. This example can also be treated via the Kondratiev and Fourier transforms.

2.2.4 Manifolds with Cylindrical Ends

Let us now describe the general setting encompassing the previous three examples. Let $\overline{M}$ be a *smooth, compact manifold with boundary* $\partial\overline{M}$ and let $r \geq 0$ be a smooth function on $\overline{M}$ such that $\partial\overline{M} = r^{-1}(0)$ and $dr \neq 0$ on $\partial\overline{M}$. Up to Lipschitz equivalence, a *manifold with cylindrical ends* is one that is isometric to $M := \overline{M} \smallsetminus \partial\overline{M}$ endowed with a metric that, near the boundary, is of the form

$$g_{cyl} := \frac{dr^2}{r^2} + h\,, \tag{1}$$

where h is a semi-definite tensor that restricts to a true metric on $\partial\overline{M}$. In local coordinates $x = (r, x_2, \dots, x_n)$ near the boundary, the Sobolev spaces $H^m(M; g_{cyl})$ associated to the metric g_{cyl} identify with the Babuška-Kondratiev (or weighted Sobolev) spaces

$$\mathcal{K}_a^m(M) := \{r^{\alpha_1 - a}\partial^\alpha u \in L^2(M),\ |\alpha| \leq m\} \tag{2}$$

with $a = \dim(\overline{M})/2$. These function spaces (for all a) arise naturally if we study the manifold with conical points $\overline{M}/\sim$, where $\sim$ collapses each component of $\partial\overline{M}$ to a point.

Let us now turn to the definition of *differential operators*. Assume, first that $\overline{M} = [0, \infty) \times \mathbb{R}^n \ni (r, x')$ and let $\alpha = (\alpha_1, \alpha_2, \ldots, \alpha_n) = (\alpha_1, \alpha') \in \mathbb{Z}_+^{n+1}$ be a generic multi-index. Then the "right differential operators" for the cylindrical ends geometry are the *totally characteristic differential operators* P of the form

$$P = \sum_{|\alpha| \leq m} a_\alpha(r, x')(r\partial_r)^{\alpha_1} \partial_{x'}^{\alpha'} . \tag{3}$$

Let us endow M with the metric $\frac{(dr)^2}{r^2} + (dx')^2$. Then, the Laplacian for this metric is a totally characteristic differential operator. Furthermore, after the Kondratiev transform $r = e^t$, this metric becomes just the usual Euclidean metric on $\mathbb{R}^{n+1} \ni (t, x')$. On the other hand, the differential operator P becomes

$$P = \sum_{|\alpha| \leq m} a_\alpha(e^t, x')\partial_t^{\alpha_1} \partial_{x'}^{\alpha'} , \tag{4}$$

which has the property that its coefficients have limits $a_\alpha(0, x')$ as $t \to -\infty$. This leads us to an important new ingredient, namely the "b-normal" (simply, "normal") operator

$$\tilde{P} := \sum_{|\alpha| \leq m} a_\alpha(0, x')\partial_t^{\alpha_1} \partial_{x'}^{\alpha'} , \tag{5}$$

obtained by "freezing the coefficients" at $t = -\infty$. The normal operator is an important instance of a "limit operator" mentioned earlier. (Notice that $r\partial_r$ was replaced with ∂_t, so we have used the Kondratiev transform implicitly.) The Fourier transform in t then yields the *indicial family* $\widehat{P}(\tau) := \sum_\alpha a_\alpha(0, x')(i\tau)^{\alpha_1}\partial_{x'}^{\alpha'}$. Similar constructions ($\tilde{P}$ and $\widehat{P}$) can be defined on any manifold with cylindrical ends by localization, and we obtain a normal operator for each connected component L of the boundary $\partial\overline{M}$ of $\overline{M}$. We have the following (almost classical) result [7, 11, 15]:

Theorem 3 *Let D be an order m, totally characteristic differential operator on $\overline{M}$. The map $D : H^s(M; E, g_{cyl}) \to H^{s-m}(M; F, g_{cyl})$ is Fredholm if, and only if,*

(i) it is elliptic and
(ii) all its normal operators are invertible $H^s(L; E) \to H^{s-m}(L; F)$.

An approach to the analysis on manifolds with cylindrical ends is via the b-calculus of Melrose [10] and Schulze [15]. See also [8].

3 Fredholm Operators on Manifolds with Nice Ends (Lie Manifolds)

The Laplace operator in cylindrical coordinates in $\mathbb{R}^3$ or the Schrödinger operators with singular potentials of the form $Z\rho^{-2\gamma}$, $\gamma \in \mathbb{R}_+ \smallsetminus \{0, 1/2, 1\}$, do not fit into the framework of manifolds with cylindrical ends. This raises the question of defining compatible pseudodifferential calculi and obtaining the associated Fredholm alternative (or Fredholm conditions) on more general manifolds. It does not seem possible to obtain convenient characterizations of Fredholm operators on general non-compact manifolds. We will thus restrict ourselves to a class of non-compact manifolds, which we will call "nice manifolds" for the purpose of this paper. This class of manifolds consists of manifolds with nice ends and are, for the most part, "Lie manifolds," a class of manifolds that we introduce next. While we do not define precisely which manifolds are nice (this is rather technical, see [3, 4]), we do discuss several examples, among which the ones mentioned in the introduction, namely manifolds with (asymptotically): cylindrical ends, hyperbolic ends, and Euclidean ends.

3.1 Lie Manifolds

We now introduce and discuss Lie manifolds, their geometry, and, most importantly, their *associated differential operators.*

3.1.1 Definition of Lie Manifolds

Assume that we are given a compact manifold with corners $\overline{M}$, whose interior is $M := \overline{M} \smallsetminus \partial\overline{M}$, and a subspace

$$\mathcal{V} \subset \mathcal{V}_b(\overline{M}) := \{X \in \mathcal{C}^\infty(\overline{M}; T\overline{M}) \mid X \text{ } \textit{tangent} \text{ to all faces of } \overline{M}\}\,. \tag{6}$$

Definition 1 ([1]) The pair $(\overline{M}, \mathcal{V})$ is a Lie manifold if:

(i) $\mathcal{V} \subset \mathcal{V}_b(\overline{M})$ is closed under the Lie bracket $[\ ,\ ]$;
(ii) $\mathcal{V}$ is a finitely-generated, projective $\mathcal{C}^\infty(\overline{M})$–module;
(iii) $\mathcal{C}_c^\infty(M; TM) \subset \mathcal{V}$ (recall that $M := \overline{M} \smallsetminus \partial\overline{M}$).

This definition is based on earlier, similar constructions due to Connes, Cordes, Kondratiev, Mazzeo, Mazya, Melrose, Plamenevskij (a relevant old article with Mazya is described in [12]), Schrohe, Schulze, Skandalis, and many others. Informally, a *Lie manifold* $(\overline{M}, \mathcal{V})$ is a manifold M with:

(i) a *compactification* $\overline{M}$ such that $M = \overline{M} \smallsetminus \partial\overline{M}$, whose role is to control the behavior at infinity of the coefficients of our differential operators, and

(ii) a *Lie algebra* of vector fields $\mathcal{V} \subset \mathcal{V}_b$ on this compactification, whose role is to define the "nice" differential operators.

Remark 1 Let $(\overline{M}, \mathcal{V})$ be a Lie manifold. Saying that $\mathcal{V}$ is a *projective* $\mathcal{C}^\infty(\overline{M})$-module means that it is stable under multiplication by functions in $\mathcal{C}^\infty(\overline{M})$ and it has a basis around each point of $\overline{M}$. In particular, $\mathcal{V}$ is a complex vector space. Saying that $\mathcal{C}_c^\infty(M; TM) \subset \mathcal{V}$ means that there are no "obstructions in the interior" for the vector fields in $\mathcal{V}$. Compact (smooth) manifolds and manifolds with cylindrical ends are Lie manifolds. These and other examples will be discussed below in Sect. 3.3.

3.1.2 Lie Manifolds and Geometry

As in [1], since $\mathcal{V}$ is a projective $\mathcal{C}^\infty(\overline{M})$-module, the Serre-Swan theorem gives the existence of a vector bundle $A \to \overline{M}$ such that

$$\mathcal{V} \simeq \Gamma(\overline{M}; A)\,. \tag{7}$$

The Lie algebra structure on the sections of A means that it is a *Lie algebroid*. The inclusion $\mathcal{C}_c^\infty(M; TM) \subset \mathcal{V}$ means that $A = TM$ in the interior $M := \overline{M} \smallsetminus \partial\overline{M}$ of $\overline{M}$. Thus, once we have chosen a metric on A, that metric will induce a metric on TM (or, which is the same thing, on M). A metric obtained in this way will be called *compatible*, and is unique up to Lipschitz equivalence. For a smooth bundle $E \to \overline{M}$, we define the Sobolev spaces $H^s(M; E)$ using the powers of the Laplacian for any compatible metric.

3.1.3 Differential Operators

Our main interest lies in the algebra $\mathrm{Diff}_{\mathcal{V}}(\overline{M})$, which consists of the differential operators generated by $\mathcal{C}^\infty(\overline{M})$ and by derivatives in $\mathcal{V}$. Given two vector bundles $E, F \to \overline{M}$, we have a similar definition of $\mathrm{Diff}_{\mathcal{V}}(\overline{M}; E, F)$. All the geometric operators on M with a compatible metric will belong to some space of the form $\mathrm{Diff}_{\mathcal{V}}(\overline{M}; E, F)$. For instance, the Levi-Civita connection on M extends to a differential operator in $\mathrm{Diff}_{\mathcal{V}}(\overline{M}; A, A^* \otimes A)$ (an A-connection). Hence, since $\overline{M}$ is compact, its interior M will have bounded curvature (together with its covariant derivatives).

3.2 *Formulation of the Main Result: Fredholm Conditions on "Nice" Manifolds*

We are now closer to giving a precise statement for the Fredholm conditions in Theorem 1. Let $(\overline{M}, \mathcal{V})$ be a Lie manifold and $D \in \mathrm{Diff}_{\mathcal{V}}(\overline{M}; E, F)$. If D has order

m, then it defines a continuous map $D : H^s(M; E) \to H^{s-m}(M; F)$. To $(\overline{M}, \mathcal{V})$ and D we associate:

(i) spaces Z_α, $\alpha \in I$, the *orbits* of $\mathcal{V}$ on $\overline{M}$ (they do not dependent on D);
(ii) isotropy groups G_α, $\alpha \in I$, (also independent of D); and
(iii) G_α-invariant differential operator D_α on $Z_\alpha \times G_\alpha$

as follows. Let $\mathcal{V}_x := \{X \in \mathcal{V} | \, X(x) = 0\}$ and $I_x := \{f \in \mathcal{C}^\infty(\overline{M}) | \, f(x) = 0\}$, for $x \in \partial\overline{M}$. Then $\mathcal{V}_x/I_x\mathcal{V}$ is a Lie algebra and, for $x \in Z_\alpha$, G_α is the simply-connected Lie group that integrates it:

$$Lie(G_\alpha) = \mathcal{V}_x/I_x\mathcal{V}\,.$$

For each $\alpha \in I$, the operator D_α is obtained by restricting D to the orbit Z_α and letting the vector fields act on G_α via the map $\mathcal{V}_x \to Lie(G_\alpha)$. The operators D_α are the *limit operators*. We can now formulate our main result [3, 4].

Theorem 4 *Let $D \in \mathrm{Diff}_{\mathcal{V}}(\overline{M}; E, F)$ be of order m and assume that $(\overline{M}, \mathcal{V})$ is "nice." We have that $D : H^s(M; E) \to H^{s-m}(M; F)$ is Fredholm if, and only if, D is elliptic and all D_α, $\alpha \in I$, are invertible.*

Our Lie manifold $(M, \mathcal{V})$ is *nice* if there exists a *Hausdorff* Lie groupoid $\mathcal{G}$ such that

- its Lie algebroid $A(\mathcal{G}) \simeq A$ of Eq. (7) (so $\Gamma(\overline{M}; A) = \mathcal{V}$) and
- $\mathcal{G}$ satisfies the Effros-Hahn conjecture (a statement about $C^*(\mathcal{G})$ that implies that the regular representations of $C^*(\mathcal{G})$ determine its invertible elements).

The proof is obtained by considering the norm closure $\overline{\Psi}^0(\mathcal{G})$ of the algebra of order zero pseudodifferential operators on $\mathcal{G}$ [3, 4]. Let $\mathcal{K}$ be the algebra of compact operators on $L^2(M)$. Recall that an operator is Fredholm if, and only if, it is invertible modulo $\mathcal{K}$. We thus want to characterize the invertible elements of $\overline{\Psi}^0(\mathcal{G})/\mathcal{K}$. Since $\overline{\Psi}^0(\mathcal{G})/C^*(\mathcal{G}) \simeq \mathcal{C}(S^*A)$ via the principal symbol and $\mathcal{G}$ satisfies the Effros-Hahn conjecture, in addition to the principal symbol, it is enough to look at the regular representations of $C^*(\mathcal{G})/\mathcal{K}$. Each of these regular representations yields a limit operator. This completes the proof. A useful property here is that nice manifolds are closed under general blow-ups [3, 4].

3.3 *Examples and Applications*

The example "zero" is that of a smooth, compact manifold without boundary M. Then M is a Lie manifold with $\overline{M} = M$ and $\mathcal{V} = \Gamma(M; TM)$ (all smooth vector fields on M). In particular, we have $A = TM$. The boundary $\partial\overline{M} := \overline{M} \smallsetminus M$ is empty, which is consistent with the fact that there are no limit operators. Then Theorem 4 becomes simply the (well known) Theorem 2. In all our following examples, $\overline{M}$ will be a smooth, compact manifold with boundary $\partial\overline{M}$ and we shall concentrate on a tubular neighborhood $U = [0, \epsilon) \times \partial\overline{M} \ni (r, y)$ of the boundary, because the interior $M := \overline{M} \smallsetminus \partial\overline{M}$ can be treated with classical tools. Here are our main three examples:

(i) Let $\mathcal{V} := \mathcal{V}_b$, the set of smooth vector fields on $\overline{M}$ that are *tangent* to the boundary $\partial\overline{M}$ of $\overline{M}$, as before. We have $\mathcal{V}_b = \mathcal{C}^\infty(\overline{M})r\partial_r + \sum \mathcal{C}^\infty(\overline{M})\partial_y$ on U, so the projectivity condition is satisfied and we obtain a Lie manifold $(\overline{M}, \mathcal{V}_b)$. Any metric of the form $g = r^{-2}(dr)^2 + h$ as in Eq. (1) will be compatible (with the Lie manifold structure) on M, and hence M is a *manifold with cylindrical ends*. The bundle $A = {}^bTM$ was considered by Melrose. The index set I consists of the connected components of $\partial\overline{M}$ and $G_\alpha = \mathbb{R}$. The differential operators $\mathrm{Diff}_{\mathcal{V}}(\overline{M})$ are the totally characteristic differential operators on $\overline{M}$ (already discussed) and the limit operators D_α are the normal operators of D defined earlier.

(ii) Let next $\mathcal{V} := \mathcal{V}_0 := r\mathcal{C}^\infty(\overline{M}; T\overline{M})$, the set of smooth vector fields on $\overline{M}$ that *vanish* at the boundary $\partial\overline{M}$ of $\overline{M}$. Then $\mathcal{V}_0 = \mathcal{C}^\infty(\overline{M})r\partial_r + \sum \mathcal{C}^\infty(\overline{M})r\partial_y$ on U, so the projectivity condition is satisfied and, again, $(\overline{M}, \mathcal{V}_0)$ is a Lie manifold. The choice of a metric on A (a compatible metric on M) makes the interior $M := \overline{M} \smallsetminus \partial\overline{M}$ of $\overline{M}$ a so called *asymptotically hyperbolic manifold*, since the metric is of the form $g = \frac{h}{r^2}$ with h a true metric on $\overline{M}$. The orbits $Z_\alpha \subset \partial\overline{M}$ are reduced to points (so $I = \partial\overline{M}$) and $G_\alpha = T_\alpha\partial\overline{M} \rtimes \mathbb{R}$, which is a *non-commutative* group (obtained as the semi-direct product from the action of $\mathbb{R}$ on $T_\alpha\partial\overline{M}$ by dilations).

(iii) To round up our podium, let $\mathcal{V} := \mathcal{V}_{sc} := r\mathcal{V}_b \subset \mathcal{V}_0$. From example (i), we see that $\mathcal{V}_{sc} := \mathcal{C}^\infty(\overline{M})r^2\partial_r + \sum \mathcal{C}^\infty(\overline{M})r\partial_y$ on U, so the projectivity condition is yet again satisfied to yield a Lie manifold $(\overline{M}, \mathcal{V}_{sc})$. Euclidean, asymptotically Euclidean, and asymptotically conical spaces are modeled by this type of Lie manifolds. The orbits $Z_\alpha = \{\alpha\}$ are again reduced to points (so again $I = \partial\overline{M}$). However, this time, $G_\alpha = T_\alpha\partial\overline{M} \times \mathbb{R}$ is *commutative*.

The Laplacian $\Delta_{\mathbb{R}^3}u = r^{-2}\big[(r\partial_r)^2u + \partial_\theta^2u + (r\partial_z)^2u\big]$ in *cylindrical coordinates* $(r, \theta, z) \in [0, \infty) \times S^1 \times \mathbb{R}$ is closely related to the example (ii) above. Ignoring the factor r^{-2}, the relevant operator is generated by the vector fields $r\partial_r$, ∂_θ, and $r\partial_z$, which are again tangent to the boundary. This example was one of the original motivations to considering a framework more general than that of manifolds with cylindrical ends.

We conclude this section with an example that goes even beyond Lie manifolds. Namely, let us we consider the Schrödinger operator with potential $V = \rho^{-2\gamma}V_0$, with V_0 *smooth in generalized spherical coordinates* (ρ, x'), $\rho \in [0, \infty)$, $x' \in S^{n-1}$, and $\gamma \geq 0$:

$$\Delta + V = \rho^{-2}\big[(\rho\partial_\rho)^2 + (n-2)\rho\partial_\rho + \Delta_{S^{n-1}} + \rho^{2-2\gamma}V_0\big] \tag{8}$$

Let us assume first that $2\gamma \in \{0, 1, 2\}$. Ignoring the factor ρ^{-2}, the resulting operator

$$(\rho\partial_\rho)^2 + (n-2)\rho\partial_\rho + \Delta_{S^{n-1}} + \rho^{2-2\gamma} V_0 \tag{9}$$

is in the b-calculus [10, 15]. This property can be used, among other applications, to study the domain of $\Delta + V$ and the regularity of its eigenfunctions. If $2\gamma \in \mathbb{R} \smallsetminus \{0, 1, 2\}$, the operator $\Delta + V$ can still be studied with a modified pseudodifferential calculus [3]. For instance, if $\gamma > 1$ and if we write $\Delta + V = \rho^{-2\gamma}\big[(\rho^\gamma\partial_\rho)^2 + (n - 1 - \gamma)\rho^{\gamma-1}\rho^\gamma\partial_\rho + \rho^{2\gamma-2}\Delta_{S^{n-1}} + V_0\big]$, then the resulting operator

$$(\rho^\gamma\partial_\rho)^2 + (n-1-\gamma)\rho^{\gamma-1}\rho^\gamma\partial_\rho + \rho^{2\gamma-2}\Delta_{S^{n-1}} + V_0 \tag{10}$$

is in the $c_{\gamma,\gamma-1}$-calculus of [3] (recall that V_0 is smooth in polar coordinates).

Other examples come from semi-Riemannian and sub-Riemannian geometry [2].

4 Pseudodifferential Operators and Problems

Work related to the algebras $\Psi^\infty(\mathcal{G})$ used in the proof of our main result was done by Androulidakis, Connes, Debord, Mazzeo, Melrose, Monthubert, Ruzhansky, Schrohe, Schulze, Skandalis, and many others. See [1, 5, 6, 9, 13, 14] for references. In that case, $\Psi^{-\infty}(\mathcal{G})$ is the algebra generated by the conormal distributions of order $-\infty$ on $A \to M$. In turn, $\Psi^m(\mathcal{G})$ is linearly generated by the conormal distributions of order m on $A \to M$ and by $\Psi^{-\infty}(\mathcal{G})$. A generalization of these algebras in contained in [3]. In the case of manifolds with cylindrical ends, we have $\mathcal{V} = \mathcal{V}_b$ and $\Psi^\infty(\mathcal{G})$ consists of the properly supported operators in the b-calculus of Melrose and Schulze (and hence, it is a dense algebra in a suitable topology). Similarly, for asymptotically hyperbolic manifolds, $\mathcal{V} = \mathcal{V}_0$ and $\Psi^\infty(\mathcal{G})$ consists of the properly supported operators in the 0-calculus of Mazzeo and Schulze. Finally, for asymptotically euclidean manifolds, $\mathcal{V} = \mathcal{V}_{sc}$ and $\Psi^\infty(\mathcal{G})$ consists of the properly supported operators in the SG-calculus of Parenti and Schrohe, the same calculus as the "scattering calculus" of Melrose.

4.1 Problems

We conclude by formulating two problems.

Problem 1 (Connes, Bohlen-Schrohe) Find the index of a Fredholm operator $D \in \Psi^m(\mathcal{G})$.

Problem 2 (Baldare-Benameur-Lesch-V.N., Ruzhansky) Let G be a compact Lie group. Let M be compact smooth connected with a G action and E be a G-bundle. Let $\overline{M}$ be the Albin-Melrose compactification of the principal orbit bundle (for the G) action and $\mathcal{G}$ be its Lie groupoid. Describe $\Psi^m(M; E)^G$ using $\Psi^m(\mathcal{G})$ and the Ruzhansky calculus.

Acknowledgments Support: authors Ivan Beschastnyi and Catarina Carvalho by FCT–Fundação para a Ciência e Tecnologia, projects UIDB/04106/2020 and UIDB/04721/2020, respectively. Author Victor Nistor by ANR grant OpART and Yu Qiao by NSFC (11971282).

References

1. B. Ammann, R. Lauter, V. Nistor, On the geometry of Riemannian manifolds with a Lie structure at infinity. Int. J. Math. Math. Sci. **1–4**, 161–193 (2004)
2. I. Beschastnyi, Closure of the Laplace-Beltrami operator on 2d almost-Riemannian manifolds and semi-Fredholm properties of differential operators on Lie manifolds. Result. Math. **78**(2), 56 (2023). Id/No 59
3. I. Beschastnyi, C. Carvalho, V. Nistor, Y. Qiao, Work in progress
4. C. Carvalho, V. Nistor, Y. Qiao, Fredholm conditions on non-compact manifolds: theory and examples, in *Operator Theory, Operator Algebras, and Matrix Theory*. Operator Theory: Advances and Applications vol. 267 (Birkhäuser/Springer, Cham, 2018), pp. 79–122
5. A. Connes, *Noncommutative Geometry* (Academic Press, San Diego, 1994)
6. C. Debord, G. Skandalis, Adiabatic groupoid, crossed product by $\mathbb{R}_+^*$ and pseudodifferential calculus. Adv. Math. **257**, 66–91 (2014)
7. V.A. Kondrat'ev, Boundary value problems for elliptic equations in domains with conical or angular points. Transl. Moscow Math. Soc. **16**, 227–313 (1967)
8. R. Lauter, J. Seiler, Pseudodifferential analysis on manifolds with boundary – a comparison of b-calculus and cone algebra, in *Approaches to Singular Analysis* (Springer, Berlin, 2001), pp. 131–166
9. R. Mazzeo, Elliptic theory of differential edge operators. I. Commun. Partial Differ. Equ. **16**(10), 1615–1664 (1991)
10. R. Melrose, *The Atiyah-Patodi-Singer Index Theorem.* Research Notes in Mathematics, vol. 4 (A. K. Peters, Wellesley, 1993), xiv, 377pp.
11. R. Melrose, G. Mendoza, Elliptic operators of totally characteristic type. MSRI Preprint
12. S.A. Nazarov, B.A. Plamenevsky, *Elliptic Problems in Domains with Piecewise Smooth Boundaries*. de Gruyter Expositions in Mathematics, vol. 13 (Walter de Gruyter, Berlin, 1994)
13. M. Ruzhansky, V. Turunen, *Pseudo-Differential Operators and Symmetries*. Pseudo-Differential Operators. Theory and Applications, vol. 2 (Birkhäuser Verlag, Basel, 2010). Background analysis and advanced topics
14. E. Schrohe, Fréchet algebra techniques for boundary value problems on noncompact manifolds: Fredholm criteria and functional calculus via spectral invariance. Math. Nachr. **199**, 145–185 (1999)
15. B.W. Schulze, *Pseudo-Differential Operators on Manifolds with Singularities*. Studies in Mathematics and its Applications, vol. 24 (North-Holland, Amsterdam, 1991)

Singular Value Decomposition for the X-Ray Transforms on the Reduced Heisenberg Group, and a Two-Radius Theorem

Steven Flynn

Abstract We give an explicit Singular Value Decomposition of the sub-Riemannian X-ray transform on the Heisenberg group with compact center. By studying the singular values, we obtain a two-radius theorem for integrals over sub-Riemannian geodesics. We also state intertwining properties of distinguished differential operators. We conclude with a description of ongoing work.

Keywords X-ray Transform · Heisenberg group · Two-radius theorem · Sub-Riemannain geometry

2020 Mathematics Subject Classification 44A12, 53C17, 43A80

1 Introduction

The X-ray transform assigns to a function its integrals over closed geodesics. The Heisenberg group with its sub-Riemannian structure provides a rich example of such integral functionals. Of crucial importance for practical applications of the X-ray Transform is its Singular Value Decomposition [6].

2 Definitions

2.1 Heisenberg Group

The Heisenberg group is $\mathbb{H} = \mathbb{C} \times \mathbb{R}$ with the multiplication law

$$(x + iy, t)(u + iv, s) = (x + u + i(y + v), t + s + \tfrac{1}{2}(xv - yu))$$

S. Flynn (✉)
Department of Mathematics, University of Padova, Padova, Italy
e-mail: stevenpatrick.flynn@unipd.it

D. Cardona et al. (eds.), *Extended Abstracts 2021/2022*, Research Perspectives Ghent Analysis and PDE Center 3, https://doi.org/10.1007/978-3-031-48579-4_2

The reduced Heisenberg group is the quotient of $\mathbb{H}$ by a discrete subgroup of the center

$$\overline{\mathbb{H}} := \mathbb{H}/\Gamma, \quad \Gamma := \{(0, k\pi) \in \mathbb{C} \times \mathbb{R} : k \in \mathbb{Z}\}.$$

Heuristically, the harmonic analysis on the Heisenberg group is controlled by the center $Z(\mathbb{H}) = \{0 + 0i\} \times \mathbb{R}$. Having compact center, analyis of the reduced Heisenberg group resembles that of the circle.

2.2 X-Ray Transform

Let $r \in \mathbb{Q}^+$ be a positive rational number. Of interest are curves on $\overline{\mathbb{H}}$ modeled by

$$\gamma_r(s) := \left(\sqrt{r}e^{is/\sqrt{r}}, \tfrac{1}{2}\sqrt{r}s\right) \in \overline{\mathbb{H}} \cong \mathbb{C} \times (\mathbb{R}/\pi\mathbb{Z}). \tag{1}$$

Take for granted that the set of closed unit speed *sub-Riemannian* geodesics [7] on $\overline{\mathbb{H}}$ is generated by left translations by $(z, t) \in \overline{\mathbb{H}}$ of the curves γ_r above, for $r \in \mathbb{Q}^+$. The requirement that r is rational is analogous to the requirement that closed geodesics on the flat torus have rational slope. If r were irrational, the curve γ_r would not be closed, so we ignore that case.

Indeed, for $r = a/b$ with $a, b \in \mathbb{N}^+$ coprime,

$$\gamma_r \text{ is a closed curve in } \overline{\mathbb{H}} \text{ with period } 2\pi\sqrt{ab}.$$

After one period of γ_r with $r = a/b$ for coprime (a, b),

(i) $a \in \mathbb{N}^+$ counts the number of times γ_r winds around the S^1 component of $\overline{\mathbb{H}}$,
(ii) $b \in \mathbb{N}^+$ counts the number of counterclockwise rotations made by γ_r around its central axis.

Call a the *vertical winding number*, and b the *horizontal winding number* of γ_r in $\overline{\mathbb{H}}$.

Definition 1 For vertical and horizontal winding numbers a and b, and $r = a/b$ the X-ray transform

$$I_r : C_c^\infty(\overline{\mathbb{H}}) \to C_c^\infty(\overline{\mathbb{H}})$$

of a compactly supported smooth function $f \in C_c^\infty(\overline{\mathbb{H}})$ is

$$I_r f(z, t) = \int_0^{2\pi\sqrt{ab}} f\left((z, t)\gamma_r(s)\right) ds, \quad (z, t) \in \overline{\mathbb{H}}. \tag{2}$$

We also write $If(r; z, t) := I_r f(z, t)$.

Remark 1 The operator I is the Heisenberg analog of the X-ray transform on a torus. It is related to the X-ray transform on the full Heisenberg group [3] by a type of torus-projection operator (as in [5]).

That I_r is well-defined is straightforward. By the Cauchy Schwartz Inequality and compactness of the domain of integration, I_r extends to a bounded operator on $L^2(\overline{\mathbb{H}})$. It is a simple matter to extend I_r to other function spaces, but we focus on the L^2 theory.

The primary concern is injectivity of the X-ray transform. Written as in Eq. (2), the X-ray transform is a special case of the Pompeiu Transform [1]. However, to the author's knowlegdge, the special case of integration over left-translates a closed sub-Riemannian geodesic has not been considered.

3 Spectral Decomposition of I_r, I_r^* and $I_r^* I_r$

The spectral decomposition of the X-ray transform I_r has a continuous part and a discrete part respecting the orthogonal decomposition

$$L^2(\overline{\mathbb{H}}) \cong L^2(\mathbb{C}) \oplus {}^0L^2(\overline{\mathbb{H}}). \tag{3}$$

The substances in the orthogonal decomposition above are

$$L^2(\mathbb{C}) \cong \{f \in L^2(\overline{\mathbb{H}}) : f(z,t) = f(z,0),\ \forall (z,t) \in \overline{\mathbb{H}}\}$$

$${}^0L^2(\overline{\mathbb{H}}) := \{f \in L^2(\overline{\mathbb{H}}) : \int_0^{\pi} f(z,t)dt = 0,\ \forall z \in \mathbb{C}\}.$$

Proposition 1 *I_r preserves the orthogonal decomposition above. i.e,*

$$I_r : L^2(\mathbb{C}) \to L^2(\mathbb{C}) \qquad I_r : {}^0L^2(\overline{\mathbb{H}}) \to {}^0L^2(\overline{\mathbb{H}}).$$

The continuous part of I_r is its restriction to $L^2(\mathbb{C})$, which is essentially the mean value transform on the plane $M^R : L^2(\mathbb{C}) \to L^2(\mathbb{C})$,

$$M^R f(z) := \frac{1}{2\pi} \int_0^{2\pi} f(z + Re^{i\theta})d\theta, \quad f \in L^2(\mathbb{C}).$$

Theorem 1 *The restriction of I_r to $L^2(\mathbb{C})$ is a scalar multiple of the mean value transform. In particular*

$$I_r f(z) = 2\pi\sqrt{ab}\, M^{\sqrt{r}} f(z), \quad f \in L^2(\mathbb{C}).$$

Therefore

$$I_r f(z) = \frac{\sqrt{ab}}{2\pi} \int_{\mathbb{R}^2} J_0(R|\xi|)\widehat{f}(\xi)e^{iz\cdot\xi}\,d\xi.$$

Here J_0 is the zeroth order Bessel function defined in (12).

Proof For $f \in L^2(\mathbb{C})$,

$$\begin{aligned}
I_r|_{L^2(\mathbb{C})} f(z,t) &= \int_0^{2\pi\sqrt{ab}} f\left((z,t)(\sqrt{r}e^{is/\sqrt{r}}, \tfrac{1}{2}\sqrt{r}s)\right) ds \\
&= \int_0^{2\pi\sqrt{ab}} f\left(z + e^{i\sqrt{r}s/\sqrt{r}}, t + \tfrac{1}{2}\sqrt{r}s + \tfrac{1}{2}\sqrt{r}\operatorname{Im}\left(\overline{z}e^{is/\sqrt{r}}\right)\right) ds \\
&= \int_0^{2\pi\sqrt{ab}} f(z + \sqrt{r}e^{is/\sqrt{r}})ds = \sqrt{r}\int_0^{2\pi b} f(z + \sqrt{r}e^{is})ds \\
&= (2\pi b)\sqrt{r}M^{\sqrt{r}} f(z) = 2\pi\sqrt{ab}M^{\sqrt{r}} f(z),
\end{aligned}$$

Furthermore

$$\begin{aligned}
\int_{\mathbb{R}^2} M^R g(z)e^{-iz\cdot\xi}dz &= \int_{\mathbb{R}^2} \frac{1}{2\pi}\int_0^{2\pi} g(z + Re^{i\theta})e^{-iz\cdot\xi}d\theta dz \\
&= \frac{1}{2\pi}\int_0^{2\pi} e^{iRe^{i\theta}\cdot\xi}d\theta \int_{\mathbb{R}^2} g(z)e^{-iz\cdot\xi}dz = J_0(R|\xi|)\widehat{g}(\xi)
\end{aligned}$$

The result follows. □

We now focus on the Singular Value Decomposition of the discrete part of I_r.

Theorem 2 (Spectral Decomposition of I_r on ${}^0L^2(\overline{\mathbb{H}})$) *Fix $r = a/b \in \mathbb{Q}^+$. Then for all $j, k = 0, 1, 2, \ldots$*

$$I_r\psi^n_{jk} = \sqrt{ab}\,\delta_{\mathbb{Z}}(rn)\,c(rn, j)\psi^n_{j+r|n|,k}$$

where

$$c(m, j) = 2\pi\sqrt{\frac{j!}{(j+m)!}}m^{m/2}e^{-(i+1)\pi m/2}L^m_j(m), \quad m \in \mathbb{N}$$

with $c(-m, j) = c(m, j)$ and

$$\delta_{\mathbb{Z}}(m) = \begin{cases} 1 & m \in \mathbb{Z} \\ 0 & m \notin \mathbb{Z} \end{cases}.$$

Here $\{\psi^n_{jk} : n \in \mathbb{Z}^*, j, k \in \mathbb{N}\}$ defined in Eq. (10), is an orthonormal Hilbert Basis of ${}^0L^2(\overline{\mathbb{H}})$. The generalized Laguerre functions are given in (11).

Proof For $n \in \mathbb{Z}^*$, $j, k \in \mathbb{N}$, $r \in \mathbb{Q}^+$, and $(z, t) \in \overline{\mathbb{H}}$,

$$
\begin{aligned}
I_r\psi^n_{jk}(z,t) &= \frac{\sqrt{|n|}}{\pi} I_r M^n_{jk}(z,t) = \frac{\sqrt{|n|}}{\pi}\int_0^{2\pi\sqrt{ab}} M^n_{jk}\left((z,t)\gamma_r(s)\right) ds \\
&= \frac{\sqrt{|n|}}{\pi}\int_0^{2\pi\sqrt{ab}} \sum_{l=0}^{\infty} M^n_{jl}\left(\gamma_r(s)\right) M^n_{lk}(z,t) ds, \qquad \text{by (9)}, \\
&= \sum_{l=0}^{\infty}\left[\int_0^{2\pi\sqrt{ab}} M^n_{jl}\left(\gamma_r(s)\right) ds\right] \psi^n_{lk}(z,t).
\end{aligned}
$$

Now

$$
\begin{aligned}
\int_0^{2\pi\sqrt{ab}} M^n_{jl}\left(\gamma_r(s)\right) ds &= \int_0^{2\pi\sqrt{ab}} M^n_{jl}\left(\sqrt{r}e^{is/\sqrt{r}}, \tfrac{1}{2}\sqrt{r}s\right) ds \\
&= \sqrt{r}\int_0^{2\pi\sqrt{abr}} M^n_{jl}\left(\sqrt{r}e^{is}, \tfrac{1}{2}rs\right) ds, \quad \text{via } s \mapsto s\sqrt{r} \\
&= \sqrt{r}\int_0^{2\pi b} e^{i((j-l)+r|n|)s} M^n_{jl}\left(\sqrt{r}, 1\right) ds, \quad \text{by (6)} \\
&= \sqrt{ab}\int_0^{2\pi} e^{i(b(j-l)+a|n|)s} ds\, M^n_{jl}\left(\sqrt{r}, 1\right), \quad s \mapsto bs \\
&= 2\pi\sqrt{ab}\,\left[\delta_0\left(b(j-l)+a|n|\right)\right] M^n_{jl}(\sqrt{r}, 0).
\end{aligned}
$$

Thus

$$
\begin{aligned}
I_r\psi^n_{jk}(z,t) &= 2\pi\sqrt{ab}\sum_{l=0}^{\infty}\delta(j-l+r|n|) M^n_{jl}(\sqrt{r}, 0)\psi^n_{lk}(z,t) \\
&= \delta_{\mathbb{Z}}(rn)(2\pi\sqrt{ab}) M^n_{j,j+|n|r}(\sqrt{r}, 0)\psi^n_{j+|n|r,k}(z,t)
\end{aligned}
$$

as desired. □

Remark 2 In contrast with [3], we see that the spectral decomposition of the X-ray transform involves the "Bessel spectrum" (corresponding ot the finite dimensional representations of $\overline{\mathbb{H}}$). On the full Heisneberg group, this part of the spectrum has Plancherel measure zero. See [4] or [8].

For fixed $r \in \mathbb{Q}^+$, the X-ray transform I_r is a convolution operator from $L^2(\overline{\mathbb{H}})$ to itself. We define the adjoint using the same measure (the Haar measure, which here is the Lebesgue measure) on the domain and target space.

Theorem 3 *The formal adjoint $I_r^* : L^2(\overline{\mathbb{H}}) \to L^2(\overline{\mathbb{H}})$ is given by*

$$I_r^* f(z,t) = \int_0^{2\pi\sqrt{ab}} f\left((z,t)\gamma_r(s)^{-1}\right) ds.$$

Remark 3 Note that $\gamma_r(s)^{-1}$ is not a geodesics for the left-invariant metric. In particular I_r is not formally self-adjoint. This fact contrasts with the case for the X-ray transform on the torus with a fixed directional parameter,

Corollary 1 (Spectral Decomposition of I_r on ${}^0L^2(\overline{\mathbb{H}})$) *Fix $r = a/b \in \mathbb{Q}^+$. Then for all $j, k = 0, 1, 2, ...$*

$$I_r^* \psi_{jk}^n = \sqrt{ab}\, \delta_{\mathbb{Z}}(rn)\, \overline{c(n, j - r|n|)} \psi_{j-r|n|,k}^n,$$

with $c(m, j)$ defined above for $j \geq 0$ and $c(m, j) = 0$ for $j < 0$.

Corollary 2 (Spectral Decomposition of N_r on ${}^0L^2(\overline{\mathbb{H}})$) *The normal operator $N_r := I_r^* I_r$ is diagonalized by the basis $\{\psi_{jk}^n\}$:*

$$N_r \psi_{jk}^n = (ab)\, \delta_{\mathbb{Z}}(rn)\, |c(rn, j)|^2 \psi_{jk}^n.$$

Corollary 3 (Singular Value Decomposition of I_r) *For $f \in {}^0L^2(\overline{\mathbb{H}})$, $I_r f = U_r \circ D_r f$ where*

$$D_r \psi_{jk}^n = s(n, j, r) \psi_{jk}^n$$

$$U_r \psi_{jk}^n = e^{-\pi r|n|/2} \sigma(r|n|, j) \psi_{j+r|n|,k}^n$$

where $\sigma(r|n|, j) = \operatorname{sgn}(L_j^{(r|n|)}(r|n|))$ and

$$s(n, j, r) = \sqrt{ab}\, \delta_{\mathbb{Z}}(rn)\, |c(rn, j)|.$$

It is illuminating to see the action of the X-ray Transform on the (reduced) Heisenberg fan, which is the set

$$R := \cup_{j \in \mathbb{N}^*} R_j \quad R_j = R_j^+ \cup R_j^-, \quad R_j^{\pm} = \{(\pm 2n, 2n(2j+1)) : n \in \mathbb{N}^+\}.$$

R is the set of pairs of eigenvalues of $\mathcal{L}$ and $-iT$ (defined in Sect. 4) corresponding to the joint eigenvectors ψ_{jk}^n (we omit the part of the Heisenberg fan corresponding to the Bessel spectrum): The operator U_r in the SVD of I_r acts on the Heisenberg fan according to the picture below.

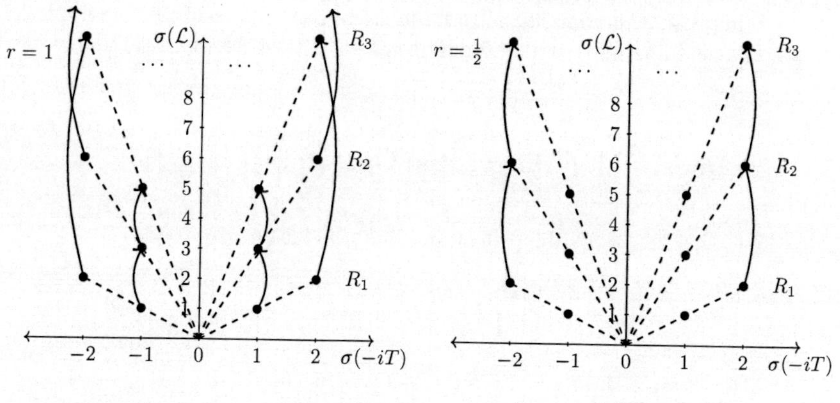

In the figures above, dashed lines are rays of the Heisenberg fan (on the full Heisenberg group). Dots are elements of the Heisenberg fan of $\overline{\mathbb{H}}$, each $(2n, 2|n|(2j+1))$ corresponding to the subspace $\{\psi^n_{jk}\}_{k\in\mathbb{N}}$. Arrows represent the action of U_r for $r = 1$ and $r = 1/2$ in the first and second graphic respectively.

The following functions appear in the expression for the singular values:

Definition 2 (Diagonal Laguerre Function) For $x \in \mathbb{R}$ and $j \in \mathbb{N}$, call

$$l_j(x) = L_j^{(x)}(x) := \frac{1}{2\pi i}\oint_C \left(\frac{e^{-z/(1-z)}}{1-z}\right)^x \frac{dz}{(1-z)z^{j+1}}$$

the *diagonal Laguerre function* of order j. Here C is a counterclockwise circle around the origin not encircling the point $z = 1$.

Next we show that any L^2 function on $\overline{\mathbb{H}}$ is determined by its integrals over geodesics of two $r_1, r_2 \in \mathbb{Q}^+$ satisfying a compatibility condition. This type of results is called a Two-Radius Theorem [2], but a more appropriate name might be a "Two-Momenta Theorem," since $\lambda_i := 1/\sqrt{r_j}$ is the momentum of the sub-Riemannian geodesic γ_{r_j} [7].

Corollary 4 (Two-Radius Theorem) *Let $f \in L^2(\overline{\mathbb{H}})$ with $I_{r_1} f = I_{r_2} f = 0$ for some $r_i \in \mathbb{Q}^+$. Then $f = 0$ provided that*

1. *r_1/r_2 is not a ratio of roots of diagonal Laguerre functions $l_j(x) := L_j^{(x)}(x)$ for any $j \in \mathbb{N}$*
2. *$\sqrt{r_1/r_2}$ is not a ratio of roots of the Bessel function J_0.*

Proof For and $n \in \mathbb{Z}^*$ let $s(n, j, r_1)$ and $s(n, j, r_2)$ be the singular values for ψ^n_{jk} corresponding to I_{r_1} and I_{r_2} respectively. Then $f = 0$ provided that $s(n, j, r_1)$ and $s(n, j, r_2)$ are not both zero. We have $s(n, j, r_1) = s(n, j, r_2) = 0$ if and only if $r_j n$ is a zero of the diagonal Laguerre function l_j for $j = 1, 2$. In this case, r_1/r_2 is a ratio of zeros of Laguerre polynomials. Similarly, $J_0(\sqrt{r_i}|\xi|)$ vanishes if and only if $r_j|\xi|$ is a zero of J_0. Then r_1/r_2 is a ratio of zeros of J_0. □

Remark 4 There is a lof of room to strengthen this result. For example, one may consider f in more general function spaces. This is the topic of a future work.

4 Intertwining Differential Operators

Note that the standard left-invariant vector fields on $\mathbb{H}$ (or on $\overline{\mathbb{H}}$) are

$$X := \partial_x - \frac{1}{2} y \partial_t \qquad Y := \partial_y + \frac{1}{2} x \partial_t \qquad T := \partial_t .$$

Also, the standard right-invariant vector fields are

$$\tilde{X} := \partial_x + \frac{1}{2} y \partial_t \qquad \tilde{Y} := \partial_y - \frac{1}{2} x \partial_t \qquad \tilde{T} := \partial_t .$$

Let $-\mathcal{L} := X^2 + Y^2$, be the left sub-laplacian, and $-\tilde{\mathcal{L}} := \tilde{X}^2 + \tilde{Y}^2$ be the right sub-laplacian. Also let $\Box_r = \mathcal{L} + rT^2$ on $\overline{\mathbb{H}}$.

Proposition 2 *For any right-invariant vector field* $\tilde{V}$ *and* $f \in S(\overline{\mathbb{H}})$ *we have*

$$I_r \left(\tilde{V} f \right) (z, t) = \tilde{V} I_r f (z, t).$$

Proof Idea I_r is a group convolution operator by a compactly supported measure on $\overline{\mathbb{H}}$:

$$I_r f = \mu_r *_{\mathbb{H}} f$$

where

$$\int_{\overline{\mathbb{H}}} f d\mu = \int_0^{2\pi\sqrt{ab}} f(\gamma_r(s))\, ds = I_r f(0).$$

And for a right-invariant vector field on $\overline{\mathbb{H}}$,

$$\tilde{V} (\mu_r *_{\mathbb{H}} f) = \mu_r *_{\mathbb{H}} \left(\tilde{V} f \right).$$

□

Corollary 5 *The X-ray transform* I_r *commutes with* T *and the right sub-laplacian:*

$$I_r (T f) = T (I_r f) \qquad I_r \left(\tilde{\mathcal{L}} f \right) = \tilde{\mathcal{L}} (I_r f) .$$

We use these facts to prove the most interesting intertwining property of I_r:

Theorem 4 *I_r intertwines the left sublaplacian $\mathcal{L}$ with the operator $\Box_r$ on $\overline{\mathbb{H}}$:*

$$I_r\,(\mathcal{L}f) = \Box_r\,(I_r f)\,, \quad f \in S(\overline{\mathbb{H}}).$$

We first note the helical symmetry of I_r.

Lemma 1 *Define the rotation map $\mathcal{R}_\theta^* f(z,t) = f(e^{i\theta}z, t)$. Then a straightforward computation gives*

$$I_r\left(\mathcal{R}_\theta^* f\right) = \mathcal{R}_\theta^* I_r f\left(z, t - \tfrac{1}{2}r\theta\right)$$

for $f \in C_c^\infty(G)$. Differentiating in θ yields

$$I_r(\partial_\theta f)(z,t) = (\partial_\theta - \tfrac{1}{2}rT)I_r f(z,t).$$

Proof of Theorem 4 A straightforward computation gives

$$\tilde{\mathcal{L}} - \mathcal{L} = 2\partial_\theta T.$$

Thus

$$I_r\left((\mathcal{L} - \tilde{\mathcal{L}})f\right)(z,t) = -2\left(\partial_\theta - \tfrac{1}{2}rT\right)T I_r f(z,t),$$

so that

$$\begin{aligned} I_r\,(\mathcal{L}f) &= I_r\left(\tilde{\mathcal{L}}f\right) + (-2\partial_\theta + r)\,T I_r f = \tilde{\mathcal{L}}\,(I_r f) + (-2\partial_\theta + r)\,T I_r f \\ &= \left(\tilde{\mathcal{L}} - 2\partial_\theta T + rT^2\right)(I_r f) = \left(\mathcal{L} + rT^2\right)(I_r f) \\ &=: \Box_r\,(I_r f) \end{aligned}$$

as desired. □

Remark 5 The functions ψ_{jk}^n are joint eigenfunctions of $\mathcal{L}$ and T (see 5). Indeed

$$\begin{cases} \mathcal{L}\psi_{jk}^n &= 2|n|(1+2j)\,\psi_{jk}^n \\ T\psi_{jk}^n &= 2in\,\psi_{jk}^n. \end{cases}$$

So

$$\Box_r \psi_{jk}^n = 2|n|\,(1 + 2(j - |n|r))\,\psi_{jk}^n.$$

What's more the functions ψ^n_{jk} are eigenfunctions of the right sublaplacian $\tilde{\mathcal{L}}$. Indeed

$$\tilde{\mathcal{L}}\psi^n_{jk} = 2|n|(1+2k)\psi^n_{jk}.$$

5 Ongoing Work

1. Explicit Inversion formulas. The two-radius theorem implies that a function $f \in L^2(\overline{\mathbb{H}})$ may be recovered form its X-ray transform If. We can therefore write down an inversion formula in terms of the singular values, but it is desirable to have a closed form inversion formula.
2. Chacterize the zeros of the diagonal Laguerre functions. The following conjecture, if true, would imply a one-radius theorem.

   Conjecture 1 The only $(j, n) \in \mathbb{N} \times \mathbb{Z}^*$ for which $l_j(n) := L^n_j(n) = 0$ is $(j, n) = (2, 2)$.

3. Stability estimates. We would like to know if the recovery of f from If is stable. That is, for a suitable choice of Sobolev scale $H^s(\overline{\mathbb{H}})$ and $H^s(\overline{\mathbb{H}} \times \mathbb{Q}^+)$ for which

   $$\|f\|_{H^s} \leq C\|If\|_{H^{s'}}$$

   We expect that, with respect to a natural choice of Sobolev scale based on the intertwining properties above, that I is one half smoothening. That is $s' = s + \frac{1}{2}$.

- Non-Euclidea metrics on the base. The geodesics in (1) are determined by lifting the Eudliean metric $g = dx^2 + dy^2$ from $\mathbb{R}^2$ to the horizontal distribution of $\overline{\mathbb{H}}$. What can we say about the X-ray transform associated to the lift of a non-Euclidean metric from $\mathbb{R}^2$ to a sub-Riemannian metric on the Heisenberg group?
- Can we obtain a similar explicit spectral decomposition for the X-ray transform on more general Carnot groups such as the Engel group of free Nilpotent groups? Can we state more general two-radius theorems in these cases?

Appendix

The proofs of the following results may be found, for example, in [4, 8]

Entry Functions on the Heisenberg Group

Consider the family of functions, called *matrix coefficients* or *entry functions* defined on the Heisenberg group:

$$M_{jk}^{h}(z,t)=\begin{cases}\sqrt{\frac{k!}{j!}}\left(+\sqrt{h}z\right)^{j-k}L_{k}^{(j-k)}\left(h|z|^{2}\right)e^{-h|z|^{2}/2}e^{2iht} & j\geq k\\ \sqrt{\frac{j!}{k!}}\left(-\sqrt{h}\overline{z}\right)^{k-j}L_{j}^{(k-j)}\left(h|z|^{2}\right)e^{-h|z|^{2}/2}e^{2iht} & j\leq k\end{cases}, \tag{4}$$

and $M_{jk}^{h}(z,t)=M_{jk}^{|h|}(\overline{z},-t)$ for $h<0$.

The matrix coefficients are joint eigenfunctions of the left sublaplacian $\mathcal{L}$ and the Reed vector field T.

$$\begin{cases}\mathcal{L}M_{jk}^{h}(z,t)=2|h|\,(1+2j)\,M_{jk}^{h}(z,t)\\ TM_{jk}^{h}(z,t)=2ihM_{jk}^{h}(z,t).\end{cases} \tag{5}$$

Lemma 2 (Useful Properties on M_{jk}^{h}) *Let $h>0$*

(1) For all $j,k=0,1,2,\ldots$

$$M_{jk}^{h}(z,t)=M_{jk}^{1}(\sqrt{h}z,ht) \tag{6}$$

$$M_{jk}^{-h}(z,t)=M_{jk}^{h}(\overline{z},-t)$$

(2) For all $j,j',k,k'=0,1,2,\ldots$

$$\int_{\mathbb{C}}M_{jk}^{h}(z,0)\overline{M}_{j'k'}^{h}(z,0)dz=\frac{\pi}{h}\delta_{jj'}\delta_{kk'}. \tag{7}$$

(3) For any $m,n\in\mathbb{Z}^{*}$ and $j,k=0,1,2,\ldots$

$$\int_{0}^{2\pi}M_{jk}^{n}(e^{ims},s/2)ds=2\pi\,\delta\left(|n|+m(j-k)\right)M_{jk}^{n}(1,0). \tag{8}$$

(4) For all $j,k=0,1,2,\ldots$

$$M_{jk}^{h}\left((w,s)(z,t)\right)=\sum_{l=0}^{\infty}M_{jl}^{h}(z,t)M_{lk}^{h}(w,s),\quad (w,s),(z,t)\in\mathbb{H}. \tag{9}$$

Definition 3 Define the rescaled functions

$$\psi_{jk}^{n}:=\frac{\sqrt{|n|}}{\pi}M_{jk}^{n} \tag{10}$$

Since $M^n_{jk}(z,t) = e^{2int} M^n_{jk}(z,0)$ we see that t may be taken mod π. Thus ψ^n_{jk} is a function on $L^2(\overline{\mathbb{H}})$.

As a consequence of Lemma 2, the set $\{\psi^n_{jk}\}$ is orthonormal. In fact, more is true [8]:

Theorem 5 *The functions* $\{\psi^n_{jk} : n \in \mathbb{Z}^*, j,k = 0,1,2,...\}$ *are an orthonormal basis for* ${}^0L^2(\overline{\mathbb{H}})$.

Special Functions

$$\sum_{j=0}^{\infty} t^j L_j^{(\alpha)}(x) = \frac{1}{(1-t)^{\alpha+1}} e^{-xt/(1-t)}, \quad |t| < 1. \tag{11}$$

$$J_0(r) = \frac{1}{2\pi} \int_0^{2\pi} e^{ir\cos\theta} d\theta \tag{12}$$

Acknowledgments Support from the Research Project Grant RPG-2020-037: *Quantum Limits for Sub-elliptic Operators* (PI: Véronique Fischer; CoPI: Clotilde Fermanian-Kammerer) is gratefully acknowledged.

References

1. M. Agranovsky, et al., Injectivity of the pompeiu transform in the heisenberg group. J. d'Analyse Math. **63**(1), 131–173 (1994). ISSN: 0021-7670, 1565-8538. https://doi.org/10.1007/BF03008422
2. J. Delsarte, K. B. Vedak, Lectures on topics in mean periodic functions and the two-radius theorem, Vol. 22 (Tata Institute of Fundamental Research, Bombay, 1961).
3. S. Flynn, Injectivity of the Heisenberg X-ray transform. J. Funct. Anal. **280**(5), 108886 (2021). ISSN: 0022-1236. https://doi.org/10.1016/j.jfa.2020.108886
4. G.B. Folland, *Harmonic Analysis in Phase Space*. The Annals of Mathematics Studies no. 122 (Princeton University Press, Princeton, 1989), 277pp. ISBN: 978-0-691-08527-2; 978-0-691-08528-9
5. J. Ilmavirta, O. Koskela, J. Railo, Torus computed tomography. SIAM J. Appl. Math. **80**(4), 1947–1976 (2020). ISSN: 0036-1399, 1095-712X. https://doi.org/10.1137/19M1268070. arXiv:1906.05046
6. R.K. Mishra, F. Monard, Range characterizations and singular value decomposition of the geodesic x-ray transform on disks of constant curvature. J. Spectr. Theory **11**(3), 1005–1041 (2021). ISSN: 1664-039X. https://doi.org/10.4171/JST/364
7. R. Montgomery, *A Tour of Subriemannian Geometries, Their Geodesics and Applications*. Mathematical Surveys and Monographs, vol. 91 (American Mathematical Society, Providence, 2006). ISBN: 978-0-8218-4165-5; 978-1-4704-1318-7. https://doi.org/10.1090/surv/091
8. S. Thangavelu, *Harmonic Analysis on the Heisenberg Group* (Birkhäuser, Boston, 1998). ISBN: 978-1-4612-7275-5; 978-1-4612-1772-5. https://doi.org/10.1007/978-1-4612-1772-5

Nonlocal Functionals with Non-standard Growth

Minhyun Kim

Abstract In this note, we review recent progress on the De Giorgi–Nash–Moser theory for nonlocal functionals with non-standard growth, which include functionals with Orlicz growth, variable exponents, double phase growth, and orthotropic structure. Some open problems are suggested.

Keywords De Giorgi–Nash–Moser theory · Nonlocal functional · Non-standard growth

2020 Mathematics Subject Classification 35B65; 35A15; 47G20

1 Introduction

In this note, we summarize recent progress and suggest some open problems in the De Giorgi–Nash–Moser theory for nonlocal functionals[1] modeled on

$$(1-s)\iint \frac{|u(x)-u(y)|^p}{|x-y|^{n+sp}}\,\mathrm{d}y\,\mathrm{d}x, \tag{1}$$

where $n \in \mathbb{N}$, $s \in (0,1)$, and $p \in (1,\infty)$, and for their Euler–Lagrange equations. Since it is known [4] that (1) converges to the local functional

$$\int |Du|^p\,\mathrm{d}x \tag{2}$$

[1] Functionals of the form (1) are called nonlocal functionals since the Euler–Lagrange equation of (1) is given by the fractional p-Laplace equation $(-\Delta_p)^s u = 0$, which is a nonlocal operator.

M. Kim (✉)
Department of Mathematics and Research Institute for Natural Sciences, Hanyang University, Seoul, South Korea
e-mail: minhyun@hanyang.ac.kr

D. Cardona et al. (eds.), *Extended Abstracts 2021/2022*, Research Perspectives Ghent Analysis and PDE Center 3, https://doi.org/10.1007/978-3-031-48579-4_3

in various senses as $s \nearrow 1$, we are particularly interested in regularity estimates that are robust in the sense that the constants in the estimates stay uniform as $s \nearrow 1$. Through this, the theories of local and nonlocal functionals are unified. Since the theory for nonlocal functionals with standard growth (1) is completed (see, for instance, [15, 18, 19, 30, 31]), we focus on the non-standard growth cases.

Let us recall that the De Giorgi–Nash–Moser theory for the functional (2) has been generalized to that for a more general class of local functionals of the form

$$\int F(x, Du)\,\mathrm{d}x, \tag{3}$$

see a survey [37]. The most fundamental example of (3) is, of course, a functional with standard growth $F(x,\xi) \sim |\xi|^p$, but there is a number of interesting examples of functionals with non-standard growth as listed below: $F(x,\xi) = G(|\xi|)$, $F(x,\xi) = |\xi|^{p(x)}$, $F(x,\xi) = |\xi|^p + a(x)|\xi|^q$, and $F(x,\xi) = \sum_{k=1}^n |\xi_k|^{p_i}$. Moreover, these local functionals have recently been extended to nonlocal functionals with non-standard growth. In the following four sections, we study the De Giorgi–Nash–Moser theory for nonlocal functionals corresponding to each of these examples.

2 Nonlocal Functionals with Orlicz Growth

Let G be an Orlicz function with $G(1) = 1$. As a nonlocal analogue of the functional

$$\int G(|Du|)\,\mathrm{d}x, \tag{4}$$

we consider

$$(1-s) \iint G\left(\frac{|u(x)-u(y)|}{|x-y|^s}\right) \frac{\mathrm{d}y\,\mathrm{d}x}{|x-y|^n}. \tag{5}$$

It is a natural extension in the sense that (5) converges (and Γ-converges) to (4), up to constants, as $s \nearrow 1$ (see [1, 26]).

The De Giorgi–Nash–Moser theory for the functional (5), and for more general functionals with measurable coefficients, is more or less complete. Let us summarize results known in the literature. Throughout this section, we fix an open set Ω in $\mathbb{R}^n$, let $0 < s_0 \leq s < 1 < p \leq q$, and assume that G satisfies

$$pG(t) \leq tg(t) \leq qG(t) \quad \text{for } t \geq 0, \tag{6}$$

where $g = G'$. The Orlicz–Sobolev space $V^{s,G}(\Omega|\mathbb{R}^n)$ is defined by

$$V^{s,G}(\Omega|\mathbb{R}^n) = \left\{ u \in L^0(\mathbb{R}^n) : u|_\Omega \in L^G(\Omega), \iint_{(\Omega^c \times \Omega^c)^c} G\left(\frac{|u(x) - u(y)|}{|x - y|^s}\right) \frac{\mathrm{d}y\,\mathrm{d}x}{|x - y|^n} < \infty \right\},$$

where $L^0(\mathbb{R}^n)$ denotes the space of all measurable functions on $\mathbb{R}^n$. The tail term in this framework is defined by

$$\mathrm{Tail}_g(u; x_0, R) = R^s g^{-1}\left((1 - s)R^s \int_{\mathbb{R}^n \setminus B_R(x_0)} g\left(\frac{|u(y)|}{|y - x_0|^s}\right) \frac{\mathrm{d}y}{|y - x_0|^{n+s}}\right).$$

We begin with the local boundedness and the weak Harnack inequality.

Theorem 1 (Local Boundedness [13, Theorem 3.1]) *If $u \in V^{s,G}(\Omega|\mathbb{R}^n)$ is a local sub-minimizer of (5), then u is locally bounded from above. Moreover, for any $B_R = B_R(x_0) \subset \Omega$ and $\delta \in (0, 1)$, it holds that*

$$\sup_{B_{R/2}} G\left(\frac{u_+}{R^s}\right) \le C \fint_{B_R} G\left(\frac{u_+}{R^s}\right) \mathrm{d}x + \delta G\left(\frac{\mathrm{Tail}_g(u_+; x_0, R/2)}{(R/2)^s}\right),$$

where $C = C(n, s_0, q, \delta) > 0$ is a constant.

We remark that we only need the second inequality of (6) as an assumption in Theorem 1. See [5, 12] for results similar to Theorem 1.

Theorem 2 (Weak Harnack Inequality [13, Theorem 4.1]) *If $u \in V^{s,G}(\Omega|\mathbb{R}^n)$ is a local super-minimizer of (5) which is nonnegative in $B_R = B_R(x_0) \subset \Omega$, then*

$$\fint_{B_{R/2}} G^\varepsilon\left(\frac{u}{R^s}\right) \mathrm{d}x \le C \inf_{B_{R/2}} G^\varepsilon\left(\frac{u}{R^s}\right) + C G^\varepsilon\left(\frac{\mathrm{Tail}_g(u_-; x_0, R)}{R^s}\right), \tag{7}$$

where $\varepsilon \in (0, 1)$ and $C > 0$ are constants depending only on n, s_0, p, and q.

In [13, Theorem 4.1], the left-hand side of (7) is given by the averaged integral over B_R. We point out, however, that the integral must be taken over a smaller ball, say $B_{R/2}$, as in (7).

Using Theorems 1 and 2, we obtain the Hölder estimate and the Harnack inequality.

Theorem 3 (Hölder Estimate [12, Theorem 1.1]) *If $u \in V^{s,G}(\Omega|\mathbb{R}^n)$ is a local minimizer of (5), then u is locally Hölder continuous in Ω. Moreover, for any $B_R = B_R(x_0) \subset \Omega$, it holds that*

$$R^\alpha [u]_{C^\alpha(\overline{B_{R/4}})} \le C\|u\|_{L^\infty(B_{R/2})} + \mathrm{Tail}_g(u; x_0, R/2),$$

where $\alpha \in (0, 1)$ and $C > 0$ are constants depending only on n, s_0, p, and q.

Theorem 4 (Harnack Inequality [13, Theorem 1.1]) *If $u \in V^{s,G}(\Omega|\mathbb{R}^n)$ is a local minimizer of (5) which is nonnegative in $B_R = B_R(x_0) \subset \Omega$, then*

$$\sup_{B_{R/2}} u \le C \inf_{B_{R/2}} u + C\mathrm{Tail}_g(u_-; x_0, R)$$

for some $C = C(n, s_0, p, q) > 0$.

Note that Theorems 1–4 are generalizations of the results in [34, 36, 38] for local functionals. See [5, 7] for results similar to Theorems 3 and 4.

Theorem 2 with a small exponent $\varepsilon \in (0, 1)$ is sufficient to prove Theorem 4, but it is of independent interest to find an optimal exponent in Theorem 2. In fact, in the standard growth case ($G(t) = t^p$), the weak Harnack inequality (up to the boundary) with an optimal exponent is useful in the study of harmonic functions, see [33]. However, it is still open in the Orlicz case. We conjecture that if $1 < p \le \min\{q, n/s\}$, then for any $\delta \in (0, \frac{n}{n-sp})$, it holds that

$$⨍_{B_{R/2}} g^\delta\left(\frac{u}{R^s}\right) \mathrm{d}x \le C \inf_{B_{R/2}} g^\delta\left(\frac{u}{R^s}\right) + Cg^\delta\left(\frac{\mathrm{Tail}_g(u_-; x_0, R)}{R^s}\right), \tag{8}$$

where C depends on δ as well. To the best of our knowledge, the Moser iteration technique in this framework has not been developed yet. To obtain (8) from (7), one may need to investigate the Moser iteration method.

3 Nonlocal Functionals with Variable Exponent

In this section, we consider a functional

$$(1-s) \iint \left(\frac{|u(x) - u(y)|}{|x-y|^s}\right)^{p(x,y)} \frac{\mathrm{d}y\,\mathrm{d}x}{|x-y|^n}, \tag{9}$$

where $p(x, y)$ is a variable exponent satisfying $1 < \inf p \le \sup p < \infty$. It is known [32, Theorem 1.1] that the functional (9) converges to

$$\int 2\pi^{\frac{n-1}{2}} \frac{\Gamma(\frac{\bar{p}(x)+1}{2})}{\Gamma(\frac{n+\bar{p}(x)}{2})} \frac{1}{\bar{p}(x)} |\nabla u(x)|^{\bar{p}(x)} \,\mathrm{d}x,$$

where $\bar{p}(x) = p(x, x)$, under a log-Hölder-type assumption on $p(x, \cdot)$, provided that u is sufficiently smooth, say $u \in C_c^2(\mathbb{R}^n)$. Interestingly, there exists a smooth variable exponent p such that the convergence may fail for $u \in W^{1,\bar{p}(\cdot)}$, see [32, Corollary 1.3]. It is worth investigating under what conditions on p the convergence might be true for all $u \in W^{1,\bar{p}(\cdot)}$, or more generally, $u \in L^{\bar{p}(\cdot)}$.

The De Giorgi–Nash–Moser theory for the functional (9), and for more general functionals with measurable coefficients, is partially established. For instance, the local boundedness and the Hölder estimate are known. In what follows, Ω denotes an open subset of $\mathbb{R}^n$ and the function space $V^{s,p(\cdot,\cdot)}(\Omega|\mathbb{R}^n)$ is defined by

$$V^{s,p(\cdot,\cdot)}(\Omega|\mathbb{R}^n) = \left\{ u \in L^0(\mathbb{R}^n) : u|_\Omega \in L^{\bar{p}(\cdot)}(\Omega), \iint_{(\Omega^c\times\Omega^c)^c} \frac{|u(x)-u(y)|^{p(x,y)}}{|x-y|^{n+sp(x,y)}}\,dy\,dx < \infty \right\}.$$

The following theorems are nonlocal analogues of the results in [24].

Theorem 5 (Local Boundedness [10, 39]) *Suppose that p is continuous and symmetric. If $u \in V^{s,p(\cdot,\cdot)}(\Omega|\mathbb{R}^n)$ satisfies*

$$\sup_{x\in\Omega} \int_{\mathbb{R}^n} \frac{u_+^{p(x,y)-1}(y)}{(1+|y|)^{n+sp(x,y)}}\,dy < \infty \tag{10}$$

and is a local sub-minimizer of (9), then u is locally bounded from above in Ω.

In [10], the authors also obtain the following uniform estimate: for each $x_0 \in \Omega$ with $p(x_0,x_0) \le n/s$, there is a radius $R \in (0,1)$ such that $B_R = B_R(x_0) \subset \Omega$, $p_+ < p_-^* := np_-/(n-\sigma p_-)$, and

$$\sup_{B_{R/2}} u \le C\left(\int_{B_R} u_+^{p_+}\,dx\right)^{\frac{1}{p_+}} + \left(\sup_{x\in B_R}\int_{\mathbb{R}^n\setminus B_{R/2}} \frac{u_+^{p(x,y)-1}(y)}{|y-x_0|^{n+sp(x,y)}}\,dy\right)^{\frac{1}{p_+-1}} + 1 \tag{11}$$

for any $\sigma \in (0,s)$, where $p_+ = \sup_{x,y\in B_R} p(x,y)$ and $p_- = \inf_{x,y\in B_R} p(x,y)$.

Theorem 6 (Hölder Estimate [39]) *Suppose that p is symmetric and satisfies*

$$\sup_{B_r(x_0)\subset\Omega}\ \sup_{x_1,x_2,y_1,y_2\in B_r(x_0)} |p(x_1,y_1)-p(x_2,y_2)| \le \frac{L}{\log(1/r)} \quad \forall r < 1/2$$

for some $L > 0$. If $u \in V^{s,p(\cdot,\cdot)}(\Omega|\mathbb{R}^n)$ satisfies (10) and is a local minimizer of (9), then u is locally Hölder continuous in Ω.

Uniform Hölder estimates are known in [10] under an additional assumption on p: assume in addition

$$\sup_{x\in B_r, y\in B_r^c} p(x,y) \le \sup_{x,y\in B_r} p(x,y) \quad \text{and} \quad \inf_{x\in B_r, y\in B_r^c} p(x,y) \le \inf_{x,y\in B_r} p(x,y)$$

for all $B_r = B_r(x_0) \subset\subset \Omega$. Then for each $x_0 \in \Omega$ with $p(x_0, x_0) \leq n/s$, there is a radius $R \in (0, 1)$ such that $B_R = B_R(x_0) \subset\subset \Omega$, $p_+ < p_-^*$, and

$$R^{\alpha}[u]_{C^{\alpha}(\overline{B_{R/2}})} \leq C\|u\|_{L^{\infty}(B_R)} + R^s + 1$$
$$+ \left(R^{s\tilde{p}_+} \sup_{x \in B_{3R/4}} \int_{\mathbb{R}^n \setminus B_R} \frac{|u(y)|^{p(x,y)-1}}{|y - x_0|^{n+sp(x,y)}} \, dy \right)^{\frac{1}{\tilde{p}_+ - 1}} \tag{12}$$

for any $\sigma \in (0, s)$, where $\tilde{p}_+ = \sup_{y \in B_R^c} p(x_0, y)$. The constants α and C depend only on n, s, σ, $\sup_{x \in \Omega, y \in \mathbb{R}^n} p(x, y)$, $\inf_{x \in \Omega, y \in \mathbb{R}^n} p(x, y)$, R, and L.

Let us make some remarks. The robustness of uniform estimates (11) and (12) as $s \nearrow 1$ is nowhere written, but the author believes that these estimates can be made robust by investigating the dependence of constants on s. Note that a more general class of functionals allowing a variable order of differentiability $s(x, y)$ instead of a constant order s is studied in [39]. Finally, we point out that the weak Harnack inequality and the Harnack inequality remain open.

4 Nonlocal Double Phase Functionals

Another example of a nonlocal functional with non-standard growth is given by

$$\iint \left((1 - s)\frac{|u(x) - u(y)|^p}{|x - y|^{n+sp}} + a(x, y)(1 - t)\frac{|u(x) - u(y)|^q}{|x - y|^{n+tq}} \right) dy \, dx, \tag{13}$$

where $0 < s \leq t < 1 < p \leq q < \infty$ and a is a measurable function satisfying $0 \leq a(x, y) = a(y, x) \leq \|a\|_{L^{\infty}(\mathbb{R}^n)} < \infty$ for all $x, y \in \mathbb{R}^n$. The convergence of (13) to the (local) double phase functional as $s = t \nearrow 1$ is partially known: see [27, Theorem 5.73] for the case $q = 2$ and [28] for the translation-invariant case, i.e., $a(x, y) = a(x + z, y + z)$ for all z.

As a generalization of [14, 16], the De Giorgi–Nash–Moser theory for the nonlocal functional (13) is studied in [6]. To state the main results in [6], let us recall the function spaces

$$\mathcal{A}(\Omega) = \left\{ u \in L^0(\mathbb{R}^n) : u|_{\Omega} \in L^p(\Omega), \right.$$
$$\left. \iint_{(\Omega^c \times \Omega^c)^c} H(x, y, |u(x) - u(y)|) \, dy \, dx < \infty \right\}$$

and

$$L^{q-1}_{sp}(\mathbb{R}^n) = \left\{ u \in L^0(\mathbb{R}^n) : \int_{\mathbb{R}^n} \frac{|u(x)|^{q-1}}{(1 + |x|)^{n+sp}} \, dx < \infty \right\},$$

where Ω denotes an open subset of $\mathbb{R}^n$ and

$$H(x, y, \tau) = \frac{\tau^p}{|x-y|^{sp}} + a(x, y)\frac{\tau^q}{|x-y|^{tq}}.$$

We refer the reader to [17, 25, 41] for the regularity theory when $t \le s$.

Theorem 7 (Local Boundedness [6, Theorem 1.1]) *Assume $q \le np/(n-sp)$ when $sp < n$. If $u \in \mathcal{A}(\Omega) \cap L^{q-1}_{sp}(\mathbb{R}^n)$ is a local minimizer of (13), then u is locally bounded in Ω.*

Theorem 8 (Hölder Estimate [6, Theorem 1.2]) *Assume that a satisfies*

$$|a(x_1, y_1) - a(x_2, y_2)| \le [a]_\alpha(|x_1 - x_2| + |y_1 - y_2|)^\alpha \quad \forall (x_1, y_1), (x_2, y_2) \in \mathbb{R}^n \times \mathbb{R}^n$$

for some $\alpha > 0$, and that $tq \le sp + \alpha$. If $u \in \mathcal{A}(\Omega) \cap L^{q-1}_{sp}(\mathbb{R}^n)$ is a local minimizer of (13) which is locally bounded in Ω, then u is locally Hölder continuous in Ω.

A couple of remarks are in order. First, the uniform regularity estimates in Theorems 7 and 8 are hidden in their proofs, and the robustness of the estimates are not investigated. It would be interesting to obtain robust estimates. Second, the weak Harnack inequality and Harnack inequality for nonlocal functional (13) are open. Note that the Harnack inequality is known for local double phase functional [2] and for mixed local and nonlocal double phase functionals [8].

5 Nonlocal Functionals with Orthotropic Structure

In this section, we consider a nonlocal functional

$$\iint |u(x) - u(y)|^p \, \mu(x, \mathrm{d}y) \, \mathrm{d}x, \tag{14}$$

where μ is given by a sum of singular measures

$$\mu(x, \mathrm{d}y) = \sum_{k=1}^n \frac{1 - s_k}{|x_k - y_k|^{1+s_k p}} \, \mathrm{d}y_k \prod_{i \ne k} \delta_{x_i}(\mathrm{d}y_i).$$

Here, $s_k \in (0, 1)$ for $k = 1, \dots, n$. Since it is the sum of one-dimensional nonlocal functionals with standard growth, it is obvious that the functional (14) with $s = s_1 = \cdots = s_n$ converges to

$$\int \sum_{k=1}^n |D_k u|^p \, \mathrm{d}x$$

as $s \nearrow 1$.

For the functional (14) and more general functionals with measurable coefficients, the De Giorgi–Nash–Moser theory is partially known [9, 11, 23]. Throughout this section, we assume that $s_1, \dots, s_n \in [s_0, 1)$ for some $s_0 \in (0, 1)$ and $p \in (1, n/\bar{s})$, where $\bar{s} = n(\sum 1/s_k)^{-1}$ is the harmonic mean of the orders of differentiability $s_1, \dots, s_n$. Let $\Omega \subset \mathbb{R}^n$ be open and define $V^{\mu}(\Omega|\mathbb{R}^n)$ by

$$V^{\mu}(\Omega|\mathbb{R}^n) = \Big\{ u \in L^0(\mathbb{R}^n) : u|_{\Omega} \in L^p(\Omega), \\ \int_{\Omega}\int_{\mathbb{R}^n} |u(x) - u(y)|^p \, \mu(x, \mathrm{d}y) \, \mathrm{d}x < \infty \Big\}.$$

We also consider rectangles

$$M_R(x) = \prod_{k=1}^{n} \left(x_k - R^{s_{\max}/s_k}, x_k + R^{s_{\max}/s_k} \right),$$

where $s_{\max} = \max\{s_1, \dots, s_n\}$, which take the anisotropy of the functional (14) into account.

Theorem 9 (Weak Harnack Inequality [11, Theorem 1.4]) *If $u \in V^{\mu}(\Omega|\mathbb{R}^n)$ is a local super-minimizer of (14) which is nonnegative in $M_R = M_R(x_0) \subset \Omega$, then*

$$\left(\fint_{M_{R/2}} u^{\varepsilon} \, \mathrm{d}x \right)^{1/\varepsilon} \le C \inf_{M_{R/4}} u + C \left(R^{s_{\max} p} \sup_{x \in M_{\frac{15R}{16}}} \int_{\mathbb{R}^n \setminus M_R} u_-^{p-1}(y) \, \mu(x, \mathrm{d}y) \right)^{\frac{1}{p-1}},$$

where $\varepsilon \in (0, 1)$ and $C > 0$ are constants depending only on n, s_0, and p.

Theorem 10 (Hölder Estimate [11, Theorem 1.5]) *If $u \in V^{\mu}(\Omega|\mathbb{R}^n)$ is a local minimizer of (14), then u is locally Hölder continuous in Ω. Moreover, for any $M_R = M_R(x_0) \subset \Omega$, it holds that*

$$R^{\alpha} [u]_{C^{\alpha}(\overline{M_{R/2}})} \le C \|u\|_{L^{\infty}(\mathbb{R}^n)}, \tag{15}$$

where $\alpha \in (0, 1)$ and $C > 0$ are constants depending only on n, s_0, and p.

The local boundedness for local minimizers of (14) is open. Moreover, there may be a room to improve the uniform estimate (15) by considering a local L^{∞}-norm and the tail term on the right-hand side instead of the global L^{∞}-norm. However, it is known that one cannot expect the Harnack inequality. See [3] for a counterexample.

In full generality, one may consider a nonlocal functional

$$\int_{\Omega} \sum_{k=1}^{n} (1 - s_k) \int_{\mathbb{R}} \frac{|u(x + re_k) - u(x)|^{p_k}}{r^{1+s_k p_k}} \, \mathrm{d}r \, \mathrm{d}x, \tag{16}$$

where $s_k \in (0, 1)$ and $p_k \in (1, \infty)$ for $k = 1, \dots, n$. We remark that the regularity theory for the functional (16) is widely open. Even for a local functional

$$\int \sum_{k=1}^{n} |D_k u|^{p_k} \, dx,$$

only a few partial results are known. See, for instance, [20–22, 29, 35, 40].

Acknowledgments The research is financially supported by the National Research Foundation of Korea (NRF) grant funded by the Korea government (MSIT) (RS-2023-00252297).

References

1. A. Alberico, A. Cianchi, L. Pick, L. Slavíková, On the limit as $s \to 1^-$ of possibly non-separable fractional Orlicz-Sobolev spaces. Atti Accad. Naz. Lincei Rend. Lincei Mat. Appl. **31**(4), 879–899 (2020)
2. P. Baroni, M. Colombo, G. Mingione, Harnack inequalities for double phase functionals. Nonlinear Anal. **121**, 206–222 (2015)
3. R.F. Bass, Z.Q. Chen, Regularity of harmonic functions for a class of singular stable-like processes. Math. Z. **266**(3), 489–503 (2010)
4. J. Bourgain, H. Brezis, P. Mironescu, Another look at Sobolev spaces, in *Optimal Control and Partial Differential Equations* (IOS, Amsterdam, 2001), pp. 439–455
5. S.S. Byun, H. Kim, J. Ok, Local Hölder continuity for fractional nonlocal equations with general growth. Math. Ann. **387**(1-2), 807–846 (2023)
6. S.S. Byun, J. Ok, K. Song, Hölder regularity for weak solutions to nonlocal double phase problems. J. Math. Pures Appl. **168**, 110–142 (2022)
7. S.S. Byun, H. Kim, K. Song, Nonlocal Harnack inequality for fractional elliptic equations with Orlicz growth. to appear in Bull. Lond. Math. Soc. (2023)
8. S.S. Byun, H.S. Lee, K. Song, Regularity results for mixed local and nonlocal double phase functionals (2023). arXiv:2301.06234
9. J. Chaker, M. Kassmann, Nonlocal operators with singular anisotropic kernels. Commun. Partial Differ. Equ. **45**(1), 1–31 (2020)
10. J. Chaker, M. Kim, Local regularity for nonlocal equations with variable exponents (2021). arXiv:2107.06043
11. J. Chaker, M. Kim, Regularity estimates for fractional orthotropic p-Laplacians of mixed order. Adv. Nonlinear Anal. **11**(1), 1307–1331 (2022)
12. J. Chaker, M. Kim, M. Weidner, Regularity for nonlocal problems with non-standard growth. Calc. Var. Partial Differ. Equ. **61**(6), paper no. 227, 31 (2022)
13. J. Chaker, M. Kim, M. Weidner, Harnack inequality for nonlocal problems with non-standard growth. Math. Ann. **386**(1–2), 533–550 (2023)
14. M. Colombo, G. Mingione, Bounded minimisers of double phase variational integrals. Arch. Ration. Mech. Anal. **218**(1), 219–273 (2015)
15. M. Cozzi, Regularity results and Harnack inequalities for minimizers and solutions of nonlocal problems: a unified approach via fractional De Giorgi classes. J. Funct. Anal. **272**(11), 4762–4837 (2017)
16. G. Cupini, P. Marcellini, E. Mascolo, Local boundedness of minimizers with limit growth conditions. J. Optim. Theory Appl. **166**(1), 1–22 (2015)

17. C. De Filippis, G. Palatucci, Hölder regularity for nonlocal double phase equations. J. Differ. Equ. **267**(1), 547–586 (2019)
18. A. Di Castro, T. Kuusi, G. Palatucci, Nonlocal Harnack inequalities. J. Funct. Anal. **267**(6), 1807–1836 (2014)
19. A. Di Castro, T. Kuusi, G. Palatucci, Local behavior of fractional p-minimizers. Ann. Inst. H. Poincaré C Anal. Non Linéaire **33**(5), 1279–1299 (2016)
20. E. Dibenedetto, U. Gianazza, V. Vespri, Remarks on local boundedness and local Hölder continuity of local weak solutions to anisotropic p-Laplacian type equations. J. Elliptic Parabol. Equ. **2**(1–2), 157–169 (2016)
21. F.G. Düzgün, P. Marcellini, V. Vespri, An alternative approach to the Hölder continuity of solutions to some elliptic equations. Nonlinear Anal. **94**, 133–141 (2014)
22. F.G. Düzgün, P. Marcellini, V. Vespri, Space expansion for a solution of an anisotropic p-Laplacian equation by using a parabolic approach. Riv. Math. Univ. Parma **5**(1), 93–111 (2014)
23. B. Dyda, M. Kassmann, Regularity estimates for elliptic nonlocal operators. Anal. PDE **13**(2), 317–370 (2020)
24. X. Fan, D. Zhao, A class of De Giorgi type and Hölder continuity. Nonlinear Anal. **36**(3), 295–318 (1999)
25. Y. Fang, C. Zhang, On weak and viscosity solutions of nonlocal double phase equations. Int. Math. Res. Not. IMRN **5**, 3746–3789 (2023)
26. J. Fernández Bonder, A.M. Salort, Fractional order Orlicz-Sobolev spaces. J. Funct. Anal. **277**(2), 333–367 (2019)
27. G.F. Foghem Gounoue, *L^2-Theory for Nonlocal Operators on Domains* (Universität Bielefeld, Bielefeld, 2020)
28. G.F. Foghem Gounoue, A remake of Bourgain-Brezis-Mironescu characterization of Sobolev spaces. Partial Differ. Equ. Appl. **4**(2), paper no. 16, 36 (2023)
29. N. Fusco, C. Sbordone, Local boundedness of minimizers in a limit case. Manuscripta Math. **69**(1), 19–25 (1990)
30. M. Kassmann, A priori estimates for integro-differential operators with measurable kernels. Calc. Var. Partial Differ. Equ. **34**(1), 1–21 (2009)
31. M. Kassmann, A new formulation of Harnack's inequality for nonlocal operators. C. R. Math. Acad. Sci. Paris **349**(11–12), 637–640 (2011)
32. M. Kim, Bourgain, Brezis and Mironescu theorem for fractional Sobolev spaces with variable exponents. Ann. Mat. Pura Appl. (4). **202**(6), 2653–2664 (2023)
33. M. Kim, K.A. Lee, S.C. Lee, The Wiener criterion for nonlocal Dirichlet problems. Commun. Math. Phys. **400**(3), 1961–2003 (2023)
34. G.M. Lieberman, The natural generalization of the natural conditions of Ladyzhenskaya and Ural'tseva for elliptic equations. Commun. Partial Differ. Equ. **16**(2–3), 311–361 (1991)
35. V. Liskevich, I.I. Skrypnik, Hölder continuity of solutions to an anisotropic elliptic equation. Nonlinear Anal. **71**(5–6), 1699–1708 (2009)
36. E. Mascolo, G. Papi, Harnack inequality for minimizers of integral functionals with general growth conditions. NoDEA Nonlinear Differ. Equ. Appl. **3**(2), 231–244 (1996)
37. G. Mingione, Regularity of minima: an invitation to the dark side of the calculus of variations. Appl. Math. **51**(4), 355–426 (2006)
38. G. Moscariello, L. Nania, Hölder continuity of minimizers of functionals with nonstandard growth conditions. Ricerche Mat. **40**(2), 259–273 (1992)
39. J. Ok, Local Hölder regularity for nonlocal equations with variable powers. Calc. Var. Partial Differ. Equ. **62**(1), paper no. 32, 31 (2023)
40. S. Piro-Vernier, F. Ragnedda, V. Vespri, Hölder regularity for bounded solutions to a class of anisotropic operators. Manuscripta Math. **158**(3–4), 421–439 (2019)
41. J.M. Scott, T. Mengesha, Self-improving inequalities for bounded weak solutions to nonlocal double phase equations. Commun. Pure Appl. Anal. **21**(1), 183–212 (2022)

A Variational Approach to the Hot Spots Conjecture

Jonathan Rohleder

Abstract We review a recent new approach to the study of critical points of Laplacian eigenfunctions. Its core novelty is a non-standard variational principle for the eigenvalues of the Laplacians with Neumann and Dirichlet boundary conditions on bounded, simply connected planar domains. This principle can be used to provide simple proofs of some previously known results on the hot spots conjecture.

Keywords Laplacian · Eigenfunctions · Hot spots conjecture · Spectral theory

2020 Mathematics Subject Classification 35P15, 35J25, 35A15

1 Introduction

The *hot spots conjecture* seems to have been formulated first by Rauch in 1974, see Section II.5 in Kawohl's book [14]. It suggests, in its strongest form, that any eigenfunction corresponding to the smallest positive eigenvalue of the Laplacian on a bounded domain with Neumann boundary conditions attains its maximum and minimum only on the boundary. This would imply that the hottest and coldest spots in an insulated body with a "generic" initial heat distribution diverge from each other for large time as much as they can, i.e. converge to the boundary; cf. [14].

Let us briefly review the major steps forward on this conjecture. First of all, the conjecture is true for simple shapes on which the eigenfunctions in question can be computed explicitly, such as balls, cubes, equilateral triangles or right isosceles triangles. Moreover, there exist by now a few non-trivial results for domains in the plane. Bañuelos and Burdzy [2] proved that the conjecture is true for obtuse triangles and sufficiently long convex domains with symmetries, and Atar and Burdzy [1] showed it for so-called lip-domains, i.e. domains enclosed by the graphs of two

J. Rohleder (✉)
Matematiska Institutionen, Stockholms Universitet, Stockholm, Sweden
e-mail: jonathan.rohleder@math.su.se

D. Cardona et al. (eds.), *Extended Abstracts 2021/2022*, Research Perspectives Ghent Analysis and PDE Center 3, https://doi.org/10.1007/978-3-031-48579-4_4

Lipschitz continuous functions with Lipschitz constants at most one. Jerison and Nadirashvili proved it for domains symmetric w.r.t. both coordinate axes [9] for which all horizontal and vertical cross sections are intervals. On the other hand Burdzy and Werner [6] and Burdzy [4] constructed counterexamples given by certain multiply-connected domains, see also the numerical study [16]. The most recent advances include a prove for general triangles by Judge and Mondal [10, 11], which was preceeded by Siudeja's partial result [22]. For further recent related results we refer the reader to, e.g., [12, 17, 18, 23, 24]. The conjecture is still open for general simply connected domains in the plane, as well as in higher dimensions, and is believed to be true at least for convex domains.

It seems that most proofs of the key results on the hot spots conjecture are either based on probabilistic methods, exploiting reflected Brownian motion [1–4, 6, 19, 23], or rely on tracing critical points of eigenfunctions under perturbations of the domain [9, 10]. The author of this note is suggesting a completely independent approach [21], inspired by the following classical observation: as is well known, in order to study nodal sets of eigenfunctions, it is convenient to make use of the fact that these eigenfunctions are optimizers of certain variational principles. For instance, the first non-trivial eigenvalue μ_2 of the Neumann Laplacian $-\Delta_{\mathrm{N}}$ on a bounded Lipschitz domain Ω is given by

$$\mu_2 = \min_{\substack{\psi \in H^1(\Omega)\setminus\{0\} \\ \int_\Omega \psi = 0}} \frac{\int_\Omega |\nabla \psi|^2}{\int_\Omega |\psi|^2}, \tag{1}$$

and a function ψ in the Sobolev space $H^1(\Omega)$ with vanishing integral is a minimizer of (1) if and only if ψ is an eigenfunction of $-\Delta_{\mathrm{N}}$ corresponding to μ_2. From this characterization it can be derived easily that ψ has precisely two nodal domains; see, e.g., [7, Chapter VI, § 6]. Therefore in order to study critical points of ψ, it seems natural to search for variational principles in which the gradient of ψ, instead of ψ itself, is a minimizer. The following theorem was obtained in [21].

Theorem 1 *Assume that $\Omega \subset \mathbb{R}^2$ is a bounded, simply connected Lipschitz domain with piecewise C^∞-smooth boundary and that all its corners, if any, are convex. Then*

$$\mu_2 = \min_{u = \binom{u_1}{u_2} \in \mathcal{H}_{\mathrm{N}}\setminus\{0\}} \frac{\int_\Omega \left(|\nabla u_1|^2 + |\nabla u_2|^2\right) - \int_{\partial\Omega} \kappa\left(|u_1|^2 + |u_2|^2\right)}{\int_\Omega \left(|u_1|^2 + |u_2|^2\right)}, \tag{2}$$

where $\mathcal{H}_{\mathrm{N}}$ consists of all vector fields with components in the Sobolev space $H^1(\Omega)$ such that their traces satisfy $\langle u, \nu\rangle = 0$ a.e. on $\partial\Omega$, and κ is the signed curvature on $\partial\Omega$ w.r.t. the outer unit normal ν, defined on all boundary points except corners. The minimizers of (2) are precisely the gradients of eigenfunctions ψ of the Neumann Laplacian corresponding to μ_2.

In the following Sect. 2 we will sketch how this variational principle is obtained; it turns out to be a particular case of a min-max principle for all eigenvalues of the Laplacians with Neumann and Dirichlet boundary conditions. In the final Sect. 3, we explain how several previously known results on the hot spots conjecture can be derived in an elementary fashion from Theorem 1.

2 A Non-standard Variational Principle

In this section we sketch the construction which leads to Theorem 1. The main idea is to construct a self-adjoint operator, acting as the negative Laplacian on vector fields, for which $\nabla\psi$ is an eigenfunction if ψ is a non-constant eigenfunction of the Neumann Laplacian.

Let us briefly fix some notation; for more details we refer the reader to [21]. On a bounded, connected Lipschitz domain $\Omega \subset \mathbb{R}^2$ we denote by $-\Delta_N$ the Neumann Laplacian, i.e.

$$-\Delta_N u = -\Delta u, \quad \mathrm{dom}(-\Delta_N) = \left\{u \in H^1(\Omega) : \Delta u \in L^2(\Omega), \partial_\nu u|_{\partial\Omega} = 0\right\},$$

and by $-\Delta_D$ the Dirichlet Laplacian,

$$-\Delta_D u = -\Delta u, \quad \mathrm{dom}(-\Delta_D) = \left\{u \in H^1(\Omega) : \Delta u \in L^2(\Omega), u|_{\partial\Omega} = 0\right\}.$$

The boundary conditions of the operators have to be understood in an appropriate weak sense; $u|_{\partial\Omega}$ denotes the trace of u on $\partial\Omega$ and $\partial_\nu u|_{\partial\Omega}$ is the derivative of u on the boundary in the direction of the outward pointing normal vector ν. Both operators are unbounded and self-adjoint in $L^2(\Omega)$, and their spectra consist of isolated eigenvalues of finite multiplicities. Let

$$0 = \mu_1 < \mu_2 \le \mu_3 \le \dots \tag{3}$$

be an enumeration of the eigenvalues of $-\Delta_N$ and let

$$\lambda_1 < \lambda_2 \le \lambda_3 \le \dots \tag{4}$$

be the eigenvalues of $-\Delta_D$, both counted according to their multiplicities. The eigenfunctions of $-\Delta_N$ corresponding to $\mu_1 = 0$ are the constant functions.

Let us now assume, in addition, that Ω has a piecewise smooth boundary. In the space $L^2(\Omega)^2$ of square-integrable two-component vector fields on Ω we define the sesquilinear form

$$\mathfrak{a}[u, v] = \int_\Omega \big(\langle\nabla u_1, \nabla v_1\rangle + \langle\nabla u_2, \nabla v_2\rangle\big) - \int_{\partial\Omega} \kappa\langle u, v\rangle, \quad u = \begin{pmatrix} u_1 \\ u_2 \end{pmatrix}, v = \begin{pmatrix} v_1 \\ v_2 \end{pmatrix},$$

with domain

$$\operatorname{dom} \mathfrak{a} := \mathcal{H}_{\mathrm{N}} := \left\{ u \in H^1(\Omega)^2 : \langle u|_{\partial\Omega}, \nu \rangle = 0 \text{ on } \partial\Omega \right\}.$$

Here we denote by κ the signed curvature function along the piecewise smooth curve $\partial\Omega$ w.r.t. the outer unit normal ν, defined on all points of $\partial\Omega$ except possible corners; in particular, κ is bounded and piecewise smooth, with possible jumps at the corners. If Ω is convex then $\kappa(x) \leq 0$ holds for almost all $x \in \partial\Omega$.

The sesquilinear form $\mathfrak{a}$ is symmetric, semi-bounded, densely defined in $L^2(\Omega)^2$ and closed. Hence, by Kato [13, VI, Theorem 2.1] there exists a unique self-adjoint operator A in $L^2(\Omega)^2$ such that $\operatorname{dom} A \subset \operatorname{dom} \mathfrak{a}$ and

$$(Au, v)_{L^2(\Omega)^2} = \mathfrak{a}[u, v], \quad u \in \operatorname{dom} A, v \in \operatorname{dom} \mathfrak{a},$$

where $(\cdot, \cdot)_{L^2(\Omega)^2}$ stands for the standard inner product in the space $L^2(\Omega)^2$. It can be computed that the operator A acts as the Laplacian,

$$Au = \begin{pmatrix} -\Delta u_1 \\ -\Delta u_2 \end{pmatrix}, \quad u \in \operatorname{dom} A,$$

and its domain consists of sufficiently regular vector fields satisfying the conditions

$$\langle u|_{\partial\Omega}, \nu \rangle = 0 \quad \text{and} \quad \partial_1 u_2 - \partial_2 u_1 = 0 \quad \text{on } \partial\Omega,$$

interpreted in a weak sense; cf. [21, Lemma 3.3 and Remark 3.4]. The operator A is intimately related to the operators $-\Delta_{\mathrm{N}}$ and $-\Delta_{\mathrm{D}}$ in the following way. We are imposing here conditions on the domain Ω which make sure that functions in the domains of $-\Delta_{\mathrm{D}}$ and $-\Delta_{\mathrm{N}}$ belong to the Sobolev space $H^2(\Omega)$. We make use of the notation $\nabla^\perp \varphi = (-\partial_2 \varphi, \partial_1 \varphi)^\top$.

Theorem 2 *Assume that $\Omega \subset \mathbb{R}^2$ is a bounded Lipschitz domain with piecewise C^∞-smooth boundary whose corners are convex. Let A be the self-adjoint operator in $L^2(\Omega)^2$ associated with the sesquilinear form $\mathfrak{a}$. Moreover, let the eigenvalues of $-\Delta_{\mathrm{N}}$ be enumerated as in (3) and let $\psi_1, \psi_2, \dots$ form an orthonormal basis of $L^2(\Omega)$ such that $-\Delta_{\mathrm{N}} \psi_k = \mu_k \psi_k$ holds for $k = 1, 2, \dots$; analogously let the eigenvalues of $-\Delta_{\mathrm{D}}$ be enumerated as in (4) and let $\varphi_1, \varphi_2, \dots$ be an orthonormal basis of $L^2(\Omega)$ consisting of corresponding eigenfunctions, $-\Delta_{\mathrm{D}} \varphi_k = \lambda_k \varphi_k$ for all k. Then the following hold.*

(1) *For each $k \geq 2$, $\nabla \psi_k$ is nontrivial, belongs to* $\operatorname{dom} A$, *and satisfies $A \nabla \psi_k = \mu_k \nabla \psi_k$. Moreover, the functions $\frac{1}{\sqrt{\mu_k}} \nabla \psi_k$ form an orthonormal basis of $\nabla H^1(\Omega)$.*
(2) *For each $k \geq 1$, $\nabla^\perp \varphi_k$ is nontrivial, belongs to* $\operatorname{dom} A$, *and satisfies $A \nabla^\perp \varphi_k = \lambda_k \nabla^\perp \varphi_k$. Moreover, the functions $\frac{1}{\sqrt{\lambda_k}} \nabla^\perp \varphi_k$ form an orthonormal basis of $\nabla^\perp H_0^1(\Omega)$.*

In particular, if Ω is simply connected then the spectrum of A coincides with the union of the positive eigenvalues of the Neumann and Dirichlet Laplacians, counted with multiplicities.

For a rigorous proof we refer the reader to [21, Theorem 3.2]. For a sketch, let ψ be any non-constant eigenfunction of $-\Delta_{\mathrm{N}}$ corresponding to an eigenvalue μ and $u = \nabla\psi$. Then u belongs to dom A: for the boundary conditions, note that

$$u \cdot \nu = \nabla\psi \cdot \nu = 0 \quad \text{on } \partial\Omega,$$

and

$$\partial_1 u_2 - \partial_2 u_1 = \partial_1\partial_2\psi - \partial_2\partial_1\psi = 0$$

by the Schwartz lemma, and this holds in particular on $\partial\Omega$. Moreover, it is clear that

$$Au = \begin{pmatrix} -\Delta\partial_1\psi \\ -\Delta\partial_2\psi \end{pmatrix} = -\nabla\Delta\psi = \mu u.$$

Similarly, if $-\Delta_{\mathrm{D}}\varphi = \lambda\varphi$ and $u = \nabla^{\perp}\varphi$, then u satisfies the boundary conditions required in dom A: firstly,

$$u \cdot \nu = \nabla^{\perp}\varphi \cdot \nu = 0 \quad \text{on } \partial\Omega$$

as this corresponds to the tangential derivative on $\partial\Omega$ and $\varphi = 0$ constantly there. Secondly,

$$\partial_1 u_2 - \partial_2 u_1 = \partial_1\partial_1\varphi + \partial_2\partial_2\varphi = \Delta\varphi = -\lambda\varphi = 0 \quad \text{on } \partial\Omega,$$

making use of the differential equation and the boundary condition for φ. It follows $A\nabla^{\perp}\varphi = \lambda\nabla^{\perp}\varphi$. Furthermore, the proof of the orthonormal basis properties mentioned in the theorem is straightforward.

Let us now assume, in addition, that Ω is simply connected. Then the famous Helmholtz decomposition reads

$$L^2(\Omega)^2 = \nabla H^1(\Omega) \oplus \nabla^{\perp} H_0^1(\Omega),$$

see, e.g., [15, Lemma 2.10]. Using this, it follows from Theorem 2 (1) and (2) that the spectrum of A consists precisely of all non-trivial eigenvalues of $-\Delta_{\mathrm{N}}$ and $-\Delta_{\mathrm{D}}$, including multiplicities. As it is well-known that $\mu_2 < \lambda_1$ (see [8, 20]), we obtain the following min-max principle.

Theorem 3 *Assume that $\Omega \subset \mathbb{R}^2$ is a bounded, simply connected domain with piecewise C^{∞}-smooth boundary and that all corners, if any, are convex. Denote by*

$$\eta_1 \leq \eta_2 \leq \ldots$$

the union of the positive eigenvalues of $-\Delta_N$ *and* $-\Delta_D$, *counted according to their multiplicities. Then*

$$\eta_j = \min_{\substack{F \subset \mathcal{H}_N \text{ subspace} \\ \dim F = j}} \max_{u = \binom{u_1}{u_2} \in F \setminus \{0\}} \frac{\int_\Omega \left(|\nabla u_1|^2 + |\nabla u_2|^2\right) - \int_{\partial\Omega} \kappa \left(|u_1|^2 + |u_2|^2\right)}{\int_\Omega \left(|u_1|^2 + |u_2|^2\right)}.$$

Especially, the first positive eigenvalue μ_2 *of* $-\Delta_N$ *is given by*

$$\mu_2 = \min_{u = \binom{u_1}{u_2} \in \mathcal{H}_N \setminus \{0\}} \frac{\int_\Omega \left(|\nabla u_1|^2 + |\nabla u_2|^2\right) - \int_{\partial\Omega} \kappa \left(|u_1|^2 + |u_2|^2\right)}{\int_\Omega \left(|u_1|^2 + |u_2|^2\right)}. \tag{5}$$

Moreover, if ψ *is an eigenfunction of* $-\Delta_N$ *corresponding to* μ_2 *then the minimum in* (5) *is attained at* $u = \nabla\psi$, *and, conversely, each minimizer* u *of* (5) *satisfies* $u = \nabla\psi$ *for some* $\psi \in \ker(-\Delta_N - \mu_2)$.

3 Application to the Hot Spots Conjecture

In this section we indicate how the variational principle of Theorem 3 can be applied to the hot spots conjecture. The first result discussed here was first proven by Atar and Burdzy [1] by probabilistic methods. It comprises the class of so-called lip domains introduced by Burdzy and Chen in [5].

Definition 1 A bounded Lipschitz domain $\Omega \subset \mathbb{R}^2$ is called *lip domain* if

$$\Omega = \left\{(x, y)^\top : f_1(x) < y < f_2(x), x \in (a, b)\right\},$$

where $f_1, f_2 : [a, b] \to \mathbb{R}$ are Lipschitz continuous functions with Lipschitz constant at most one such that $f_1(x) < f_2(x)$ for all $x \in (a, b)$, $f_1(a) = f_2(a)$ and $f_1(b) = f_2(b)$.

Figure 1 shows a typical lip domain. Other examples include right and obtuse triangles (in contrast to acute triangles) or right and obtuse trapezoids (in contrast to acute trapezoids). Clearly, lip domains are simply connected.

Theorem 4 *Assume that* Ω *is a lip domain with piecewise* C^∞*-smooth boundary whose corners are convex. Then the following assertions hold.*

(1) *If* Ω *is not a square then the first positive eigenvalue* μ_2 *of the Neumann Laplacian on* Ω *is simple, i.e. the corresponding eigenfunction* ψ *is unique up to scalar multiples.*
(2) *If* Ω *is not a rectangle then* ψ *may be chosen such that its directional derivatives in both directions* $\boldsymbol{e}_1 + \boldsymbol{e}_2$ *and* $\boldsymbol{e}_1 - \boldsymbol{e}_2$ *are positive inside* Ω, *where* $\boldsymbol{e}_1$ *and* $\boldsymbol{e}_2$ *are the standard basis vectors in* $\mathbb{R}^2$. *In particular,* ψ *does not have any critical point inside* Ω *and, hence, takes its maximum and minimum on* $\partial\Omega$ *only.*

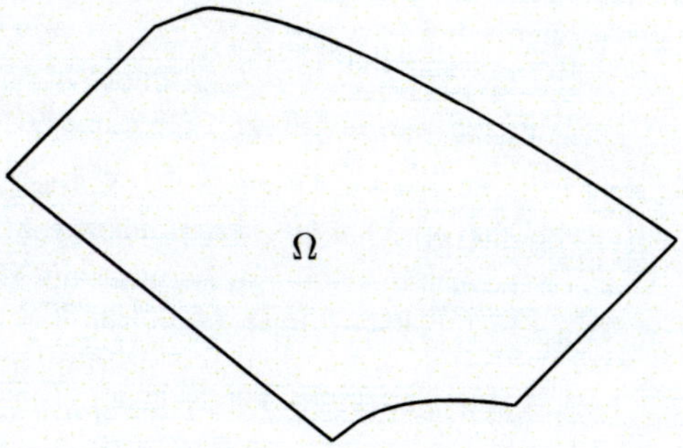

Fig. 1 A lip domain

The proof relies on the fact that rotating any lip domain by $\pi/4$ in positive direction leads to a domain for which the outer unit normal vector field ν on the boundary satisfies

$$\nu(x) \in \overline{Q_2} \cup \overline{Q_4}, \tag{6}$$

that is, $\nu(x)$ belongs to the union of the closed second and fourth quadrants in the plane, for almost all $x \in \partial\Omega$. We will now sketch the proof of Theorem 4 assuming that Ω is rotated such that (6) holds for almost all $x \in \partial\Omega$.

Let ψ be any eigenfunction of $-\Delta_{\mathrm{N}}$ corresponding to μ_2. Then $u = \nabla\psi$ minimizes (5). Take $v = (|u_1|, |u_2|)^\top$. Then

$$\|v\|_{L^2(\Omega)^2} = \|u\|_{L^2(\Omega)^2} \quad \text{and} \quad \mathfrak{a}[v] \leq \mathfrak{a}[u].$$

Moreover, the condition (6) guarantees that $v|_{\partial\Omega} \cdot \nu = 0$ holds, since the components of $\nu(x)$ have opposite signs for almost all $x \in \partial\Omega$. Hence, $v \in \mathcal{H}_{\mathrm{N}}$ and v is another minimizer of (5), whose components are non-negative everywhere in Ω. Thus $v \in \ker(A - \mu_2)$ and there exists an eigenfunction ψ' of $-\Delta_{\mathrm{N}}$ corresponding to the eigenvalue μ_2 such that $\nabla\psi' = v$. However,

$$\Delta\partial_j\psi' = -\mu_2|u_j| \leq 0, \quad j = 1, 2,$$

and, thus, the minimum principle for superharmonic functions yields that $v_j = \partial_j\psi'$ is either constantly zero or strictly positive in Ω, $j = 1, 2$. As ψ' being constant in one direction is only possible if Ω is a rectangle, the assertion (2) follows. For assertion (1), if ψ and ψ' are linearly independent and Ω is not a square, one can construct a linear combination of $\nabla\psi$ and $\nabla\psi'$ vanishing at some point in Ω, contradicting the reasoning in the proof of assertion (2). A complete proof can be found in [21, Section 4].

Theorem 3 can be employed for a more careful analysis of the nodal lines of the components of $\nabla\psi$, for an eigenfunction ψ of $-\Delta_{\mathrm{N}}$. This can be used, for instance, to reprove the following theorem of Jerison and Nadirashvili [9, Theorem 1.1].

Theorem 5 *Assume that $\Omega \subset \mathbb{R}^2$ has a piecewise C^∞-smooth boundary without non-convex corners and is symmetric with respect to both coordinate axes. Furthermore, assume that all vertical and horizontal cross sections of Ω are intervals and that Ω is not a rectangle. Then the following hold.*

(1) *For any eigenfunction ψ corresponding to μ_2 that is odd w.r.t. x and even w.r.t. y, $\partial_x \psi$ does not have any zero in Ω; if ψ is chosen such that $\psi > 0$ if $x > 0$, then $\partial_x \psi > 0$ in Ω. Moreover, $\partial_y \psi$ vanishes exactly on the axes.*
(2) *For any eigenfunction ψ corresponding to μ_2 that is even w.r.t. x and odd w.r.t. y, $\partial_y \psi$ does not have any zero in Ω; if ψ is chosen such that $\psi > 0$ if $y > 0$, then $\partial_y \psi > 0$ in Ω. Moreover, $\partial_x \psi$ vanishes exactly on the axes.*

A proof of Theorem 5 based on Theorem 3 will be published elsewhere.

It should, finally, be pointed out that the regularity assumptions on $\partial\Omega$ in the above theorems are stronger than those in [1, 9].

Acknowledgments This research is financially supported by grant no. 2022-03342 of the Swedish Research Council (VR).

References

1. R. Atar, K. Burdzy, On Neumann eigenfunctions in lip domains. J. Am. Math. Soc. **17**(2), 243–265 (2004)
2. R. Bañuelos, K. Burdzy, On the "hot spots" conjecture of J. Rauch. J. Funct. Anal. **164**, 1–33 (1999)
3. R.F. Bass, K. Burdzy, Fiber Brownian motion and the "hot spots" problem. Duke Math. J. **105**(1), 25–58 (2000)
4. K. Burdzy, The hot spots problem in planar domains with one hole. Duke Math. J. **129**(3), 481–502 (2005)
5. K. Burdzy, Z.Q. Chen, Weak convergence of reflecting Brownian motions. Electron. Commun. Probab. **3**, 29–33 (1998)
6. K. Burdzy, W. Werner, A counterexample to the "hot spots" conjecture. Ann. Math. **149**, 309–317 (1999)
7. R. Courant, D. Hilbert, *Methods of Mathematical Physics*, vol. I (Interscience Publishers, New York, 1953), xv+561pp.
8. N. Filonov, On an inequality between dirichlet and neumann eigenvalues for the laplace operator (in Russian). Algebra i Analiz **16**, 172–176 (2004); translation in St. Petersburg Math. Journal 16 (2005), 413–416
9. D. Jerison, N. Nadirashvili, The "hot spots" conjecture for domains with two axes of symmetry. J. Am. Math. Soc. **13**(4), 741–772 (2000)
10. C. Judge, S. Mondal, Euclidean triangles have no hot spots. Ann. Math. **191**, 167–211 (2020)
11. C. Judge, S. Mondal, Erratum: euclidean triangles have no hot spots. Ann. Math. **195**, 337–362 (2022)
12. C. Judge, S. Mondal, Critical points of Laplace eigenfunctions on polygons. Commun. Partial Differ. Equ. **47**, 1559–1590 (2022)
13. T. Kato, *Perturbation Theory for Linear Operators* (Springer, Berlin, 1995)
14. B. Kawohl, *Rearrangements and Convexity of Level Sets in PDE*. Lecture Notes in Mathematics, vol. 1150 (Springer, Berlin, 1985)

15. J.P. Kelliher, Eigenvalues of the Stokes operator versus the Dirichlet Laplacian in the plane. Pac. J. Math. **244**(1), 99–132 (2010)
16. A. Kleefeld, The hot spots conjecture can be false: some numerical examples. Adv. Comput. Math. **47**, paper no. 85, 31pp. (2021)
17. D. Krejčiřík, M. Tušek, Location of hot spots in thin curved strips. J. Differ. Equ. **266**, 2953–2977 (2019)
18. P. Mariano, H. Panzo, J. Wang, Improved upper bounds for the Hot Spots constant of Lipschitz domains (2022). arXiv:2110.03672
19. M.N. Pascu, Scaling coupling of reflecting Brownian motions and the hot spots problem. Trans. Am. Math. Soc. **354**(11), 4681–4702 (2002)
20. G. Pólya, Remarks on the foregoing paper. J. Math. Phys. **31**, 55–57 (1952)
21. J. Rohleder, A new approach to the hot spots conjecture (2021). arXiv:2106.05224
22. B. Siudeja, Hot spots conjecture for a class of acute triangles. Math. Z. **280**, 783–806 (2015)
23. S. Steinerberger, Hot spots in convex domains are in the tips (up to an Inradius). Commun. Partial Differ. Equ. **45**, 641–654 (2020)
24. S. Steinerberger, An upper bound on the Hot Spots constant (2022). arXiv:2106.03677

Endpoint Sobolev Inequalities for Vector Fields and Cancelling Operators

Jean Van Schaftingen

Abstract The injectively elliptic vector differential operators $A(\mathrm{D})$ from V to E on $\mathbb{R}^n$ such that the estimate

$$\|\mathrm{D}^{\ell} u\|_{L^{n/(n-(k-\ell))}(\mathbb{R}^n)} \leq \|A(\mathrm{D})u\|_{L^1(\mathbb{R}^n)}$$

holds can be characterized as the operators satisfying a cancellation condition

$$\bigcap_{\xi \in \mathbb{R}^n \setminus \{0\}} A(\xi)[V] = \{0\} .$$

These estimates unify existing endpoint Sobolev inequalities for the gradient of scalar functions (Gagliardo and Nirenberg), the deformation operator (Korn–Sobolev inequality by M.J. Strauss) and the Hodge complex (Bourgain and Brezis). Their proof is based on the fact that $A(\mathrm{D})u$ lies in the kernel of a cocancelling differential operator.

Keywords Sobolev embedding · Vector differential operators · Injective ellipticity · Bourgain-Brezis estimates

2020 Mathematics Subject Classification 35A23, 26D15, 35E05, 42B30, 42B35, 46E35

J. Van Schaftingen (✉)
Institute de Recherche en Mathématique et Physique (IRMP), Université Catholique de Louvain (UCLouvain), Louvain-la-Neuve, Belgium
e-mail: Jean.VanSchaftingen@UCLouvain.be

D. Cardona et al. (eds.), *Extended Abstracts 2021/2022*, Research Perspectives Ghent Analysis and PDE Center 3, https://doi.org/10.1007/978-3-031-48579-4_5

1 Sobolev Inequalities for Vector Fields

The classical *Sobolev inequality* [12, 16, 20] states that given $n, k \in \mathbb{N} \setminus \{0\}$ and $p \in [1, \frac{n}{k-\ell})$ there exists a constant $C_1 \in (0, \infty)$ such that each function $u \in C_c^\infty(\mathbb{R}^n, \mathbb{R})$ satisfies the inequality

$$\Big(\int_{\mathbb{R}^n} |\mathrm{D}^\ell u|^{\frac{np}{n-(k-\ell)p}}\Big)^{1-\frac{(k-\ell)p}{n}} \le C_1 \int_{\mathbb{R}^n} |\mathrm{D}^k u|^p \,. \tag{1}$$

Given linear spaces V and E and a linear differential operator $A(\mathrm{D})$ of order $k \in \mathbb{N}\setminus\{0\}$ from V to E on $\mathbb{R}^n$ defined for $u \in C^\infty(\mathbb{R}^n, V)$ at each $x \in \mathbb{R}^n$ by $A(\mathrm{D})u(x) := A(\mathrm{D}u(x))$, where $A \in \mathrm{Lin}(\mathrm{Lin}^k_{\mathrm{sym}}(\mathbb{R}^n, V), E)$, or equivalently, by

$$A(\mathrm{D})u(x) = \sum_{\substack{\alpha\in\mathbb{N}^n\\ |\alpha|=k}} A_\alpha[\partial^\alpha u(x)] = \sum_{\substack{\alpha\in\mathbb{N}^n\\ |\alpha|=k}} \partial^\alpha (A_\alpha[u])(x) \,,$$

with $A_\alpha \in \mathrm{Lin}(V, E)$ for $\alpha = (\alpha_1, \dots, \alpha_n) \in \mathbb{N}^n$ satisfying $|\alpha| := \alpha_1 + \dots + \alpha_n = k$, the goal of the present work is to determine whether every vector field $u \in C^\infty(\mathbb{R}^n, V)$ satisfies a vector Sobolev inequality

$$\Big(\int_{\mathbb{R}^n} |\mathrm{D}^\ell u|^{\frac{np}{n-(k-\ell)p}}\Big)^{1-\frac{(k-\ell)p}{n}} \le C_2 \int_{\mathbb{R}^n} |A(\mathrm{D})u|^p \,. \tag{2}$$

When $p \in (1, \infty)$, the injective ellipticity is the key notion to have (2).

Definition 1 Given $n \in \mathbb{N} \setminus \{0\}$, finite-dimensional vector spaces V and E, a homogeneous constant coefficient differential operator $A(\mathrm{D})$ of order $k \in \mathbb{N} \setminus \{0\}$ from V to E on $\mathbb{R}^n$ is *injectively elliptic* whenever for every $\xi \in \mathbb{R}^n \setminus \{0\}$, one has $\ker A(\xi) = \{0\}$.

If the operator $A(\mathrm{D})$ is injectively elliptic, we can write in the Fourier domain for every $\xi \in \mathbb{R}^n \setminus \{0\}$

$$\mathcal{F}(\mathrm{D}^k u)(\xi) = \xi^{\otimes k} \otimes A(\xi)^\dagger[\mathcal{F}(A(\mathrm{D})u)(\xi)] \,, \tag{3}$$

where the Fourier transform $\mathcal{F}u : \mathbb{R}^n \to V + iV$ of a Schwartz test function $u \in \mathcal{S}(\mathbb{R}^n, V)$ is defined for every $\xi \in \mathbb{R}^n$ by the integral formula

$$(\mathcal{F}u)(\xi) := \int_{\mathbb{R}^n} e^{-2\pi i\, \xi\cdot x}\, u(x)\,\mathrm{d}x$$

and where for each $\xi \in \mathbb{R}^n \setminus \{0\}$, $A(\xi)^\dagger$ is the Moore–Penrose generalized inverse of $A(\xi)$:

$$A(\xi)^\dagger := \big(A(\xi)^* A(\xi)\big)^{-1} A(\xi)^* \,, \tag{4}$$

with $A(\xi)^* \in \operatorname{Lin}(E, V)$ the adjoint of $A(\xi)$. Applying the Parseval identity when $p = 2$ and a classical multiplier theorem (see for example [21, ch. IV th. 3]) when $p \in (1, \infty) \setminus \{2\}$, we obtain from (3) the estimate

$$\int_{\mathbb{R}^n} |\mathrm{D}^k u|^p \le C_3 \int_{\mathbb{R}^n} |A(\mathrm{D})u|^p . \tag{5}$$

As a consequence of the inequalities (5) and (1), we get that when $p \in (1, \frac{n}{k-\ell})$, the estimate (2) holds when the operator $A(\mathrm{D})$ is injectively elliptic.

The proof outlined above does not work at all in the endpoint $p = 1$. More dramatically, Ornstein [17] has proved that if $B(\mathrm{D})$ is a homogeneous constant coefficient differential operator of order k from V to a linear space F on $\mathbb{R}^n$ and if for every $u \in C^\infty_c(\mathbb{R}^n, V)$ the estimate

$$\int_{\mathbb{R}^n} |B(\mathrm{D})u| \le C_4 \int_{\mathbb{R}^n} |A(\mathrm{D})u| , \tag{6}$$

holds, then one can write $B(\mathrm{D}) = LA(\mathrm{D})$, with $L : E \to F$ being a constant-coefficient linear mapping.

2 Sobolev Estimates for Cancelling Operators

Continuing our investigation of the Sobolev-type inequality (2), we will examine how badly $A(\mathrm{D})u$ does not control u beyond Ornstein's non-estimate (6). In order to do this, we can try to have $A(\mathrm{D})u$ as singular as possible, that is, close to a Dirac measure and thus to construct some $u : \mathbb{R}^n \to V$ such that, for some fixed vector $e \in E$,

$$A(\mathrm{D})u = e\delta_0 \tag{7}$$

on $\mathbb{R}^n$ in the sense of distributions. Taking the Fourier transform on both sides of (7), we get for every $\xi \in \mathbb{R}^n$

$$(2\pi i)^k A(\xi)[\mathcal{F}u(\xi)] = e . \tag{8}$$

The Eq. (8) will have a solution for every $\xi \in \mathbb{R}^n$ if and only if the operator $A(\mathrm{D})$ does not satisfy the following cancellation condition, introduced by the author [27].

Definition 2 The homogeneous differential operator $A(\mathrm{D})$ is *cancelling* whenever

$$\bigcap_{\xi \in \mathbb{R}^n \setminus \{0\}} A(\xi)[V] = \{0\} .$$

Thanks to the injective ellipticity of $A(\mathrm{D})$, $A(\xi)^\dagger$ can be defined by (4) and is homogeneous of degree $-k$; thanks to a construction [7, lem. 2.1]; [18] based on classical constructions in distribution theory and Fourier analysis [13, th. 3.2.3, 3.2.4, 7.1.18, th. 7.1.16] (see also [29, prop. 2] for a direct self-contained proof), one can construct a representation kernel $G_A \in C^\infty(\mathbb{R}^n \setminus \{0\}, \mathrm{Lin}(E, V))$ such that for every $\xi \in \mathbb{R}^n$,

$$\mathcal{F}G_A(\xi) = (2\pi i)^{-k}\, A^\dagger(\xi)$$

and such that for every $x \in \mathbb{R}^n \setminus \{0\}$ and every $t \in \mathbb{R} \setminus \{0\}$,

$$G_A(tx) = t^{n-k}\Big(G_A(x) - \ln|t|\, P_A(x)\Big)\,,$$

where the function $P_A : \mathbb{R}^n \to \mathrm{Lin}(E, V)$ is a homogeneous polynomial of degree $k - n$ when $k - n \geq 0$ is even, and is 0 otherwise.

Thanks to Theorem 2, we can now state our main result characterizing endpoint Sobolev inequalities [27, prop. 4.6 and 5.5].

Theorem 1 *Let $n \in \mathbb{N} \setminus \{0\}$, let V and E be finite-dimensional vector spaces and let $A(\mathrm{D})$ be a homogeneous constant coefficient differential operator of order $k \in \mathbb{N} \setminus \{0\}$ from V to E on $\mathbb{R}^n$. If $A(\mathrm{D})$ is injectively elliptic and if $\ell \in \mathbb{N} \setminus \{0\}$ satisfies $0 < k - \ell < n$, then there exists a constant $C_5 \in (0, \infty)$ such that for each $u \in C^\infty_c(\mathbb{R}^n, V)$*

$$\Big(\int_{\mathbb{R}^n} |\mathrm{D}^\ell u|^{\frac{n}{n-(k-\ell)}}\Big)^{1-\frac{k-\ell}{n}} \leq C_5 \int_{\mathbb{R}^n} |A(\mathrm{D})[u]|\,, \tag{9}$$

if and only if $A(\mathrm{D})$ is cancelling.

The necessity of the cancellation follows essentially by observing that if $e \in \bigcap_{\xi \in \mathbb{R}^n \setminus \{0\}} A(\xi)[V] \setminus \{0\}$, then a suitable approximation of $G_A[e]$ by smooth functions prevents (9) from holding.

A first consequence of Theorem 1, is the endpoint Sobolev inequality of Bourgain and Brezis [3, 4, cor. 17] (see also [14]): given $m \in \{1, \dots, n-1\}$, for every $u \in C^\infty_c(\mathbb{R}^n, \bigwedge^m \mathbb{R}^n)$,

$$\Big(\int_{\mathbb{R}^n} |u|^{\frac{n}{n-1}}\Big)^{1-\frac{1}{n}} \leq C \int_{\mathbb{R}^n} |\mathrm{d}u| + |\mathrm{d}^* u| \tag{10}$$

holds if and only $m \notin \{1, n-1\}$. Here $\mathrm{d}u \in C^\infty_c(\mathbb{R}^n, \bigwedge^{m+1} \mathbb{R}^n)$ and $\mathrm{d}^* u \in C^\infty_c(\mathbb{R}^n, \bigwedge^{m-1} \mathbb{R}^n)$ denote respectively the exterior differential and codifferential of the differential form u.

As a second consequence, we have Strauss's endpoint Korn–Sobolev inequality [22]: for every $u \in C_c^\infty(\mathbb{R}^n, \mathbb{R}^n)$, one has

$$\left(\int_{\mathbb{R}^n} |u|^{\frac{n}{n-1}}\right)^{1-\frac{1}{n}} \le C_6 \int_{\mathbb{R}^n} |\mathrm{D}_{\mathrm{sym}} u| , \tag{11}$$

where $\mathrm{D}_{\mathrm{sym}} u := (\mathrm{D}u + (\mathrm{D}u)^*)/2$ is the *symmetric derivative,* also known in elasticity as the *deformation operator.*

3 Duality Estimates for Cocancelling Operators

The proof of the sufficiency of the cancellation (1) is based on the crucial fact that $A(\mathrm{D})u$ on the right-hand side is not any function, but is constrained to satisfy some *compatibility conditions*.

Proposition 1 *Let $n \in \mathbb{N} \setminus \{0\}$, let V and E be finite-dimensional vector spaces, and let $A(\mathrm{D})$ be a homogeneous constant coefficient differential operator of order $k \in \mathbb{N} \setminus \{0\}$ from V to E on $\mathbb{R}^n$. If $A(\mathrm{D})$ is injectively elliptic, then there exists a homogeneous constant coefficient differential operator $L(\mathrm{D})$ from E to E on $\mathbb{R}^n$ such that for every $\xi \in \mathbb{R}^n \setminus \{0\}$,*

$$A(\xi)[V] = \ker L(\xi) . \tag{12}$$

The proof of Theorem 1 is based on the definition of $L(\mathrm{D})$ by requiring that for each $\xi \in \mathbb{R}^n \setminus \{0\}$,

$$L(\xi) := \det\big(A(\xi)^* A(\xi)\big) \Big(\mathrm{id}_E - A(\xi)(A(\xi)^* A(\xi))^{-1} A(\xi)^*\Big) . \tag{13}$$

Theorem 1 can be seen as a generalization of the *symmetry of second-order derivatives*

$$\partial_j(\partial_i u) = \partial_i(\partial_j u) \tag{14}$$

and of the *Saint-Venant compatibility conditions* for the symmetric derivative

$$\partial_{k\ell}(\partial_i u^j + \partial_j u^i) + \partial_{ij}(\partial_k u^\ell + \partial_\ell u^k) = \partial_{kj}(\partial_i u^\ell + \partial_\ell u^i) + \partial_{i\ell}(\partial_k u^j + \partial_j u^k) , \tag{15}$$

although the construction of Theorem 1 gives a more complicated operator than what appears in (14) and (15).

The definition of cancelling operator (Theorem 2) and the construction of compatibility conditions (Theorem 1) suggest the definition of cocancelling operator.

Definition 3 Let $n \in \mathbb{N} \setminus \{0\}$ and let E and F be finite-dimensional vector spaces. A homogeneous constant coefficient differential operator $L(\mathrm{D})$ from E to F on $\mathbb{R}^n$ is *cocancelling* whenever

$$\bigcap_{\xi \in \mathbb{R}^n \setminus \{0\}} \ker L(\xi) = \{0\} .$$

The cocancellation condition characterizes the operators for which there is a duality estimate with critical Sobolev spaces [26, 27] (see also previous results [3–5, 23–25]).

Theorem 2 *Let $n \in \mathbb{N} \setminus \{0, 1\}$, let V and E be finite-dimensional vector spaces, let $L(\mathrm{D})$ be a homogeneous constant coefficient differential operator from E to F on $\mathbb{R}^n$ and let $\ell \in \{1, \dots, n-1\}$. There exists a constant $C_7 \in (0, \infty)$ such that for every $f \in L^1(\mathbb{R}^n, E)$ that satisfies $L(\mathrm{D}) f = 0$ in $\mathbb{R}^n$ in the sense of distributions and every $\varphi \in C_c^\infty(\mathbb{R}^n, E)$ one has*

$$\left| \int_{\mathbb{R}^n} \langle f, \varphi \rangle \right| \le C_7 \int_{\mathbb{R}^n} |f| \left(\int_{\mathbb{R}^n} |\mathrm{D}^\ell \varphi|^{\frac{n}{\ell}} \right)^{\frac{\ell}{n}} \tag{16}$$

if and only if the operator $L(\mathrm{D})$ is cocancelling.

Theorem 2 states somehow that with regards to integration against vector fields that are in the kernel functions in the homogeneous Sobolev space $\dot{W}^{\ell,n/\ell}(\mathbb{R}^n, E)$ behave as if they were bounded—which is well-known not to be the case.

The necessity of the cancellation can be seen by noting that if $L(\mathrm{D})$ was not cancelling, then one would have $L(\mathrm{D})(\delta_0 e) = 0$ for some $e \in E \setminus \{0\}$; approximating the measure $\delta_0 e$ by smooth functions one would deduce from (16) that the Sobolev space $\dot{W}^{\ell,n/\ell}(\mathbb{R}^n, \mathbb{R})$ would be continuously embedded in $L^\infty(\mathbb{R}^n, \mathbb{R})$, which is not the case when $\ell \in \{1, \dots, n-1\}$.

Assuming that Theorem 2, Theorem 1 can be proved as follows. Noting that the operator $L(\mathrm{D})$ given by Theorem 2 is cocancelling, one gets by Theorem 2 that $\|A(D)u\|_{W^{-\ell,n/(n-\ell)}} \le C_8 \|A(D)u\|_{L^1}$ so that a classical multiplier theorem brings the conclusion.

Bourgain and Brezis's original proof [4] of estimates of the type of theorem 2 was based on an approximation property for critical Sobolev functions through a Littlewood–Paley decomposition, generalizing a similar results in the study of the divergence equation [1, 2]; the advantage of their proof compared to the one presented below is that it provides much stronger estimates of the form

$$\left| \int_{\mathbb{R}^n} f \cdot \varphi \right| \le C_9 \|f\|_{L^1(\mathbb{R}^n) + \dot{W}^{-1,n/(n-1)}(\mathbb{R}^n)} \left(\int_{\mathbb{R}^n} |\mathrm{D}\varphi|^n \right)^{\frac{1}{n}} . \tag{17}$$

Let us now explain how Theorem 2 can be proved in the case where $L(\mathrm{D})$ is the divergence operator, following [24]. (The reader is referred to [26, 27, 29, 30] for the general case.) Without loss of generality, we are going to estimate the integral

$$\int_{\mathbb{R}^n} f \cdot e_n \phi\,, \tag{18}$$

for $\phi \in C_c^\infty(\mathbb{R}^n, \mathbb{R})$, where e_n is the nth vector in the canonical basis of $\mathbb{R}^n$. By Fubini's theorem we have

$$\int_{\mathbb{R}^n} f \cdot e_n \phi = \int_{\mathbb{R}} \Big(\int_{\mathbb{R}^{n-1}} f(\cdot, x_n) \cdot e_n \phi(\cdot, x_n) \Big) \,\mathrm{d}x_n\,. \tag{19}$$

We are now going to estimate the inner integral on the right-hand side of (19). First, we immediately have for each $\psi \in C^\infty(\mathbb{R}^{n-1}, \mathbb{R})$,

$$\Big| \int_{\mathbb{R}^{n-1}} f(\cdot, x_n) \cdot e_n \psi \Big| \le \|\psi\|_{L^\infty(\mathbb{R}^{n-1})} \int_{\mathbb{R}^{n-1}} |f(\cdot, x_n)|\,. \tag{20}$$

On the other hand, by the Gauss–Ostrogradsky divergence theorem, we also have

$$\Big| \int_{\mathbb{R}^{n-1}} f(\cdot, x_n) \cdot e_n \psi \Big| = \Big| \int_{\mathbb{R}^{n-1} \times (x_n, \infty)} \operatorname{div}(f \Psi) \Big| \le \|\mathrm{D}\psi\|_{L^\infty(\mathbb{R}^{n-1})} \int_{\mathbb{R}^n} |f|\,. \tag{21}$$

since $\operatorname{div} f = 0$. Interpolating between the estimates (20) and (21) and applying the Morrey–Sobolev embedding on $\mathbb{R}^{n-1}$, we get

$$\begin{aligned} \Big| \int_{\mathbb{R}^{n-1}} f(\cdot, x_n) \cdot e_n \psi \Big| &\le C_{10} \Big(\int_{\mathbb{R}^{n-1}} |f(\cdot, x_n)| \Big)^{1-1/n} \Big(\int_{\mathbb{R}^n} |f| \Big)^{1/n} |\psi|_{C^{0,1/n}(\mathbb{R}^{n-1})} \\ &\le C_{11} \Big(\int_{\mathbb{R}^{n-1}} |f(\cdot, x_n)| \Big)^{1-\frac{1}{n}} \Big(\int_{\mathbb{R}^n} |f| \Big)^{\frac{1}{n}} \Big(\int_{\mathbb{R}^{n-1}} |\mathrm{D}\psi|^n \Big)^{\frac{1}{n}}\,. \end{aligned} \tag{22}$$

Combining (19) and (22), we deduce in view of Hölder's inequality that

$$\begin{aligned} \Big| \int_{\mathbb{R}^n} f \cdot e_n \phi \Big| &\le C_{11} \Big(\int_{\mathbb{R}^n} |f| \Big)^{\frac{1}{n}} \int_{\mathbb{R}} \Big(\int_{\mathbb{R}^{n-1}} |f(\cdot, x_n)| \Big)^{1-\frac{1}{n}} \Big(\int_{\mathbb{R}^{n-1}} |\mathrm{D}\phi(\cdot, x_n)|^n \Big)^{\frac{1}{n}} \,\mathrm{d}x_n \\ &\le C_{11} \int_{\mathbb{R}^n} |f| \Big(\int_{\mathbb{R}^n} |\mathrm{D}\phi|^n \Big)^{\frac{1}{n}}\,. \end{aligned} \tag{23}$$

Therefore for any vector $\nu \in \mathbb{R}^n$, it follows from (23) that we have proved

$$\Big| \int_{\mathbb{R}^n} f \cdot \nu \phi \Big| \le C_{11} |\nu| \Big(\int_{\mathbb{R}^n} |f| \Big) \Big(\int_{\mathbb{R}^n} |\mathrm{D}\phi|^n \Big)^{\frac{1}{n}}\,. \tag{24}$$

Decomposing $\varphi = \sum_{j=1}^n e_i \phi_i$ with $\phi_i := e_i \cdot \varphi$, we finally obtain (16) from (24).

Theorem 2 when $L(\mathrm{D})$ is the divergence is equivalent, thanks to Smirnov's result on the approximation of divergence-free measures [19], to the Bourgain, Brezis and Mironescu's estimate on circulation integrals [5] (see also [23]): if $\Gamma \subset \mathbb{R}^n$ is a closed curve with tangent vector t and length $|\Gamma|$, then for every vector field $\varphi \in C_c^\infty(\mathbb{R}^n, \mathbb{R}^n)$, one has

$$\Big|\int_\Gamma \langle \varphi, t\rangle\Big| \le C_2 |\Gamma| \Big(\int_{\mathbb{R}^n} |\mathrm{D}\varphi|^n\Big)^{\frac{1}{n}} \; ; \tag{25}$$

the geometric flavour of (25) raises several natural open questions on sharp constants in higher dimensions $n \ge 3$ [8].

4 Further Results

The cancellation condition can also be proved to be a necessary and sufficient condition for other estimates.

In the scale of fractional Sobolev spaces, for any injectively elliptic operator $A(\mathrm{D})$ and assuming that $k, \ell \in \mathbb{N}$, $p \in (1, \infty)$ and $\sigma \in (0, 1)$ satisfy $k - n = \ell + \sigma - \frac{n}{p}$, the estimate

$$\Big(\int_{\mathbb{R}^n}\int_{\mathbb{R}^n} \frac{|\mathrm{D}^\ell u(y) - \mathrm{D}^\ell u(x)|^p}{|y - x|^{n+\sigma p}} \,\mathrm{d}y \,\mathrm{d}x\Big)^{\frac{1}{p}} \le C_{13} \int_{\mathbb{R}^n} |A(\mathrm{D})[u]| \; , \tag{26}$$

holds for every $u \in C_c^\infty(\mathbb{R}^n, V)$ if and only if if and only if the operator $A(\mathrm{D})$ is cancelling [27] (see also [29]).

Similarly, for any injectively elliptic operator $A(\mathrm{D})$, the Hardy inequality

$$\int_{\mathbb{R}^n} \frac{|\mathrm{D}^\ell u(x)|}{|x|^{k-\ell}} \,\mathrm{d}x \le C_{14} \int_{\mathbb{R}^n} |A(\mathrm{D})[u]| \; , \tag{27}$$

holds for every $u \in C_c^\infty(\mathbb{R}^n, V)$ if and only if if and only if $A(\mathrm{D})$ is cancelling [27] (see also [29] for the proof); this results originates in Maz'ya's work [15] (see also [6]).

Finally, Raiţă [18] has proved that for any injectively elliptic operator $A(\mathrm{D})$ the uniform estimate

$$\|\mathrm{D}^{n-k} u\|_{L^\infty(\mathbb{R}^n)} \le C_{15} \int_{\mathbb{R}^n} |A(\mathrm{D})[u]| \; , \tag{28}$$

is equivalent to the *weak cancellation* property that for every $e \in E$

$$\int_{\mathbb{S}^{n-1}} \xi^{\otimes k-n} A^{-1}(\xi)[e] \,\mathrm{d}\xi = 0 \tag{29}$$

(see also [29, §5.4]).

Endpoint estimates similar to Theorems 1 and 2 can also be obtained on stratified homogeneous groups [9, 31], on the hyperbolic plane [10] and on symmetric spaces of noncompact type [11].

For a more detailed exposition on endpoint Sobolev inequalities and cancelling operators, we refer the reader to the quite formal lecture notes [29], their somehow more informal counterpart [30] and to the survey article [28].

References

1. J. Bourgain, H. Brezis, Sur l'équation div $u = f$. C. R. Math. Acad. Sci. Paris **334**(11), 973–976 (2002)
2. J. Bourgain, H. Brezis, On the equation div $Y = f$ and application to control of phases. J. Am. Math. Soc. **16**(2), 393–426 (2003). https://doi.org/10.1090/S0894-0347-02-00411-3
3. J. Bourgain, H. Brezis, New estimates for the Laplacian, the div-curl, and related Hodge systems. C. R. Math. Acad. Sci. Paris **338**(7), 539–543 (2004). https://doi.org/10.1016/j.crma.2003.12.031
4. J. Bourgain, H. Brezis, New estimates for elliptic equations and Hodge type systems. J. Eur. Math. Soc. **9**(2), 277–315 (2007). https://doi.org/10.4171/JEMS/80
5. J. Bourgain, H. Brezis, P. Mironescu, $H^{1/2}$ maps with values into the circle: minimal connections, lifting, and the Ginzburg–Landau equation. Publ. Math. Inst. Hautes Études Sci. **99**, 1–115 (2004). https://doi.org/10.1007/s10240-004-0019-5
6. P. Bousquet, P. Mironescu, An elementary proof of an inequality of Maz'ya involving L^1 *vector fields*, in *Nonlinear Elliptic Partial Differential Equations*. Contemporary Mathematics, vol. 540 (American Mathematical Society, Providence, 2011), pp. 59–63. https://doi.org/10.1090/conm/540/10659
7. P. Bousquet, J. Van Schaftingen, Hardy-Sobolev inequalities for vector fields and canceling differential operators. Ind. Univ. Math. J. **63**(5), 1419–1445 (2014). https://doi.org/10.1512/iumj.2014.63.5395
8. H. Brezis, J. Van Schaftingen, Circulation integrals and critical Sobolev spaces: problems of optimal constants, in *Perspectives in Partial Differential Equations, Harmonic Analysis and Applications*. Proceedings of Symposia in Pure Mathematics, vol. 79 (American Mathematical Society, Providence, 2008), pp. 33–47. https://doi.org/10.1090/pspum/079/2500488
9. S. Chanillo, J. Van Schaftingen, Subelliptic Bourgain-Brezis estimates on groups. Math. Res. Lett. **16**(3), 487–501 (2009). https://doi.org/10.4310/MRL.2009.v16.n3.a9
10. S. Chanillo, J. Van Schaftingen, P.L. Yung, Variations on a proof of a borderline Bourgain-Brezis Sobolev embedding theorem. Chin. Ann. Math. Ser. B **38**(1), 235–252 (2017). https://doi.org/10.1007/s11401-016-1069-y
11. S. Chanillo, J. Van Schaftingen, P.L. Yung, Bourgain-Brezis inequalities on symmetric spaces of non-compact type. J. Funct. Anal. **273**(4), 1504–1547 (2017). https://doi.org/10.1016/j.jfa.2017.05.005
12. E. Gagliardo, Proprietà di alcune classi di funzioni in più variabili. Ricerche Mat. **7**, 102–137 (1958)
13. L. Hörmander, *The analysis of linear partial differential operators I: Distribution Theory and Fourier Analysis*. Grundlehren der Mathematischen Wissenschaften, vol. 256, 2nd edn. (Springer, Berlin, 1990)
14. L. Lanzani, E.M. Stein, A note on div curl inequalities. Math. Res. Lett. **12**(1), 57–61 (2005). https://doi.org/10.4310/MRL.2005.v12.n1.a6
15. V.G. Maz'ya, Estimates for differential operators of vector analysis involving L^1-norm. J. Eur. Math. Soc. **12**(1), 221–240 (2010). https://doi.org/10.4171/JEMS/195

16. L. Nirenberg, On elliptic partial differential equations. Ann. Scuola Norm. Sup. Pisa **13**, 115–162 (1959)
17. D. Ornstein, A non-equality for differential operators in the L_1 norm. Arch. Rational Mech. Anal. **11**, 40–49 (1962). https://doi.org/10.1007/BF00253928
18. B. Raiţă, Critical L^p-differentiability of BV^A-maps and canceling operators. Trans. Am. Math. Soc. **372**(10), 7297–7326 (2019). https://doi.org/10.1090/tran/7878
19. S.K. Smirnov, Decomposition of solenoidal vector charges into elementary solenoids, and the structure of normal one-dimensional flows (in Russian). Algebra i Analiz **5**(4), 206–238 (1993); English transl., St. Petersburg Math. J. **5** (1994), no. 4, 841–867
20. S. Sobolev, Sur un théorème d'analyse fonctionnelle. Rec. Math. Moscou n. Ser. **4**, 471–497 (1938)
21. E.M. Stein, *Singular Integrals and Differentiability Properties of Functions*. Princeton Mathematical Series, vol. 30 (Princeton University Press, Princeton, 1970)
22. M.J. Strauss, *Variations of Korn's and Sobolev's Equalities: Partial Differential Equations* (University of California, Berkeley, 1971). Proc. Sympos. Pure Math., vol. XXIII, Amer. Math. Soc., Providence, R.I., 1973, pp. 207–214
23. J. Van Schaftingen, A simple proof of an inequality of Bourgain, Brezis and Mironescu. C. R. Math. Acad. Sci. Paris **338**(1), 23–26 (2004). https://doi.org/10.1016/j.crma.2003.10.036
24. J. Van Schaftingen, Estimates for L^1-vector fields. C. R. Math. Acad. Sci. Paris **339**(3), 181–186 (2004). https://doi.org/10.1016/j.crma.2004.05.013
25. J. Van Schaftingen, Estimates for L^1 vector fields with a second order condition. Acad. Roy. Belg. Bull. Cl. Sci. **15**(1–6), 103–112 (2004)
26. J. Van Schaftingen, Estimates for L^1 vector fields under higher-order differential conditions, J. Eur. Math. Soc. (JEMS) **10** (2008), no. 4, 867–882, https://doi.org/10.4171/JEMS/133
27. J. Van Schaftingen, Limiting Sobolev inequalities for vector fields and canceling linear differential operators. J. Eur. Math. Soc. **15**(3), 877–921 (2013). https://doi.org/10.4171/JEMS/380
28. J. Van Schaftingen, Limiting Bourgain-Brezis estimates for systems of linear differential equations: theme and variations. J. Fixed Point Theory Appl. **15**(2), 273–297 (2014). https://doi.org/10.1007/s11784-014-0177-0
29. J. Van Schaftingen, Injective ellipticity, cancelling operators, and endpoint Gagliardo-Nirenberg-Sobolev inequalities for vector fields. Lecture Notes for the CIME Summer School "Geometric and Analytic Aspects of Functional Variational Principles" June 27–July 1, 2022 (2022)
30. J. Van Schaftingen, Limiting Sobolev estimates for vector fields and cancelling differential operators. Lecture notes for the Spring School on Analysis 2023 "Function Spaces and Applications XII", Paseky nad Jizerou, May 28–June 3 (2023)
31. J. Van Schaftingen, P.L. Yung, Limiting Sobolev and Hardy inequalities on stratified homogeneous groups. Ann. Fenn. Math. **47**(2), 1065–1098 (2022)

Scattering of Maxwell Potentials on Curved Spacetimes

Grigalius Taujanskas

Abstract We report on the recent construction of a scattering theory for Maxwell potentials on curved spacetimes (Nicolas and Taujanskas, Conformal scattering of Maxwell potentials (2022). arXiv:2211.14579).

Keywords Scattering · Massless fields · Maxwell potentials · Conformal geometry · Asymptotic analysis

2020 Mathematics Subject Classification 35P25, 53A30, 34K25

1 Introduction

The study of the asymptotic structure of isolated systems in general relativity has been a rich area of research since at least the 1960s. A number[1] of landmark results [5, 7, 9, 12] have been established, however many important questions, particularly regarding the fine asymptotic properties of fields and the rigorous analytic formulations of scattering theories, remain. In particular, it is of interest to study the far-field regime of massless fields—such as gravity or electromagnetism—on curved spacetimes, where they are scattered by background curvature. Since massless fields enjoy an essential conformal invariance, Penrose's conformal method [12] provides an excellent conceptual framework to study their scattering and asymptotics.

In [11] the author and J.-P. Nicolas construct a complete scattering theory for Maxwell potentials on a class of curved, non-stationary spacetimes. The scattering construction of [11] in principle allows for reasonably general backgrounds: they

[1] Too many to list here, see for example [9] for a more complete bibliography.

G. Taujanskas (✉)
Trinity Hall, Trinity Lane, Cambridge, UK
e-mail: taujanskas@dpmms.cam.ac.uk

D. Cardona et al. (eds.), *Extended Abstracts 2021/2022*, Research Perspectives Ghent Analysis and PDE Center 3, https://doi.org/10.1007/978-3-031-48579-4_6

may contain matter fields,[2] as long as the conformal boundary is suitably smooth and the spacetime is sufficiently close to Minkowski space.[3] In the case of *vacuum*,[4] a concrete subclass of such spacetimes—to which we refer as Corvino–Schoen–Chruściel–Delay (CSCD) spacetimes—may be constructed using the initial data gluing theorems of [2–4, 6] for the vacuum Einstein equations, and Friedrich's theorem for the semi-global stability of Minkowski space [7, 8]. This produces an infinite-dimensional family of vacuum spacetimes which have good conformal compactification properties: they are asymptotically simple in the sense of Penrose [12], their null and timelike infinities can be ensured to be C^k for any integer k, and they are exactly Schwarzschildean, or Kerrian, in a neighbourhood of spatial infinity. For simplicity, we work with the case of exactly Schwarzschildean spacetimes near i^0. CSCD spacetimes are described in more detail below.

2 Background Spacetimes and Field Equations

2.1 CSCD Spacetimes

We work on spacetimes $(\mathcal{M}, g_{ab})$ which are four dimensional, globally hyperbolic, asymptotically flat, and arise as developments of the vacuum Einstein equations

$$\mathrm{Ric}(g) = 0$$

from initial data (h_{ab}, κ_{ab}) on a Cauchy hypersurface $\Sigma \simeq \mathbb{R}^3$ such that:

(i) outside a given compact set $K \subset \Sigma$, the data $(h_{ab}, \kappa_{ab}) = (g_{ab}^{\mathrm{Schw}}(t = 0), 0)$ is exactly Schwarzschild at $t = 0$, so that the development $(\mathcal{M}, g_{ab})$ is exactly Schwarzschild in a neighbourhood of spatial infinity i^0,
(ii) the data (h_{ab}, κ_{ab}) is sufficiently close to Minkowskian data in the sense required by the theorems of, say, [2],
(iii) the initial metric h_{ab} satisfies the condition that $\|r^2 \mathrm{Ric}(h)\|_{L^\infty(\Sigma)}$ is not too large,[5] for r an appropriately defined radial coordinate on Σ which coincides with the standard Schwarzschildean radial coordinate on $\Sigma \setminus K$.

---

[2] In principle our scattering construction allows for matter fields provided they decay sufficiently fast at infinity, specifically that $\hat{g}_{ab}\hat{\Box}\Omega \approx 4\hat{\nabla}_a\hat{\nabla}_b\Omega$ and $\hat{\Psi}_{0,1,2,3,4} \approx 0$, where $\approx$ denotes equality on $\mathscr{I}$. For simplicity, here we report on the vacuum case.

[3] Proximity to Minkowski space allows one to construct a concrete and fairly large class of spacetimes on which the scattering theory of [11] holds. However, this is likely not strictly necessary provided the spacetime has the correct Penrose diagram, and there may be examples of 'large' allowable background spacetimes.

[4] In fact, the authors of [2] comment that their constructions should apply to certain non-vacuum constraint equations, e.g. the Einstein–Maxwell system.

[5] Condition (iii) is not part of the assumptions of the theorems of Corvino, Schoen, Chruściel and Delay, but comes from the construction of the initial potential. See Sect. 4.2.

With these conditions, $(\mathcal{M}, g_{ab})$ is then asymptotically simple with a C^k conformal compactification (for some k sufficiently large) at $\mathscr{I}^{\pm}$ and $i^{\pm}$. Being exactly Schwarzschild in $D^{+}(\Sigma \setminus K)$, at i^0 the spacetime is conformally singular. We denote the conformally rescaled metric by $\hat{g}_{ab} = \Omega^2 g_{ab}$, where Ω is the corresponding conformal factor.

2.2 Field Equations

Maxwell's equations are conformally invariant and are given by

$$\nabla^a F_{ab} = 0 = \nabla_{[a} F_{bc]} \iff \hat{\nabla}^a \hat{F}_{ab} = 0 = \hat{\nabla}_{[a} \hat{F}_{bc]},$$

where ∇ is the Levi-Civita connection of g_{ab} and $\hat{\nabla}$ is the Levi-Civita connection of $\hat{g}_{ab}$, and $\hat{F}_{ab} = F_{ab}$, i.e. F_{ab} has conformal weight zero. In terms of the potential the equations read

$$\Box A_a - \nabla_b(\nabla_a A^a) + \mathrm{R}_{ab} A^a = 0, \tag{1}$$

and, without a choice of gauge, are also conformally invariant provided A_a is chosen to have conformal weight zero. We choose an NP tetrad $(l^a, m^a, \bar{m}^a, n^a)$ on $\mathcal{M}$ and the conformal scaling $(\hat{l}^a, \hat{m}^a, \hat{\bar{m}}^a, \hat{n}^a) = (\Omega^{-2} l^a, \Omega^{-1} m^a, \Omega^{-1} \bar{m}^a, n^a)$ so that on $\mathscr{I}^+$ the vector field n^a becomes a generator of $\mathscr{I}^+$, and define the components of A_a and F_{ab}

$$\begin{pmatrix} A_0 & A_1 & A_2 \\ F_0 & F_1 & F_2 \end{pmatrix} = \begin{pmatrix} A_a l^a & A_a n^a & A_a m^a \\ F_{ab} l^a m^b & \frac{1}{2} F_{ab}(l^a n^b + \bar{m}^a m^b) & F_{ab} \bar{m}^a n^b \end{pmatrix},$$

with the associated conformal weights inherited from the scaling of the tetrad. Moreover, we choose a uniformly timelike vector field T^a, which in the case of Minkowski space is exactly $T^a = \partial_t$ and in the general case coincides with the Schwarzschildean Killing vector field ∂_t in a neighbourhood of i^0; we denote by $\mathbf{A}$ the projection of A_a to hypersurfaces orthogonal to T^a, and write $\mathfrak{a} = T^a A_a$.

3 Main Results

The main results of [11] can be summarised in the following theorems. Our tetrad is adapted to future null infinity, so the following results are explicitly stated only in the case of $\mathscr{I}^+$. The analogous gauge conditions and function spaces on $\mathscr{I}^-$ can be obtained by interchanging the vector fields l^a and n^a.

Theorem 1 *Let* $(\mathcal{M}, g_{ab}) = (\mathbb{R}^4, \eta_{ab})$ *be the Minkowski spacetime. Then a finite energy solution to* (1) *admits the gauge*

$$\nabla_a A^a = \nabla \cdot \mathbf{A} = \mathfrak{a} = 0, \tag{2}$$

and there exist bounded, invertible linear operators

$$\mathfrak{T}_K^{\pm} : \dot{H}^1_C(\Sigma) \oplus L^2_C(\Sigma) \longrightarrow \dot{\mathcal{H}}^1(\mathscr{I}^{\pm})$$
$$(\mathbf{A}, \dot{\mathbf{A}})|_{\Sigma} \longmapsto (\hat{A}_0^{\pm}, \hat{A}_1^{\pm}, \hat{A}_2^{\pm}),$$

corresponding to the future/past development according to (1) *in the gauge* (2) *on* $\mathcal{M}$, *which map finite energy Maxwell potential initial data on* Σ *to finite energy Maxwell potential characteristic data on* $\mathscr{I}^{\pm}$. *The function spaces above are given by*

$$\dot{H}^1_C(\Sigma) = \{\mathbf{A} \in \dot{H}^1(\Sigma; \mathbb{R}^3) \,:\, \nabla \cdot \mathbf{A} = 0\},$$
$$L^2_C(\Sigma) = \{\dot{\mathbf{A}} \in L^2(\Sigma; \mathbb{R}^3) \,:\, \nabla \cdot \dot{\mathbf{A}} = 0\},$$

and

$$\dot{\mathcal{H}}^1(\mathscr{I}^+) = \left\{ (\hat{A}_0^+, \hat{A}_1^+, \hat{A}_2^+) \,:\, \hat{A}_0^+ = \int_{-\infty}^{u} 2\,\mathrm{Re}\,\eth\bar{\hat{A}}_2^+ \,\mathrm{d}u,\ \hat{A}_1^+ = 0, \right.$$
$$\left. \int_{\mathscr{I}^+} |\partial_u \hat{A}_2^+|^2 \,\mathrm{d}u \wedge \mathrm{dv}_{\mathbb{S}^2} < \infty \right\}$$
$$\simeq \dot{H}^1(\mathbb{R}; L^2(\mathbb{S}^2)),$$

and analogously for $\dot{\mathcal{H}}(\mathscr{I}^-)$. *Consequently, there exists a bounded, invertible linear scattering operator*

$$\mathscr{S}_K = \mathfrak{T}_K^+ \circ (\mathfrak{T}_K^-)^{-1} : \dot{\mathcal{H}}^1(\mathscr{I}^-) \longrightarrow \dot{\mathcal{H}}^1(\mathscr{I}^+),$$
$$(\hat{A}_0^-, \hat{A}_1^-, \hat{A}_2^-) \longmapsto (\hat{A}_0^+, \hat{A}_1^+, \hat{A}_2^+)$$

which corresponds to the development according to (1) *in the gauge* (2) *of* $(\hat{A}_0^-, \hat{A}_1^-, \hat{A}_2^-)$ *from* $\mathscr{I}^-$. *The subscript* K *in the above refers to the standard timelike Killing field* $K = \partial_t$.

Moreover, the Morawetz vector field

$$K_0 = (t^2 + r^2)\partial_t + 2tr\partial_r,$$

gives rise to a stronger scattering theory given by bounded, invertible linear operators $\mathfrak{T}^{\pm}_{K_0}$, *where*

$$\mathfrak{T}^{+}_{K_0} : r^{-1}\dot{H}^1_C(\Sigma)^{\text{curl}} \oplus r^{-1}L^2_C(\Sigma) \longrightarrow u^{-1}\dot{\mathcal{H}}^1(\mathscr{I}^+)$$

and similarly for $\mathfrak{T}^{-}_{K_0}$, *where*

$$r^{-1}\dot{H}^1_C(\Sigma)^{\text{curl}} = \{\mathbf{A} \in \dot{H}^1(\Sigma;\mathbb{R}^3) \,:\, \nabla\cdot\mathbf{A} = 0,\ r(\nabla\times\mathbf{A}) \in L^2(\Sigma;\mathbb{R}^3)\},$$
$$r^{-1}L^2_C(\Sigma) = \{\dot{\mathbf{A}} \in L^2(\Sigma;\mathbb{R}^3) \,:\, \nabla\cdot\dot{\mathbf{A}} = 0,\ r\dot{\mathbf{A}} \in L^2(\Sigma;\mathbb{R}^3)\},$$

and

$$u^{-1}\dot{\mathcal{H}}^1(\mathscr{I}^+) = \left\{(\hat{A}^+_0, \hat{A}^+_1, \hat{A}^+_2) \,:\, \hat{A}^+_0 = \int_{-\infty}^{u} 2\,\mathrm{Re}\,\eth\bar{\hat{A}}^+_2\,\mathrm{d}u,\ \hat{A}^+_1 = 0,\right.$$
$$\left.\int_{\mathscr{I}^+}\left(u^2|\partial_u\hat{A}^+_2|^2 + |\eth\bar{\hat{A}}^+_2|^2\right)\mathrm{d}u\wedge\mathrm{dv}_{\mathbb{S}^2} < \infty\right\},$$

and similarly for $v^{-1}\dot{\mathcal{H}}^1(\mathscr{I}^-)$. *The resulting scattering operator*

$$\mathscr{S}_{K_0} = \mathfrak{T}^{+}_{K_0} \circ (\mathfrak{T}^{-}_{K_0})^{-1} : v^{-1}\dot{\mathcal{H}}^1(\mathscr{I}^-) \longrightarrow u^{-1}\dot{\mathcal{H}}^1(\mathscr{I}^+)$$

is linear, bounded, invertible, and maps past asymptotic data $(\hat{A}^-_0, \hat{A}^-_1, \hat{A}^-_2)$ *to future asymptotic data* $(\hat{A}^+_0, \hat{A}^+_1, \hat{A}^+_2)$ *through a development according to* (1) *in the gauge* (2).

Theorem 2 *Let* $(\mathcal{M}, g_{ab})$ *be a CSCD spacetime as described in Sect. 2.1. Then a finite energy solution to* (1) *admits a gauge which satisfies the conditions*

(i) $\nabla_a A^a = 0$ *in a neighbourhood of* Σ *and a neighbourhood of* $\mathscr{I}^+$,
(ii) $\mathfrak{a}|_\Sigma = 0 = \nabla\cdot\mathbf{A}|_\Sigma$, *and*
(iii) $\hat{A}^{[1]}_1|_{\mathscr{I}^+} = 0$,

where $\hat{A}^{[1]}_1 = \Omega^{-1}\hat{A}_1$, *and there exist bounded, invertible linear operators*

$$\mathfrak{T}^{\pm} : \dot{H}^1_C(\Sigma)^{\text{curl}} \oplus L^2(\Sigma) \longrightarrow \dot{\mathcal{H}}^1(\mathscr{I}^\pm),$$
$$(\mathbf{A}, \nabla_T\mathbf{A})|_\Sigma \longmapsto (\hat{A}^\pm_0, \hat{A}^\pm_1, \hat{A}^\pm_2),$$

corresponding to the future/past development according to (1) *on* $\mathcal{M}$ *in the above gauge, which map finite energy Maxwell potential initial data on* Σ *to finite energy Maxwell potential characteristic data on* $\mathscr{I}^\pm$. *The function spaces above are given by*

$$\dot{H}^1_C(\Sigma)^{\text{curl}} = \{\mathbf{A} \in \dot{H}^1(\Sigma) \,:\, \nabla\cdot\mathbf{A} = 0,\ \nabla\times\mathbf{A} \in L^2(\Sigma)\}$$

and $\dot{\mathcal{H}}^1(\mathscr{I}^{\pm})$ *as in Theorem 1. Consequently, there exists a bounded, invertible linear scattering operator*

$$\mathscr{S} = \mathfrak{T}^+ \circ (\mathfrak{T}^-)^{-1} : \dot{\mathcal{H}}^1(\mathscr{I}^-) \longrightarrow \dot{\mathcal{H}}^1(\mathscr{I}^+)$$
$$(\hat{A}_0^-, \hat{A}_1^-, \hat{A}_2^-) \longmapsto (\hat{A}_0^+, \hat{A}_1^+, \hat{A}_2^+)$$

which corresponds to the development of $(\hat{A}_0^-, \hat{A}_1^-, \hat{A}_2^-)$ *from* $\mathscr{I}^-$ *according to* (1) *on* $\mathcal{M}$ *in the above gauge.*

4 Remarks

4.1 Conformal Scale

The construction of the gauge and the spaces of characteristic data rely on the existence of a conformal scale which satisfies a number of conditions. In effect, we construct a conformal scale in which $\mathscr{I}^+$ is almost as 'flat' as in the case of Minkowski space, which in general is permitted by the smoothness of the conformal boundary and the rapid decay of background matter fields at infinity. In this scale we have that the spin coefficients (cf. [13]) $\hat{\lambda}$, $\hat{\pi}$, $\hat{\mu}$, $\hat{\tau}$ and $\hat{\gamma}$ vanish on $\mathscr{I}^+$, $\hat{\nu}$ vanishes in a neighbourhood of $\mathscr{I}^+$, and $\hat{\mu}$ is real in a neighbourhood of $\mathscr{I}^+$. Moreover, the components $\hat{\Phi}_{21}$ and $\hat{\Phi}_{22}$ of the trace-free Ricci tensor of $\hat{g}_{ab}$ vanish on $\mathscr{I}^+$, as does the full rescaled Weyl tensor. This conformal scale is the analogue of the conformal factor $\Omega = r^{-1}$ in Minkowski space.

4.2 Spaces of Data

The spaces of initial and characteristic data are derived from the (conformally covariant) Maxwell stress-energy tensor and a choice of timelike conformal Killing field, together with various gauge conditions on the potential. For the space of characteristic data (on $\mathscr{I}^+$), the expression for the transverse component $\hat{A}_0^+$ comes from the reduction of our gauge to $\mathscr{I}^+$. Precisely, the Lorenz gauge in the physical spacetime reduces to the condition $\hat{A}_1 = 0$ on $\mathscr{I}^+$ at first order in Ω, and to the condition

$$-f\hat{A}_1^{[1]} + \hat{\text{þ}}'\hat{A}_0 - 2\,\mathrm{Re}\,\hat{\eth}\bar{\hat{A}}_2 = 0$$

on $\mathscr{I}^+$ at second order in Ω, for a smooth function f. This becomes an ODE for $\hat{A}_0$ on $\mathscr{I}^+$ if we impose the additional gauge condition that $\hat{A}_1^{[1]} = 0$ on $\mathscr{I}^+$, which we then solve by integrating in the Bondi parameter u, $\hat{\text{þ}}' = \partial_u$, on $\mathscr{I}^+$. Note that we set $\hat{A}_0^+$ to vanish at i^0. If the free data $\hat{A}_2^+$ is smooth and compactly supported, for example, then the formula for $\hat{A}_0^+$, being an integral along $\mathscr{I}^+$ of a function which decays towards both i^+ and i^0, means that $\hat{A}_0^+$ can be chosen to vanish at either i^0 or i^+, but not both. We expect the difference $\hat{A}_0^+|_{i^+} - \hat{A}_0^+|_{i^0}$ to be related to the electromagnetic memory effect. This will be explored elsewhere.

The conditions on the space of initial data are reasonably self-explanatory. What is not immediately obvious, however, is that the condition on the *background spacetime*

$$\|r^2 \operatorname{Ric}(h)\|_{L^\infty(\Sigma)} < C^{-1} \tag{3}$$

for some constant C in fact arises from the construction of the initial data. This is due to the following reason. On a general (e.g. CSCD) spacetime, even in the gauge (2) on Σ, the canonical energy on Σ does not define a norm on the potential due to the presence of the Ricci curvature of h. One must therefore show that there is a one-to-one correspondence between finite energy fields $\mathbf{E}$, $\mathbf{B} \in L^2(\Sigma)$ and potentials in a suitable space by some other means. Essentially, this amounts to solving the elliptic system

$$\boldsymbol{\Delta}\mathbf{A}_i + \mathbf{R}_{ij}\mathbf{A}^j = -(\boldsymbol{\nabla} \times \mathbf{B})_i$$

on Σ. However, the Ricci curvature $\mathbf{R}_{ij} = \operatorname{Ric}(h)_{ij}$ on Σ is in general not positive-definite, and so standard elliptic theory fails here. Indeed, it is not clear what the kernel of the operator from $\dot{H}^1(\Sigma; \mathbb{R}^3)$ into $\dot{H}^{-1}(\Sigma; \mathbb{R}^3)$, as defined by the left-hand side of the above equation, is in general. If the assumption (3) on $\operatorname{Ric}(h)$ is made, however, then it is possible to ensure, using Hardy's inequality on Σ, that $\|\mathbf{A}\|_{\dot{H}^1(\Sigma)} \lesssim \|\mathbf{B}\|_{L^2(\Sigma)}$, which is then sufficient to control the regularity of the initial data.

4.3 Goursat Problem

The invertibility of the operators $\mathfrak{T}^\pm$ is equivalent to the well-posedness of the characteristic initial value problem (or *Goursat* problem) for (1) from $\mathscr{I}^\pm$ with finite energy characteristic data. The main analytic tool that enables us to solve the Goursat problem is Bär and Wafo's extension (see Theorem 23 in [1]) of a theorem due to Hörmander [10], which ensures that this can be done from compactly supported data on $\mathscr{I}^+$. Some care is required, however, since the component $\hat{A}_0^+$ is not compactly supported even if $\hat{A}_2^+ \in C_c^\infty(\mathscr{I}^+)$. This leads us to solve the Goursat problem near i^+ separately, where the solution is pure gauge and we first solve a wave equation for the Maxwell field $\hat{F}_{ab}$ instead. We then recover the potential near i^+ using our gauge conditions.

4.4 Role of Timelike Conformal Symmetry

As stated in Theorem 1, in the case of Minkowski space one has, in addition to the standard timelike Killing field $K = \partial_t$, the *conformal* timelike Killing field $K_0 = (t^2 + r^2)\partial_t + 2tr\partial_r$, known as the Morawetz vector field, the generator of 'inverted time translations' on $\mathcal{M}$. Since the Maxwell stress-energy tensor is traceless, K_0 also provides a conserved energy which carries different weights, $\mathbf{E}, \mathbf{B} \in r^{-1}L^2(\Sigma; \mathbb{R}^3)$. At null infinity, while K becomes tangent to $\mathscr{I}^+$, K_0 is transverse to $\mathscr{I}^+$, so the energy on $\mathscr{I}^+$ picks up angular derivatives (see the definition of the space of scattering data $u^{-1}\dot{\mathcal{H}}(\mathscr{I}^+)$ in Theorem 1). The overall result is that the spaces of initial and scattering data with respect to K_0 are strictly smaller than with respect to K, resulting in a stronger scattering theory. Loosely speaking, $\mathscr{S}_K$ therefore decomposes into a 'direct sum' which contains $\mathscr{S}_{K_0}$ as a factor.

Acknowledgments This work was partly supported by the EPSRC grant [EP/L05811/1]. We thank Gustav Holzegel and Juan Valiente Kroon for useful discussions.

References

1. C. Bär, R.T. Wafo, Initial value problems for wave equations on manifolds. Math. Phys. Anal. Geom. **18**, 7 (2015)
2. P.T. Chruściel, E. Delay, On mapping properties of the general relativistic constraints operator in weighted function spaces, with applications. Mém. Soc. Math. France **94**, 109 (2003)
3. J. Corvino, Scalar curvature deformation and a gluing construction for the Einstein constraint equations. Commun. Math. Phys. **214**(1), 137–189 (2000)
4. J. Corvino, R.M. Schoen, On the asymptotics for the vacuum Einstein constraint equations. J. Differ. Geom. **73**(2), 185–217 (2006)
5. D. Christodoulou, S. Klainerman, *The Global Nonlinear Stability of Minkowski Space, PMS-41* (Princeton University Press, Princeton, 1994)
6. P.T. Chruściel, E. Delay, Existence of non-trivial, vacuum, asymptotically simple spacetimes. Class. Quant. Gravity **19**(12), 3389 (2002)
7. H. Friedrich, On the existence of n–geodesically complete or future complete solutions of Einstein's field equations with smooth asymptotic structure. Commun. Math. Phys. **107**, 587–609 (1986)
8. H. Friedrich, B.G. Schmidt, Conformal geodesics in general relativity. Proc. Roy. Soc. Lond. Ser. A **414**, 171–195 (1987)
9. P. Hintz, A. Vasy, Stability of Minkowski spaces and polyhomogeneity of the metric. Ann. PDE **6**, 2 (2020)
10. L. Hörmander, A remark on the characteristic Cauchy problem. J. Funct. Anal. **93**(2), 270–277 (1990)
11. J.P. Nicolas, G. Taujanskas, Conformal scattering of Maxwell potentials (2022). arXiv:2211.14579
12. R. Penrose, Zero rest-mass fields including gravitation: asymptotic behaviour. Proc. Roy. Soc. Lond. Ser. A **284**(1397), 159–203 (1965)
13. R. Penrose, W. Rindler, *Spinors and Space-Time Vol. 2: Spinor and Twistor Methods in Space-Time Geometry* (Cambridge University Press, Cambridge, 1986)

Part II
Analysis and PDEs

Remark on the Ill-Posedness for KdV-Burgers Equation in Fourier Amalgam Spaces

Divyang G. Bhimani and Saikatul Haque

Abstract We have established (a weak form of) ill-posedness for the KdV-Burgers equation on a real line in Fourier amalgam spaces $\widehat{w}_s^{p,q}$ with $s < -1$. The particular case $p = q = 2$ recovers the result in Molinet and Ribaud (Int. Math. Res. Not. 2002:1979–2005 (2002)). The result is new even in Fourier Lebesgue space $\mathcal{F}L_s^q$ which corresponds to the case $p = q(\neq 2)$ and in modulation space $M_s^{2,q}$ which corresponds to the case $p = 2, q \neq 2$.

Keywords Korteweg-de Vries-Burgers (KdV-B) equation · Ill-posedness · Fourier amalgam spaces · Fourier-Lebesgue spaces · Modulsation spaces

2000 Mathematics Subject Classification 35Q53, 42B35

1 Introduction

In this paper we consider the Korteweg-de Vries–Burgers (KdV-B) equation posed on the real line:

$$\begin{cases} u_t + u_{xxx} - u_{xx} + uu_x = 0 \\ u(0, x) = u_0 \end{cases} \quad (t, x) \in \mathbb{R} \times \mathbb{R}, \tag{1}$$

where $u = u(t, x) \in \mathbb{R}$.

The KdV-B (1) was derived as a model for the propagation of long, weakly nonlinear dispersive waves in certain physical contexts when dissipative effects

D. G. Bhimani (✉)
Department of Mathematics, Indian Institute of Science Education and Research, Pune, India
e-mail: divyang.bhimani@iiserpune.ac.in

S. Haque
Harish-Chandra Research Institute, Prayagraj, India
e-mail: saikatulhaque@hri.res.in

D. Cardona et al. (eds.), *Extended Abstracts 2021/2022*, Research Perspectives Ghent Analysis and PDE Center 3, https://doi.org/10.1007/978-3-031-48579-4_7

occur (see [6]). Many authors have studied the short and long time behaviours of solutions of KdV, KdV-B and several of its variants in the context of Sobolev spaces, see e.g. [8, 9] and the references therein. In recent years, there has been a great deal of interest in studying dispersive PDEs with Cauchy data in low regularity spaces. See e.g. survey article [14]. Recently, in [17] authors have studied KdV in modulation spaces, and in [2–4] authors have studied ill-poedesness for wave, BBM and NLS in Fourier amalgam spaces, see also [13, 16]. We refer to [1] by Bejenaru-Tao for abstract well-posedness and ill-posedness theory. However, we note that there are no known well/ill-posedness results for (1) in modulation and Fourier amalgam spaces.

In this note, we would like to initiate the study of ill-posedness for (1) in the realm of **Fourier amalgam spaces. We now briefly recall these spaces. In order to study** well-posednes for 1D cubic nonlinear Schrödinger equations, in [11, 12], Oh and Forlano have introduced the Fourier amalgam space $\widehat{w}_s^{p,q}$ $(1 \leq p, q \leq \infty, s \in \mathbb{R})$:

$$\widehat{w}_s^{p,q}(\mathbb{R}) = \left\{ f \in \mathcal{S}'(\mathbb{R}) : \|f\|_{\widehat{w}_s^{p,q}} = \left\| \|\chi_{n+Q}(\xi)\mathcal{F}f(\xi)\|_{L^p_\xi(\mathbb{R})} \langle n \rangle^s \right\|_{\ell^q_n(\mathbb{Z})} < \infty \right\}$$

where $Q = (-1/2, 1/2]$ and $\mathcal{F}$ denote the Fourier transform. These spaces recapture several known spaces:

$$\widehat{w}_s^{p,q}(\mathbb{R}) = \begin{cases} \mathcal{F}L_s^q(\mathbb{R}) \text{ (Fourier-Lebesgue space)} & \textit{if} \quad p = q \\ M_s^{2,q}(\mathbb{R}) \text{ (modulation space)} & \textit{if} \quad p = 2 \\ H^s(\mathbb{R}) \text{ (Sobolev space)} & \textit{if} \quad p = q = 2. \end{cases}$$

See also Remarks 2, 3, 4. We now state our main result.

Theorem 1 *Let* $s < -1$. *Then there does not exist any* $T > 0$ *such that* (1) *admits a unique local solution defined on the interval* $[0, T]$ *and such that the flow-map* $u_0 \mapsto u(t),\ t \in [0, T]$ *is* C^2*-differentiable at zero from* $\widehat{w}_s^{p,q}(\mathbb{R})$ *to* $C([0, T]; \widehat{w}_s^{p,q}(\mathbb{R}))$.

Theorem 1 is new for $p = q \neq 2$ and recovers the result of Molinet-Ribaud in [8, Theorem 1.2]. The method of proof for Theorem 1 rely on showing the "unboundednes" of the second Picard iterate associated with 1- which was initiated by Bourgain [7] for KdV and later for mKdV by N. Tzvetkov in [15]. See also [5, Section 4] and the references there in for the further comments.

We plan to address the norm-inflation (the stronger phenomenon than the mere ill-poseness) and even the worst situation of norm inflation with infinite loss of regularity for (1) in our future works. We note that recently similar questions we have already addressed for Hartree, nonlinear Schrödinger, BBM and wave equations in [2–4]. We also expect to develop well-posedness theory for (1) in $\widehat{w}^{p,q}$ space in the future.

We conclude this section by following remarks.

Remark 2 Recall that the Fourier Lebesgue space $\mathcal{F}L^q_s$ is defined by $\{f \in \mathcal{S}'(\mathbb{R}^d) : \|f\|_{\mathcal{F}L^q_s} := \|\langle\xi\rangle^s \mathcal{F}f\|_{L^q} < \infty\}$.

Remark 3 For any given function f which is locally in B (Banach space) (i.e, $gf \in B, \forall g \in C_0^\infty(\mathbb{R}^d)$), we set $f_B(x) = \|fg(\cdot - x)\|_B$. The Feichtinger's [10] Wiener amalgam space $W(B, C)$ endowed with the norm $\|f\|_{W(B,C)} = \|f_B\|_C$. The Fourier amalgam spaces is a Fourier image of particular Wiener amalgam spaces, specifically, $\mathcal{F}W(L^p, \ell^q_s) = \widehat{w}^{p,q}_s$.

Remark 4 Let $\rho \in \mathcal{S}(\mathbb{R}^d)$, $\rho : \mathbb{R}^d \to [0, 1]$ be a smooth function satisfying $\rho(\xi) = 1$ if $|\xi|_\infty = \max(|\xi_1|, ..., |\xi_d|) \leq \frac{1}{2}$ and $\rho(\xi) = 0$ if $|\xi|_\infty \geq 1$. Let ρ_n be a translation of ρ, that is, $\rho_n(\xi) = \rho(\xi - n), n \in \mathbb{Z}^d$ and denote $\sigma_n(\xi) = \frac{\rho_n(\xi)}{\sum_{\ell\in\mathbb{Z}^d} \rho_\ell(\xi)}, n \in \mathbb{Z}^d$. Then the frequency-uniform decomposition operators can be defined by

$$\Box_n = \mathcal{F}^{-1}\sigma_n\mathcal{F}, \qquad n \in \mathbb{Z}^d.$$

The modulation $M^{p,q}_s(\mathbb{R}^d)$ (with $1 \leq p, q \leq \infty, s \in \mathbb{R}$) is defined by the norm:

$$M^{p,q}_s(\mathbb{R}^d) = \left\{ f \in \mathcal{S}'(\mathbb{R}^d) : \|f\|_{M^{p,q}_s(\mathbb{R})} := \left\| \|\Box_n f\|_{L^p_x(\mathbb{R})} \langle n\rangle^s \right\|_{\ell^q_n(\mathbb{Z}^d)} \right\}.$$

2 Proof of Theorem 1

The integral version of (1) is given by

$$u(t) = S(t)u_0 - \frac{1}{2}\int_0^t S(t-\tau)\partial_x[u(\tau)]^2 d\tau \tag{2}$$

where $\{S(t)\}_{t\geq 0}$ given by

$$\mathcal{F}S(t)u_0 = e^{-t\xi^2 + it\xi^3}\mathcal{F}u_0, \qquad t \geq 0 \tag{3}$$

is the semi-group associated to the linear part of (1). The proof of Theorem 1 follows from the fact that the second Picard iterate (given by (4)) associated to (2) is not continuous at zero from $\widehat{w}^{p,q}_s(\mathbb{R})$ to $C([0, T]; \widehat{w}^{p,q}_s(\mathbb{R}))$. We refer to the proof of Theorem 1.10 in [5] and the reference therein for detail.

Proof of Theorem 1 We define the sequence of initial data $\{\phi_N\}_{N\geq 1}$ by

$$\mathcal{F}\phi_N = N\left(\chi_{I_N} + \chi_{I_N}(-\xi)\right)$$

where $I_N = [N, N+2]$ and $\mathcal{F}\phi_N$ denotes the space Fourier transform of ϕ_N.

Let us compute $\|\phi_N\|_{\widehat{w}_s^{p,q}}$. Note that with $\Omega = I_N \cup (-I_N)$

$$\begin{aligned}\|\phi_N\|_{\widehat{w}_s^{p,q}} &= \left\| \|\chi_{n+Q}(\xi)\mathcal{F}\phi_N\|_{L^p_\xi(\mathbb{R})} \langle n\rangle^s \right\|_{\ell^q_n(\mathbb{Z})} \\ &= N \left\| \|\chi_{n+Q}(\xi)\chi_\Omega(\xi)\|_{L^p_\xi(\mathbb{R})} \langle n\rangle^s \right\|_{\ell^q_n(\mathbb{Z})}\end{aligned}$$

Now $\|\chi_{n+Q}(\xi)\chi_\Omega(\xi)\|_{L^p_\xi(\mathbb{R})}$ survives only if $n \in \mathcal{G} := \{m \in \mathbb{Z} : (m+Q) \cap \Omega \neq \emptyset\}$, and for these n's one must have $|n| \sim N$. Since $\#(\mathcal{G}) \sim 1$ and $\|\chi_{n+Q}(\xi)\chi_\Omega(\xi)\|_{L^p_\xi(\mathbb{R})} = \|\chi_{n+Q}(\xi)\|_{L^p_\xi(\mathbb{R})} = 1$ for almost all $n \in \mathcal{G}$, we conclude

$$\begin{aligned}\|\phi_{0,N}\|_{\widehat{w}_s^{p,q}} &= N \left\| \|\chi_{n+Q}(\xi)\chi_\Omega(\xi)\|_{L^p_\xi(\mathbb{R})} \langle n\rangle^s \right\|_{\ell^q_n(\mathbb{Z})} \\ &= N \left(\sum_{n\in\mathcal{G}} \|\chi_{n+Q}(\xi)\chi_\Omega(\xi)\|^q_{L^p_\xi(\mathbb{R})} \langle n\rangle^{sq} \right)^{1/q} \\ &\sim N(N^{sq})^{1/q} = N^{1+s}.\end{aligned}$$

Therefore $\|\phi_N\|_{\widehat{w}_{-1}^{p,q}} \sim 1$ and $\phi_N \to 0$ in $\widehat{w}_s^{p,q}(\mathbb{R})$ for $s < -1$. This sequence yields a counter example to the continuity of the second iteration of the Picard Scheme in $\widehat{w}_s^{p,q}(\mathbb{R})$ for $s < -1$ that is given by

$$A_2(t,h,h) = \int_0^t S(t-\tau)\partial_x[S(\tau)h]^2 d\tau \tag{4}$$

where $\{S(t)\}_{t\geq 0}$ is defined in (3). Indeed, computing the space Fourier transform we get

$$\begin{aligned}&\mathcal{F}(A_2(t,\phi_N,\phi_N))(\xi) \\ &= \int_0^t e^{-(t-\tau)\xi^2+i(t-\tau)\xi^3}(i\xi)[\mathcal{F}S(\tau)\phi_N * \mathcal{F}S(\tau)\phi_N](\xi)d\tau \\ &= \int_0^t e^{-(t-\tau)\xi^2+i(t-\tau)\xi^3}(i\xi)\int_{\mathbb{R}}[\mathcal{F}S(\tau)\phi_N](\xi-\xi_1)[\mathcal{F}S(\tau)\phi_N](\xi_1)d\xi_1 d\tau \\ &= \int_0^t e^{-(t-\tau)\xi^2+i(t-\tau)\xi^3}(i\xi)\int_{\mathbb{R}} e^{-\tau(\xi-\xi_1)^2+i\tau(\xi_x i_1)^3}\mathcal{F}\phi_N(\xi-\xi_1) \\ &\qquad\qquad \times e^{-\tau\xi_1^2+i\tau\xi_1^3}\mathcal{F}\phi_N(\xi_1)d\xi_1 d\tau \\ &= \int_{\mathbb{R}} e^{-t\xi^2}e^{it\xi^3}\mathcal{F}\phi_N(\xi_1)\mathcal{F}\phi_N(\xi-\xi_1)(i\xi)\end{aligned}$$

$$\times \int_0^t e^{-(\xi_1^2+(\xi-\xi_1)^2-\xi^2)\tau} e^{i(\xi_1^3+(\xi-\xi_1)^3-\xi^3)\tau} d\tau d\xi_1$$

$$= e^{-t\xi^2} e^{it\xi^3}(i\xi) \int_{\mathbb{R}} \mathcal{F}\phi_N(\xi_1)\mathcal{F}\phi_N(\xi-\xi_1) \frac{e^{-(\xi_1^2+(\xi-\xi_1)^2-\xi^2)t} e^{i3\xi\xi_1(\xi-\xi_1)t} - 1}{-2\xi_1(\xi-\xi_1) + i3\xi\xi_1(\xi-\xi_1)} d\xi_1.$$

We note that

$$\begin{aligned} &|\mathcal{F}(A_2(t,\phi_N,\phi_N))(\xi)| \\ &= N^2|\xi| \left| \int_{K_\xi} \frac{e^{-(\xi_1^2+(\xi-\xi_1)^2)t} e^{i3\xi\xi_1(\xi-\xi_1)t} - e^{-\xi^2 t}}{-2\xi_1(\xi-\xi_1) + i3\xi\xi_1(\xi-\xi_1)} d\xi_1 \right| \end{aligned} \tag{5}$$

where

$$K_\xi = \{\xi_1 : \xi - \xi_1 \in I_N, \xi_1 \in -I_N\} \cup \{\xi_1 : \xi_1 \in I_N, \xi - \xi_1 \in -I_N\}.$$

Note that for any $\xi \in [-1/2, 1/2]$, one has $|K_\xi| \geq 1$ and

$$\begin{cases} 3\xi\xi_1(\xi-\xi_1) \leq cN^2 \\ 2\xi_1(\xi-\xi_1) \sim N^2 \end{cases} \quad \forall \xi_1 \in K_\xi.$$

Therefore, fixing $0 < t < 1$, for $\xi \in [-1/2, 1/2]$, we have

$$\operatorname{Re}\left(e^{-(\xi_1^2+(\xi-\xi_1)^2)t)} e^{i3\xi\xi_1(\xi-\xi_1)t} - e^{-\xi^2 t}\right) \leq -e^{-t/4} + e^{-2(N+2)^2 t}$$

which leads for $N = N(t) > 0$ large enough (so that $e^{-2(N+2)^2 t} \leq \frac{1}{2}e^{-t/4}$) to

$$\left| \int_{K_\xi} \frac{e^{-(\xi_1^2+(\xi-\xi_1)^2)t} e^{i3\xi\xi_1(\xi-\xi_1)t} - e^{-\xi^2 t}}{-2\xi_1(\xi-\xi_1) + i3\xi\xi_1(\xi-\xi_1)} d\xi_1 \right| \geq c\frac{e^{-t/4}}{N^2}. \tag{6}$$

Now using the fact $\|a_n\|_{\ell_n^q(\mathbb{Z})} \geq a_0$ we have

$$\begin{aligned} \|A_2(t,\phi_N,\phi_N)\|_{\widehat{w}_s^{p,q}} &= \left\| \left\|\chi_{n+Q}(\xi)\mathcal{F}A_2(t,\phi_N,\phi_N)(\xi)\right\|_{L_\xi^p(\mathbb{R})} \langle n\rangle^s \right\|_{\ell_n^q(\mathbb{Z})} \\ &\geq \left\|\chi_{0+Q}(\xi)\mathcal{F}A_2(t,\phi_N,\phi_N)(\xi)\right\|_{L_\xi^p(\mathbb{R})} \\ &= \left(\int_{(-\frac{1}{2},\frac{1}{2}]} |\mathcal{F}A_2(t,\phi_N,\phi_N)(\xi)|^p d\xi\right)^{1/p}. \end{aligned}$$

Using (5) and (6) we have

$$\|A_2(t, \phi_N, \phi_N)\|_{\widehat{w}_s^{p,q}} \geq cN^2 \frac{e^{-t/4}}{N^2} \left(\int_{(-\frac{1}{2}, \frac{1}{2}]} |\xi|^p d\xi \right)^{1/p} \geq c_0$$

for some positive constant c_0 independent of N and t.

Since $\phi_N \to 0$ in $\widehat{w}_s^{p,q}(\mathbb{R})$, for $s < -1$, this ensure that, for any fixed $t > 0$, the map $u_0 \mapsto A_2(t, u_0, u_0)$ is not continuous at zero from $\widehat{w}_s^{p,q}(\mathbb{R})$ into $\widehat{w}_s^{p,q}(\mathbb{R})$. □

Acknowledgments S. Haque is thankful to DST-INSPIRE (DST/INSPIRE/04/2022/001457) for financial support. S. Haque is also thankful to Harish-Chandra Research Institute for the excellent research facilities.

References

1. I. Bejenaru, T. Tao, Sharp well-posedness and ill-posedness results for a quadratic non-linear Schrödinger equation. J. Funct. Anal. **233**, 228–259 (2006)
2. D.G. Bhimani, S. Haque, Strong ill-posedness for fractional Hartree and cubic NLS equations (2021). arXiv:2101.03991
3. D.G. Bhimani, S. Haque, Norm inflation for Benjamin–Bona–Mahony equation in Fourier amalgam and Wiener amalgam spaces with negative regularity. Mathematics **9**, 23, 3145 (2021)
4. D.G. Bhimani, S. Haque, Norm inflation with infinite loss of regularity at general initial data for nonlinear wave equations in Wiener amalgam and Fourier amalgam spaces. Nonlinear Anal. **223**, 113076 (2022)
5. D.G. Bhimani, S. Haque, The Hartree and Hartree-Fock equations in lebesgue $\widehat{L}^p$ and Fourier-Lebesgue $\widehat{L}^p$ spaces. Ann. Henri Poincaré **24**(3), 1005–1049 (2023)
6. D.R. Van Dooren, Comments on: "Stability of solitary waves in shallow water. Phys. Fluids **19**(6), 771–777 (1976)
7. J. Bourgain, Periodic korteweg de vries equation with measures as initial data. Sel. Math. New Ser. **3**, 115–159 (1997)
8. L. Molinet, F. Ribaud, On the low regularity of the Korteweg-de Vries-Burgers equation. Int. Math. Res. Not. **2002**, 1979–2005 (2002)
9. L. Molinet, S. Vento, Sharp ill-posedness and well-posedness results for the KdV-Burgers equation: the real line case. Ann. Sc. Norm. Super. Pisa Cl. Sci. **10**, 531–560 (2011)
10. H.G. Feichtinger, Banach convolution algebras of Wiener type, in *Functions, Series, Operators*, vols. I–II (Budapest, 1980). Colloquia Mathematica Societatis János Bolyai, vol. 35 (North-Holland, Amsterdam, 1983), pp. 509–524
11. J. Forlano, On the deterministic and probabilistic Cauchy problem of nonlinear dispersive partial differential equations. Ph.D. Thesis, Heriot-Watt University (2020)
12. J. Forlano, T. Oh, Normal form approach to the one-dimensional cubic nonlinear Schrödinger equation in Fourier-amalgam spaces (preprint)
13. N. Kishimoto, A remark on norm inflation for nonlinear Schrödinger equations. Commun. Pure Appl. Anal. **18**, 1375 (2019)
14. M. Ruzhansky, M. Sugimoto, B. Wang, Modulation spaces and nonlinear evolution equations, in *Evolution Equations of Hyperbolic and Schrödinger Type*. Progress in Mathematics, vol. 301 (Birkhäuser/Springer, Basel, 2012), pp. 267–283

15. N. Tzvetkov, Remark on the local ill-posedness for kdv equation. Comptes Rendus de l'Acad. Sci. Ser. I-Math. **329**, 1043–1047 (1999)
16. T. Oh, A remark on norm inflation with general initial data for the cubic nonlinear Schrödinger equations in negative Sobolev spaces. Funkcialaj Ekvacioj **60**, 259–277 (2017)
17. T. Oh, Y. Wang, On global well-posedness of the modified KdV equation in modulation spaces. Discrete Contin. Dyn. Syst. **41**(6), 2971–2992 (2021)

Γ-Convergence for the Bi-Laplace-Beltrami Equation on Hypersurfaces

T. Buchukuri, R. Duduchava, G. Tephnadze, and M. Tsaava

Abstract A mixed boundary value problem for the bi-Laplacian equation in a thin layer around a surface $\mathcal{C}$ with the boundary is investigated. We track what happens in Γ-limit when the thickness of the layer converges to zero. It is shown how the mixed type boundary value problem (BVP) for the bi-Laplace equation in the initial thin layer transforms in the Γ-limit into an appropriate Dirichlet BVP for the bi-Laplace-Beltrami equation on the surface. For this we apply the variational formulation and the calculus of Günter's tangential differential operators on a hypersurface and layers. This approach allow global representation of basic differential operators and of corresponding BVPs in terms of the standard cartesian coordinates of the ambient Euclidean space $\mathbb{R}^n$.

Keywords Boundary value problem · Γ-convergence · Bi-Laplace-Beltrami equation

2020 Mathematics Subject Classification 35J57, 45E10, 47B35

T. Buchukuri (✉)
A. Razmadze Mathematical Institute, I. Javakhishvili State University, Tbilisi, Georgia

R. Duduchava
Institute of Mathematics, University of Georgia, Tbilisi, Georgia

A. Razmadze Mathematical Institute, I. Javakhishvili State University, Tbilisi, Georgia
e-mail: r.duduchava@ug.edu.ge

G. Tephnadze · M. Tsaava
Institute of Mathematics, University of Georgia, Tbilisi, Georgia
e-mail: g.tephnadze@ug.edu.ge; m.tsaava@ug.edu.ge

D. Cardona et al. (eds.), *Extended Abstracts 2021/2022*, Research Perspectives
Ghent Analysis and PDE Center 3, https://doi.org/10.1007/978-3-031-48579-4_8

1 Introduction

In the paper is demonstrated what happens with a mixed type boundary value problem (BVP) for the bi-Laplace equation in a thin layer Ω^ε around a surface $\mathcal{C}$ in $\mathbb{R}^3$ when the thickness of the layer Ω^ε tends to zero: $\varepsilon \to 0$. The described problem is reformulated in the variational form and the limit of associated functionals is understood in the sense of Γ-convergence. The main tool is the representation of differential operators with the help of Gunter's derivatives-the system of tangential derivatives $\mathcal{D}_j := \partial_j - \nu_j \partial_{\boldsymbol{\nu}}$, $j = 1, 2, 3$ on the surface and the normal derivative $\partial_{\boldsymbol{\nu}} := \sum_{j=1}^{3} \nu_j \partial_j$, where $\boldsymbol{\nu} = (\nu_1, \nu_2, \nu_3)^\top$ is the unit normal vector field on the mid-surface $\mathcal{C}$. The first-order differential operator $\mathcal{D}_j$ is the directional derivative along $\pi_{\mathcal{C}} e^j$, where $\pi_{\mathcal{C}} : \mathbb{R}^3 \to T\mathcal{C}$ is the orthogonal projection onto the tangent plane to surface $\mathcal{C}$ and $e^1, \ldots, e^n$ is the canonical basis in the Euclidean space $e^j = (\delta_{jk})_{1 \le k \le 3} \in \mathbb{R}^3$, with δ_{jk} denoting the Kronecker symbol (cf. [7, 10, 12]).

Calculus of Gunter's derivatives on a hypersurface allows one to represent the most basic partial differential operators (PDO's), as well as their associated boundary value problems on a hypersurface $\mathcal{C}$ globally by means of the standard spatial coordinates in $\mathbb{R}^n$. Such BVPs arise in a variety of situations and have many practical applications. See, for example, [11, §72] for the heat conduction by surfaces, [2, §10] for the equations of surface flow, [1, 6] for the vacuum Einstein equations describing gravitational fields, [14] for the Navier-Stokes equations on spherical domains, as well as the references therein.

A hypersurface $\mathcal{C}$ in $\mathbb{R}^3$ has the natural structure of a 2-dimensional Riemannian manifold and the aforementioned PDE's are not the immediate analogues of the ones corresponding to the flat, Euclidean case, since they have to take into consideration geometric characteristics of $\mathcal{C}$ such as curvature. Inherently, these PDE's are originally written in local coordinates, intrinsic to the manifold structure of $\mathcal{C}$.

The surface gradient $\mathcal{D}_{\mathcal{C}} := (\mathcal{D}_1, \mathcal{D}_2, \mathcal{D}_3)^\top$ is defined on $\mathcal{C}$, and has a relatively simple structure. In terms of $\mathcal{D}_{\mathcal{C}}$, the Laplace-Beltrami operator on $\mathcal{C}$ simply becomes (see [13, pp. 8])

$$\Delta_{\mathcal{C}} = \mathcal{D}_{\mathcal{C}}^{*} \mathcal{D}_{\mathcal{C}} \quad \text{on} \quad \mathcal{C} \tag{1}$$

and the bi-Laplace-Beltrami operator has a following form

$$\Delta_{\mathcal{C}}^2 = \Delta_{\mathcal{C}}\, \Delta_{\mathcal{C}} \quad \text{on} \quad \mathcal{C}.$$

Friesecke et al. [9] derived a hierarchy of Plate Models from nonlinear elasticity by Γ-Convergence.

In [5] was consider the mixed BVP with zero Dirichlet but non-zero Neumann data:

$$
\begin{aligned}
&\Delta_{\Omega^\varepsilon}\widetilde{T}(\mathcal{X},t)=f(\mathcal{X},t), &&(\mathcal{X},t)\in\mathcal{C}\times(-\varepsilon,\varepsilon),\\
&\widetilde{T}^+(\mathcal{X},t)=0, &&(\mathcal{X},t)\in\partial\mathcal{C}\times(-\varepsilon,\varepsilon),\\
&(\partial_t\widetilde{T})^+(\mathcal{X},\pm\varepsilon)=q^\pm_\varepsilon(\mathcal{X}),\ \mathcal{X}\in\mathcal{C}.
\end{aligned}
\tag{2}
$$

where $\pm\partial_t = \partial_{\boldsymbol{\nu}}$ represent the normal derivatives on the surfaces $\mathcal{C}\times\{\pm\varepsilon\}$. Here $\mathcal{C}\subset\mathcal{S}$ is a smooth subsurface of a closed hypersurface $\mathcal{S}$ with smooth nonempty boundary $\partial\mathcal{C}$. In the investigation we apply that the Laplace operator $\Delta_{\Omega^\varepsilon}=\partial_1^2+\partial_2^2+\partial_3^2$ is represented up to the first order differential operator as the sum of the Laplace-Beltrami operator on the mid-surface and the square of the transversal derivative

$$\Delta_{\Omega^\varepsilon}\widetilde{T}=\Delta_{\mathcal{C}}\widetilde{T}+\partial_t^2\widetilde{T}+2\mathcal{H}_{\mathcal{C}}\partial_t\widetilde{T}.$$

The Laplace-Beltrami operator $\Delta_{\mathcal{C}}$ is defined in (1) and mean curvature $\mathcal{H}_{\mathcal{C}}(\mathcal{X}) := \frac{1}{2}\sum_{k=1}^{3}\mathcal{D}_k\mathcal{N}_k(\mathcal{X})$ of the surface are extended properly from $\mathcal{C}$ (see the forthcoming Lemma 5 below).

Introducing the function $G(\mathcal{X},t)$ which has the same Dirichlet and Neumann traces as T on the $\partial\mathcal{C}\times(-\varepsilon,\varepsilon)$ and on the upper and lower surfaces $\mathcal{C}\times\{\pm\varepsilon\}$ respectively

$$G(\mathcal{X},t)=\frac{1}{4\varepsilon}(t+\varepsilon)^2q^+_\varepsilon(\mathcal{X})-\frac{1}{4\varepsilon}(t-\varepsilon)^2q^-_\varepsilon(\mathcal{X}),\qquad q^\pm_\varepsilon\in\widetilde{\mathbb{H}}^1(\mathcal{C}) \tag{3}$$

we can reduce the problem (16) to the following boundary value problem with respect to unknown function $T=\widetilde{T}-G$

$$\Delta_{\Omega^\varepsilon}T(\mathcal{X},t)=F(\mathcal{X},t),\quad(\mathcal{X},t)\in\mathcal{C}\times(-\varepsilon,\varepsilon), \tag{4}$$

$$T^+(\mathcal{X},t)=0,\qquad(\mathcal{X},t)\in\partial\mathcal{C}\times(-\varepsilon,\varepsilon), \tag{5}$$

$$(\partial_tT)^+(\mathcal{X},\pm\varepsilon)=0,\quad\mathcal{X}\in\mathcal{C}. \tag{6}$$

where

$$F(\mathcal{X},t):=f(\mathcal{X},t)-\frac{1}{4\varepsilon}\Big((t+\varepsilon)^2\Delta_{\mathcal{C}}q^+_\varepsilon(\mathcal{X})-(t-\varepsilon)^2\Delta_{\mathcal{C}}q^-_\varepsilon(\mathcal{X})\Big) \tag{7}$$

$$-\frac{\mathcal{H}^0_{\mathcal{C}}(\mathcal{X})}{2\varepsilon}\Big((t+\varepsilon)q^+_\varepsilon(\mathcal{X})-(t-\varepsilon)q^-_\varepsilon(\mathcal{X})\Big)-\frac{1}{2\varepsilon}\big(q^+_\varepsilon(\mathcal{X})-q^-_\varepsilon(\mathcal{X})\big),$$

$(\mathcal{X},t)\in\mathcal{C}\times(-\varepsilon,\varepsilon)$.

The BVP (4)–(6) is reformulated as the minimization problem for the functional which, after scaling (stretching the variable $t = \varepsilon\tau$ and dividing the entire functional by ε) has the following form

$$\begin{aligned} E_\varepsilon(T_\varepsilon) : = & \int_{-1}^{1}\int_{\mathcal{C}} \Big[\frac{1}{2}(\mathcal{D}_\mathcal{C} T_\varepsilon)^2(\mathcal{X},\tau) + \frac{1}{2\varepsilon^2}(\partial_\tau T_\varepsilon)^2(\mathcal{X},\tau) \\ & + F_\varepsilon(\mathcal{X},\tau) T_\varepsilon(\mathcal{X},\tau)\Big] d\sigma d\tau \qquad (8) \\ F_\varepsilon(\mathcal{X},t) := & \, F(\mathcal{X},\varepsilon t), \quad (\mathcal{X},t)\in\mathcal{C}\times(-\varepsilon,\varepsilon), \\ T_\varepsilon(\mathcal{X},\tau) := & \, T(\mathcal{X},\varepsilon\tau), \; T_\varepsilon \in \widetilde{\mathbb{H}}^1(\Omega^1, \partial\mathcal{C}\times(-1,1)), \\ & F_\varepsilon \in \widetilde{\mathbb{H}}^{-1}(\Omega^1), \; q_\varepsilon^\pm \in \widetilde{\mathbb{H}}^2(\mathcal{C}). \end{aligned}$$

In [5] it was proved that, if

$$\mathcal{P}(\mathcal{C}) := \Big\{T\in\mathbb{H}^1(\Omega^1) : T(\mathcal{X},\tau) = T_\mathcal{C}(\mathcal{X}), \quad T_\mathcal{C}\in\widetilde{\mathbb{H}}^1(\mathcal{C}), \quad \tau\in[-1,1]\Big\}, \qquad (9)$$

$$f_\varepsilon(\mathcal{X},t) := f(\mathcal{X},\varepsilon t) \underset{\varepsilon\to 0}{\to} f^0(\mathcal{X}) \quad in \quad \mathbb{L}_2(\Omega^1),$$

$q_\varepsilon^\pm \in \widetilde{\mathbb{H}}^2(\mathcal{C})$ is uniformly bounded (with respect to ε) in $\mathbb{H}^2(\mathcal{C})$, and

$$\lim_{\varepsilon\to 0} q_\varepsilon^+ = \lim_{\varepsilon\to 0} q_\varepsilon^- = q_0, \quad q_0\in\mathbb{L}_2(\mathcal{C}), \quad \frac{1}{2\varepsilon}(q_\varepsilon^+ - q_\varepsilon^-) \underset{\varepsilon\to 0}{\to} q_1 \quad in \quad \mathbb{L}_2(\mathcal{C}),$$

then the functional $E_\varepsilon(T_\varepsilon)$ in (8) Γ-converges to the functional

$$E^{(0)}(T) \qquad (10)$$

$$= \begin{cases} \displaystyle\int_\mathcal{C} \Big[\langle \mathcal{D}_\mathcal{C} T_\mathcal{C}(\mathcal{X}), \mathcal{D}_\mathcal{C} T_\mathcal{C}(\mathcal{X})\rangle + 2\big(f^0(\mathcal{X}) - \mathcal{H}_\mathcal{C}^0 q_0(\mathcal{X}) & \\ \quad - q_1(\mathcal{X})\big) T_\mathcal{C}(\mathcal{X})\Big] d\sigma, & T\in\mathcal{P}(\mathcal{C}); \\ +\infty, & T\notin\mathcal{P}(\mathcal{C}). \end{cases}$$

In particular, the following Dirichlet boundary value problem on the mid-surface $\mathcal{C}$

$$\begin{aligned} & \Delta_\mathcal{C} T(\mathcal{X}) = f^0(\mathcal{X}) - \mathcal{H}_\mathcal{C}^0 q_0(\mathcal{X}) - q_1(\mathcal{X}), \quad \mathcal{X}\in\mathcal{C}, \\ & T^+(\mathcal{X}) = 0, \qquad \mathcal{X}\in\partial\mathcal{C}, \quad T\in\mathbb{H}^1(\mathcal{C}), \quad f^0, q_0, q_1 \in\mathbb{L}_2(\mathcal{C}), \end{aligned}$$

is an equivalent reformulation of the minimization problem with the energy functional (10).

In [4] by using the calculus of G*ü*nter's tangential differential operators on hypersurfaces was established Finite Element Method for the considered boundary value problem and was found approximate solution in explicit form.

In the present paper we investigate similar problem for a mixed boundary value problem for the bi-Laplacian equation in a thin layer around a surface $\mathcal{C}$ with the boundary. We trace what happens in Γ-limit when the thickness of the layer converges to zero. It is shown how the mixed type boundary value problem (BVP) for the bi-Laplace equation in the initial thin layer transforms in the Γ-limit sense into an explicit Dirichlet BVP for the bi-Laplace-Beltrami equation on the surface.

2 Auxiliary Materials

Definition 1 Let $k \geq 1$ and $\omega \subset \mathbb{R}^n$ be a compact domain. An C^k-smooth hypersurface in $\mathbb{R}^n$ is implicitly defined as the set $\mathcal{S} = \Big\{\mathcal{X} \in \omega \ : \ \Psi_{\mathcal{S}}(\mathcal{X}) = 0\Big\}$ where $\Psi_{\mathcal{S}} : \omega \to \mathbb{R}$ is C^k-smooth (or Lipschitz smooth) and is regular $\nabla\,\Psi(\mathcal{X}) \neq 0$.

$\boldsymbol{\nu}_\Gamma(t)$ is the outer unit normal vector field to the boundary $\Gamma = \partial\mathcal{S}$, which is tangent to $\mathcal{S}$ and $\boldsymbol{\nu}(\mathcal{X})$ is the outer unit normal vector field to $\mathcal{S}$. By using implicit surface functions gradient we can write the unit normal vector field on the surface in explicit form:

$$\boldsymbol{\nu}(y) := \lim_{x \to t} \frac{(\nabla\Psi_{\mathcal{S}})(x)}{|(\nabla\Psi_{\mathcal{S}})(x)|}, \qquad t \in \mathcal{S}. \tag{11}$$

In applications it is necessary to extend the vector field $\boldsymbol{\nu}(t)$ in a neighborhood of $\mathcal{S}$, preserving some important features. Here is the precise definition of extension.

Definition 2 Let $\mathcal{S}$ be a surface in $\mathbb{R}^n$ with unit normal $\boldsymbol{\nu}$. A vector filed $\mathcal{N} \in C^1(\Omega^\varepsilon)$ in a neighborhood Ω^ε of $\mathcal{S}$, will be referred to as a proper extension if $\mathcal{N}\Big|_{\mathcal{S}} = \boldsymbol{\nu}$, it is a unit vector $|\mathcal{N}| = 1$ in Ω^ε and it is a gradient vector field in the neighborhood: $\partial_j\mathcal{N}_k(x) = \partial_k\mathcal{N}_j(x)$ for all $x \in \Omega^\varepsilon$, $j, k = 1, \ldots, n$.

In [8] it was proved that the "naive" extension (cf. (11)) $\boldsymbol{\nu}(t) := \frac{(\nabla\Psi_{\mathcal{S}})(x)}{|(\nabla\Psi_{\mathcal{S}})(x)|}$, $x \in \Omega^\varepsilon$ is not proper.

Corollary 3 *For any proper extension $\mathcal{N}(x)$, $x \in \Omega^\varepsilon \subset \mathbb{R}^n$ of the unit normal vector field $\boldsymbol{\nu}$ to the surface $\mathcal{S} \subset \Omega^\varepsilon$ the equality $\partial_{\mathcal{N}}\mathcal{N}(x) = 0$ holds for all $x \in \Omega^\varepsilon$.*

In particular, for the derivatives $\mathcal{D}_k = \partial_k - \mathcal{N}_k\partial_{\mathcal{N}}$, $k = 1, \ldots, n$, which are extension into the domain Ω^ε of Günter's derivatives $\mathcal{D}_k = \partial_k - \nu_k\partial_{\boldsymbol{\nu}}$ on the surface $\mathcal{S}$, the equalities hold: $\mathcal{D}_k\mathcal{N}_j = \partial_k\mathcal{N}_j - \mathcal{N}_k\partial_{\mathcal{N}} = \partial_k\mathcal{N}_j$, $\mathcal{D}_k\mathcal{N}_j = \mathcal{D}_j\mathcal{N}_k$, for all $j, k = 1, \ldots, n$.

Let us consider the system of $(n+1)$-vectors

$$\boldsymbol{d}^{\,j} := \boldsymbol{e}^j - \mathcal{N}_j\mathcal{N}, \qquad j = 1, \dots, n \quad \text{and} \quad \boldsymbol{d}^{\,n+1} := \mathcal{N}, \tag{12}$$

where $\boldsymbol{e}^1, \dots, \boldsymbol{e}^n$ is the Cartesian basis in $\mathbb{R}^n$; the first n vectors $\boldsymbol{d}^{\,1}, \dots, \boldsymbol{d}^{\,n}$ are tangent to the surface $\mathcal{C}$, while the last one $\boldsymbol{d}^{\,n+1} = \mathcal{N}$ is orthogonal to all $\boldsymbol{d}^{\,1}, \dots, \boldsymbol{d}^{\,n}$.

Definition 4 For a function $\varphi \in \mathbb{H}^1(\Omega^\varepsilon)$ the extended gradient is

$$\mathcal{D}_{\Omega^\varepsilon}\,\varphi = \left\{\mathcal{D}_1\varphi, \dots, \mathcal{D}_n\varphi, \mathcal{D}_{n+1}\varphi\right\}^\top = \sum_{j=1}^{n+1}(\mathcal{D}_j\varphi)\boldsymbol{d}^j, \qquad \mathcal{D}_{n+1}\varphi := \partial_{\mathcal{N}}\varphi \tag{13}$$

and for a smooth vector field $\boldsymbol{U} = \sum_{j=1}^{n+1} U_j^0\boldsymbol{d}^{\,j} \in \mathcal{W}(\Omega^\varepsilon)$ the extended divergence is

$$\mathrm{div}_{\Omega^\varepsilon}\,\boldsymbol{U} := \sum_{j=1}^{n+1}\mathcal{D}_jU_j^0 + \mathcal{H}_{\mathcal{C}}^0\langle\mathcal{N}, \boldsymbol{U}\rangle = -\nabla_{\Omega^\varepsilon}^*\boldsymbol{U}, \tag{14}$$

where $\mathcal{H}_{\mathcal{C}}^0(\mathcal{X}) = (n-1)\mathcal{H}_{\mathcal{C}}(\mathcal{X})$, $\mathcal{H}_{\mathcal{C}}(\mathcal{X})$ is the extended mean curvature and

$$\mathcal{H}_{\Omega^\varepsilon}^0(x) := \sum_{j=1}^{n}\partial_j\mathcal{N}_j(x) = \sum_{j=1}^{n+1}\mathcal{D}_j\mathcal{N}_j(x)$$
$$= \sum_{j=1}^{n}\mathcal{D}_j\nu_j(\mathcal{X}) = \mathcal{H}_{\mathcal{C}}^0(\mathcal{X}), \; x \in \Omega^\varepsilon, \quad \mathcal{X} = \pi_{\mathcal{C}}x.$$

Lemma 5 *The classical gradient* $\nabla\varphi := \left\{\partial_1\varphi, \dots, \partial_n\varphi\right\}^\top$, *written in the full system of vectors* $\left\{\boldsymbol{d}^{\,j}\right\}_{j=1}^{n+1}$ *in* (12) *coincides with the extended gradient* $\mathcal{D}_{\Omega^\varepsilon}\,\varphi$ *in* (13).

Similarly, the classical divergence $\mathrm{div}\,\boldsymbol{U} := \sum_{j=1}^{n}\partial_jU_j$ *of a vector field* $\boldsymbol{U} := \sum_{j=1}^{n}U_j\boldsymbol{e}^j$, *written in the full system* (12), *coincides with the extended divergence* $\mathrm{div}\,\boldsymbol{U} = \mathrm{div}_{\Omega^\varepsilon}\,\boldsymbol{U}$ *in* (14). *The extended gradient and the negative extended divergence are dual to* $\mathcal{D}_{\Omega^\varepsilon}^* = -\mathrm{div}_{\Omega^\varepsilon}$ *and* $\mathrm{div}_{\Omega^\varepsilon}^* = -\mathcal{D}_{\Omega^\varepsilon}$ *respectively.*

The Laplace-Beltrami operator $\Delta_{\Omega^\varepsilon} := \mathrm{div}_{\Omega^\varepsilon}\mathcal{D}_{\Omega^\varepsilon}\,\varphi = -\mathcal{D}^*_{\Omega^\varepsilon}\left(\mathcal{D}_{\Omega^\varepsilon}\varphi\right)$ *on* Ω^ε*, written in the full system* (12)*, has the following form*

$$\begin{aligned}\Delta_{\Omega^\varepsilon}\varphi &= \sum_{j=1}^{n}\mathcal{D}_j^2\varphi + \partial^2_{\mathcal{N}}\varphi + \mathcal{H}^0_{\mathcal{C}}\partial_{\mathcal{N}}\varphi\\ &= \sum_{j=1}^{n+1}\mathcal{D}_j^2\varphi + \mathcal{H}^0_{\mathcal{C}}\mathcal{D}_{n+1}\varphi\,,\ \varphi\in\mathbb{H}^2(\Omega^\varepsilon)\end{aligned} \tag{15}$$

hence the bi-Laplace-Beltrami operator acquires the following form

$$\begin{aligned}\Delta^2\varphi = &\sum_{k,j=1}^{n+1}\mathcal{D}_k^2\mathcal{D}_j^2\varphi + \sum_{k=1}^{n+1}\mathcal{D}_k^2(\mathcal{H}^0_{\mathcal{C}}\mathcal{D}_{n+1}\varphi)\\ &+\sum_{j=1}^{n+1}(\mathcal{H}^0_{\mathcal{C}}\mathcal{D}_{n+1}\mathcal{D}_j^2\varphi + \mathcal{H}^0_{\mathcal{C}}\mathcal{D}_{n+1}(\mathcal{H}^0_{\mathcal{C}}\mathcal{D}_{n+1}\varphi)).\end{aligned}$$

Definition and basic properties of Γ-convergence can be found in [3]:

Definition 6 (Γ-Convergence) We say that a sequence $f_j : X \to \overline{\mathbb{R}}$ Γ-converges in X to $f_\infty : X \to \overline{\mathbb{R}}$ if for all $x \in X$ we have

(i) (lim inf inequality) For every sequence (x_j) converging to x,

$$f_\infty(x) \leqslant \liminf_j f_j(x_j);$$

(ii) (Recovery sequence) There exists a sequence $\left(x_j^0\right)$ converging to x, such that

$$f_\infty(x) \geq \limsup_j f_j\left(x_j^0\right);$$

The function f_∞ is called the Γ-limit of (f_j) and we write $f_\infty = \Gamma\text{-}\lim_j f_j$.

Γ-convergence has some "strange" properties, but it is perfectly adapted to the problems of mathematical physics. In particular, if ω is a bounded open subset of $\mathbb{R}^{n-1}$, and if $f : \mathbb{R}^n \to \mathbb{R}$ is a strictly-convex function with quadratic growth; that is, $c_1|z|^2 - c_2 \leq f(z) \leq c_3\left(1+|z|^2\right)$, for all $z \in \mathbb{R}^n$, and all $\varepsilon > 0$ the energy functional

$$E_\varepsilon(u) = \int_{\omega\times(0,\varepsilon)} f(Du)\,dx \quad \text{defined on} \quad W^{1,2}(\omega\times(0,\varepsilon)),$$

Γ-converges to

$$\Gamma \to \lim_{\varepsilon \to 0+} \frac{1}{\varepsilon} \int_{\omega \times (0,\varepsilon)} \bar{f}(Du)\, dx = \int_{\omega} \bar{f}\left(\hat{D}u\right) d\hat{x}.$$

Here $\hat{x} = (x_1, ..., x_{n-1})$, $x = \left(\hat{x}, x_n\right)$, $\hat{D}u = (D_1 u, ..., D_{n-1}u)$, $Du = \left(\hat{D}u, D_n u\right)$ and $\bar{f}(z) = \min\{f(z, b) : b \in R\}$, upon identifying $\omega \times (0, \varepsilon)$ with $\omega \times (0, 1)$ by scaling in the n-th variable and $W^{1,2}(\omega)$ with the functions in $\omega \times (0, 1)$ independent of the n-th variable.

3 Variational Formulation of Model Problem and Γ Convergence

We consider the bi-Laplace equation in "isotropic" medium, with the classical mixed boundary conditions on the boundary in the layer domain $\Omega^\varepsilon := \mathcal{C} \times (-\varepsilon, \varepsilon)$ of thickness 2ε, where $\mathcal{C} \subset \mathcal{S}$ is a smooth subsurface of a closed hypersurface $\mathcal{S}$ with smooth nonempty boundary $\partial\mathcal{C}$. We will investigate the following BVP:

$$\begin{aligned}
&\Delta^2_{\Omega^\varepsilon} v(\mathcal{X}, t) = f(\mathcal{X}, t), && (\mathcal{X}, t) \in \mathcal{C} \times (-\varepsilon, \varepsilon), \\
&v^+(\mathcal{X}, t) = 0, && (\mathcal{X}, t) \in \partial\mathcal{C} \times (-\varepsilon, \varepsilon), \\
&(\partial_t v)^+(\mathcal{X}, \pm\varepsilon) = 0, && \mathcal{X} \in \mathcal{C} \times \{\pm\varepsilon\}, \\
&(\Delta v)^+(\mathcal{X}, t) = 0 && (\mathcal{X}, t) \in \partial\mathcal{C} \times (-\varepsilon, \varepsilon), \\
&(\partial_t \Delta v)^+(\mathcal{X}, \pm\varepsilon) = h^\pm_\varepsilon && (\mathcal{X}, t) \in \mathcal{C} \times \{\pm\varepsilon\},
\end{aligned} \tag{16}$$

where $v \in \mathbb{H}^2_p(\Omega^\varepsilon)$, $h^\pm_\varepsilon \in \mathbb{H}^{-3/2}_p(\mathcal{C})$.

Results on the uniqueness and solvability of the Bi-Laplace equation in a classical setting can be found in [15].

For the investigation we assume that the bi-Laplace operator $\Delta^2_{\Omega^\varepsilon}$ is represented by the following sum:

$$\begin{aligned}
\Delta^2_{\Omega^\varepsilon} \widetilde{u} &= \Delta^2_{\mathcal{C}} \widetilde{u} + 2\partial_t^2 \Delta_{\mathcal{C}} \widetilde{u} + 2\Delta_{\mathcal{C}} \mathcal{H}_{\mathcal{C}} \partial_t \widetilde{u} \\
&\quad + 4\mathcal{H}_{\mathcal{C}} \partial_t^3 \widetilde{u} + 2\mathcal{H}_{\mathcal{C}} \Delta_{\mathcal{C}} \partial_t \widetilde{u} + 4\mathcal{H}_{\mathcal{C}} \partial_t^2 \widetilde{u} + \partial_t^4 \widetilde{u},
\end{aligned}$$

where $\mathcal{D}_4 = \partial_t$. The bi-Laplace-Beltrami operator $\Delta^2_{\mathcal{C}}$ and the mean curvature $\mathcal{H}_{\mathcal{C}}(\mathcal{X})$ of the surface are extended properly from $\mathcal{C}$.

Consider the following auxiliary BVP associated with the BVP (16)

$$\begin{aligned}
&\Delta_{\Omega^\varepsilon} G(\mathcal{X}, t) = \frac{1}{4\varepsilon}(t+\varepsilon)^2 h_\varepsilon^+ - \frac{1}{4\varepsilon}(t-\varepsilon)^2 h_\varepsilon^-, && (\mathcal{X}, t) \in \mathcal{C} \times (-\varepsilon, \varepsilon),\\
&G^\pm(\mathcal{X}, t) = 0, && (\mathcal{X}, t) \in \partial\mathcal{C} \times (-\varepsilon, \varepsilon),\\
&(\partial_t G)^\pm(\mathcal{X}, \pm\varepsilon) = 0, && \mathcal{X} \in \mathcal{C} \times \{\pm\varepsilon\},
\end{aligned} \tag{17}$$

We can reduce the problem (16) to the following boundary value problem with an unknown function $u = v - G$

$$\begin{aligned}
&\Delta^2_{\Omega^\varepsilon} u(\mathcal{X}, t) = \Delta^2_{\Omega^\varepsilon} v(\mathcal{X}, t) - \Delta_{\Omega^\varepsilon}\Delta_{\Omega^\varepsilon} G := F(\mathcal{X}, t), \quad (\mathcal{X}, t) \in \mathcal{C} \times (-\varepsilon, \varepsilon),\\
&u^+(\mathcal{X}, t) = 0, \qquad (\mathcal{X}, t) \in \partial\mathcal{C} \times (-\varepsilon, \varepsilon),\\
&(\partial_t u)^+(\mathcal{X}, \pm\varepsilon) = 0, \qquad \mathcal{X} \in \mathcal{C} \times \{\pm\varepsilon\},\\
&(\Delta u)^+(\mathcal{X}, t) = 0, \qquad (\mathcal{X}, t) \in \partial\mathcal{C} \times (-\varepsilon, \varepsilon),\\
&(\partial_t \Delta u)^+(\mathcal{X}, \pm\varepsilon) = 0, \qquad \mathcal{X} \in \mathcal{C} \times \{\pm\varepsilon\},
\end{aligned} \tag{18}$$

where $u \in \mathbb{H}^2_p(\Omega^\varepsilon)$, $h_\varepsilon^\pm \in \widetilde{\mathbb{H}}^2(\mathcal{C})$ and

$$\begin{aligned}
F(\mathcal{X}, t) = f(\mathcal{X}, t) &- \frac{1}{4\varepsilon}\Big((t+\varepsilon)^2 \Delta_{\mathcal{C}} h_\varepsilon^+(\mathcal{X}) - (t-\varepsilon)^2 \Delta_{\mathcal{C}} h_\varepsilon^-(\mathcal{X})\Big) \qquad (19)\\
&- \frac{\mathcal{H}^0_{\mathcal{C}}(\mathcal{X})}{2\varepsilon}\big((t+\varepsilon) h_\varepsilon^+(\mathcal{X}) - (t-\varepsilon) h_\varepsilon^-(\mathcal{X})\big) - \frac{1}{2\varepsilon}(h_\varepsilon^+(\mathcal{X})\\
&- h_\varepsilon^-(\mathcal{X})), \ (\mathcal{X}, t) \in \mathcal{C} \times (-\varepsilon, \varepsilon).
\end{aligned}$$

The BVP (18) is reformulated as the minimization problem for the functional which, after scaling (stretching the variable $t = \varepsilon\tau$ and dividing the entire functional by ε) has the following form

$$\begin{aligned}
E_\varepsilon(u_\varepsilon) := \int_{-1}^{1}\int_{\mathcal{C}} \Bigg[\frac{1}{2}\left(\sum_{j=1}^{3} \mathcal{D}_j^2 u_\varepsilon(\mathcal{X}, \tau) + \frac{1}{\varepsilon^2}\partial_\tau^2 u_\varepsilon(\mathcal{X}, \tau) + \frac{1}{\varepsilon}\mathcal{H}^0_{\mathcal{C}}\partial_\tau u_\varepsilon(\mathcal{X}, \tau)\right)^2\\
- F_\varepsilon(\mathcal{X}, \tau) u_\varepsilon(\mathcal{X}, \tau)\Bigg] d\sigma\, d\tau, \qquad (20)
\end{aligned}$$

$$F_\varepsilon(\mathcal{X}, t) = F(\mathcal{X}, \varepsilon t) = f(\mathcal{X}, \varepsilon t) - \frac{\varepsilon}{4}\Big((t+1)^2 \Delta_{\mathcal{C}} h_\varepsilon^+(\mathcal{X}) - (t-1)^2 \Delta_{\mathcal{C}} h_\varepsilon^-(\mathcal{X})\Big)$$

$$-\frac{\mathcal{H}^0_C(\mathcal{X})}{2}\big((t+1)h^+_\varepsilon(\mathcal{X})-(t-1)h^-_\varepsilon(\mathcal{X})\big)-\frac{1}{2\varepsilon}(h^+_\varepsilon(\mathcal{X})-h^-_\varepsilon(\mathcal{X})),$$
$$u_\varepsilon\in\widetilde{\mathbb{H}}^2(\Omega^1,\partial\mathcal{C}\times(-1,1)),\quad F_\varepsilon\in\widetilde{\mathbb{H}}^{-2}(\Omega^1),\quad h^\pm_\varepsilon\in\widetilde{\mathbb{H}}^2(\mathcal{C}).$$

The problem is: Do these energies defined on thin n-dimensional domains Ω_ε "converge" to an energy defined on the $n-1$ dimensional hypersurface $\mathcal{C}$ (the mid-surface of Ω_ε) when the domain Ω_ε is "squeezed" infinitely in the transversal direction to $\mathcal{C}$?

The main result of the present investigation is the following Theorem 7.

Theorem 7 *Let $\mathcal{P}(\mathcal{C})$ is defined in (9) and*

$$f_\varepsilon(\mathcal{X},t):=f(\mathcal{X},\varepsilon t)\underset{\varepsilon\to 0}{\to}f^0(\mathcal{X})\quad in\quad \mathbb{L}_2(\Omega^1),$$

$h^\pm_\varepsilon\in\widetilde{\mathbb{H}}^2(\mathcal{C})$ be uniformly bounded (with respect to ε) in $\mathbb{H}^2(\mathcal{C})$, and

$$\lim_{\varepsilon\to 0}h^+_\varepsilon=\lim_{\varepsilon\to 0}h^-_\varepsilon=h_0,\quad h_0\in\mathbb{L}_2(\mathcal{C}),$$
$$\frac{1}{2\varepsilon}(h^+_\varepsilon-h^-_\varepsilon)\underset{\varepsilon\to 0}{\to}h_1\quad in\quad \mathbb{L}_2(\mathcal{C}).$$

Then the functional $E_\varepsilon(u_\varepsilon)$ in (20) Γ-converges to the functional

$$E^{(0)}(T)=\begin{cases}\displaystyle\int_{\mathcal{C}}\Bigg[\bigg(\sum_{j=1}^3\mathcal{D}_j^2u_{\mathcal{C}}(\mathcal{X})\bigg)^2+2\big(f^0(\mathcal{X})-\mathcal{H}^0_{\mathcal{C}}h_0(\mathcal{X})\\ \qquad -h_1(\mathcal{X})\big)u_{\mathcal{C}}(\mathcal{X})\Bigg]d\sigma, & u\in\mathcal{P}(\mathcal{C});\\ +\infty, & u\notin\mathcal{P}(\mathcal{C}).\end{cases}\tag{21}$$

The following Dirichlet boundary value problem on the mid-surface $\mathcal{C}$

$$\begin{aligned}&\Delta^2_{\mathcal{C}}u(\mathcal{X})=f^0(\mathcal{X})-\mathcal{H}^0_{\mathcal{C}}h_0(\mathcal{X})-h_1(\mathcal{X}),\quad \mathcal{X}\in\mathcal{C},\\ &u^+(\mathcal{X})=0,\qquad \mathcal{X}\in\partial\mathcal{C},\\ &\delta^+_{\mathcal{C}}(\chi)=0,\quad \chi\in\partial\mathcal{C}\\ &u\in\mathbb{H}^2(\mathcal{C}),\quad f^0,h_0,h_1\in\mathbb{L}_2(\mathcal{C}),\end{aligned}\tag{22}$$

is an equivalent reformulation of the minimization problem with the energy functional (21).

Acknowledgments The research was supported by Shota Rustaveli National Science Foundation grant FR-21-2844.

References

1. L. Andersson, P.T. Chrusciel, Cauchy data for vacuum Einstein equations and obstructions to smoothness of null infinity. Phys. Rev. Lett. **70**, 2829–2832 (1993)
2. R. Aris, *Vectors, Tensors, and the Basic Equations of Fluid Mechanics* (Prentice-Hall, Englewood Cliffs, 1962)
3. A. Braides *Gamma-Convergence for Beginners*. Oxford Lecture Series in Mathematics and its Applications (Oxford University Press, New York, 2007)
4. T. Buchukuri, R. Duduchava, G. Tephnadze, Laplace-Beltrami equation on hypersurfaces and Γ-convergence. Trans. Razmadze Math. Inst. **170**(3), 300–307 (2016)
5. T. Buchukuri, R. Duduchava, G. Tephnadze, Laplace-Beltrami equation on hypersurfaces and Γ-convergence. Math. Methods Appl. Sci. **40**(13), 4637–4657 (2017)
6. P.G. Ciarlet, *Introduction to Linear Shell Theory*. Series in Applied Mathematics, vol. 1 (Gauthier-Villars, Éditions Scientifiques et Médicales Elsevier, Paris, 1998)
7. R. Duduchava, On multidimensional singular integral operators, I-II: the case of compact manifolds. J. Oper. Theory **11**, 41–76 (1984). 199-214
8. R. Duduchava, E. Shargorodsky, G. Tephnadze, Extension of the unit normal vector field from a hypersurface. Georgian Math. J. **22**(5), 355–359 (2015)
9. G. Friesecke, R.D. James, S. Müller, A theorem on geometric rigidity and the derivation of nonlinear plate theory from three dimensional elasticity. Commun. Pure Appl. Math. **55**(11), 1461–1506 (2002)
10. N. Günter, *Potential Theory and its Application to the Basic Problems of Mathematical Physics* (Fizmatgiz, Moscow, 1953). (Russian. Translation in French: Gauthier–Villars, Paris 1994)
11. W. Haack, *Elementare Differentialgeometrie* (in German). Lehrbücher und Monographien aus dem Gebiete der exakten Wissenschaften, vol. 20 (Birkhäuser Verlag, Basel, VIII, 1955)
12. V. Kupradze, T. Gegelia, M. Basheleishvili, T. Burchuladze, *Three-Dimensional Problems of the Mathematical Theory of Elasticity and Thermoelasticity* (North-Holland, Amsterdam, 1979). (Russian edition: Nauka, Moscow 1976)
13. U. Massari, M. Miranda, *Minimal Surfaces of Codimension One*. North-Holland Mathematics Studies, vol. 91. Notas de Matemtica, 95 (North-Holland, Amsterdam, 1984)
14. R. Temam, M. Ziane, *Navier-Stokes Equations in Thin Spherical Domains*. Optimization Methods in Partial Differential Equations, Contemporaty Mathematics, vol. 209 (AMS, Providence), pp. 281–314
15. M. Tsaava, The boundary value problems for the bi-laplace-beltrami equation. Mem. Differ. Equ. Math. Phys. **77**, 2019, 93–103

Bounded Weak Solutions with Orlicz Space Data: An Overview

David Cruz-Uribe

Abstract It is well known that non-negative solutions to the Dirichlet problem $\Delta u = f$ in a bounded domain Ω, where $f \in L^q(\Omega)$, $q > \frac{n}{2}$, satisfy $\|u\|_{L^\infty(\Omega)} \leq C\|f\|_{L^q(\Omega)}$. We generalize this result by replacing the Laplacian with a degenerate elliptic operator, and we show that we can take the data f in an Orlicz space $L^A(\Omega)$ that, in the classical case, lies strictly between $L^{\frac{n}{2}}(\Omega)$ and $L^q(\Omega)$, $q > \frac{n}{2}$.

Keywords Orlicz spaces · Degenerate elliptic equations · Bounded solutions · A priori estimates

2000 Mathematics Subject Classification 35B45, 35D30, 35J25, 46E30

1 Introduction: Uniformly Elliptic Operators

In this note we survey recent results from [7], done jointly with Scott Rodney. Let Ω be a bounded, open, and connected subset of $\mathbb{R}^n$, let $Q = Q(x)$ be an $n \times n$ positive semi-definite, self-adjoint measurable matrix function, and let v be a non-negative, measurable function. (Hereafter, we refer to v as a weight.) We are interested in studying solutions of the Dirichlet problem for the elliptic PDE

$$\begin{cases} -\operatorname{Div}(Q\nabla u) = vf & \text{for } x \in \Omega \\ u = 0 & \text{for } x \in \partial\Omega \end{cases} \tag{1}$$

Provided that $v(x) > 0$ a.e. we can define the operator

$$Lu = -v^{-1}\operatorname{Div}(Q\nabla u),$$

D. Cruz-Uribe (✉)
Department of Mathematics, University of Alabama, Tuscaloosa, AL, USA
e-mail: dcruzuribe@ua.edu

D. Cardona et al. (eds.), *Extended Abstracts 2021/2022*, Research Perspectives Ghent Analysis and PDE Center 3, https://doi.org/10.1007/978-3-031-48579-4_9

and while this does not make sense if $v(x) = 0$ on a set of positive measure, we will abuse notation and say that we are interested in solutions of the equation $Lu = f$. As we will see, there is a close interaction between the matrix Q and the weight v; in turn, much of the work on this equation is informed by the theory of weighted norm inequalities in harmonic analysis.

Our work is motivated by earlier results by Fabes et al. [11], Chanillo and Wheeden [3], Franchi et al. [12], and Sawyer and Wheeden [18–20]. Our previous work with Rodney and Rosta [9, 10] is also relevant.

Our starting point is the following classical result for uniformly elliptic operators due to Trudinger [22] (see also Maz′ya [15, 16] and Stampacchia [21]).

Theorem 1 *Let $f \in L^q(\Omega)$, $q > \frac{n}{2}$, Q uniformly elliptic, and $v = 1$. If u is a non-negative weak solution of* (1), *then*

$$\|u\|_{L^\infty(\Omega)} \leq C\|f\|_{L^q(\Omega)}.$$

The standard proof of this result using Moser iteration, and the classical Sobolev inequality

$$\left(\int_\Omega |\psi(x)|^{\frac{2n}{n-2}}\,dx\right)^{\frac{n-2}{2n}} \leq C\left(\int_\Omega |\nabla\psi(x)|^2\,dx\right)^{\frac{1}{2}}.$$

The lower bound for q is closely related to the fact that

$$\frac{n}{2} = \left(\frac{n}{n-2}\right)'$$

is the dual exponent of the "gain" in the Sobolev inequality. The bound on q in Theorem 1 is sharp: if we take Q to be the identity (so the operator becomes the Laplacian) and $\Omega = B(0,1)$, and if we let

$$f(x) = \frac{1}{|x|^2\log(e+|x|^{-1})},$$

then $f \in L^{\frac{n}{2}}(\Omega)$ but the solution to $\Delta u = f$ is unbounded at the origin.

Our first result generalizes Theorem 1 by showing that we can get closer to the endpoint by passing to a finer scale of spaces. Recall that a Young function $A : [0,\infty) \to [0,\infty)$, is an increasing, convex function that satisfies $A(0) = 0$ and $A(t)/t \to \infty$ as $t \to \infty$. We define the Orlicz space $L^A(\Omega)$ to consist of all measurable functions f such that

$$\|f\|_{L^A(\Omega)} = \inf\left\{\lambda > 0 : \int_\Omega A\left(\frac{|f(x)|}{\lambda}\right)dx \leq 1\right\} < \infty.$$

If we take $A(t) = t^{\frac{n}{2}} \log(e+t)^q$, $q > 0$, then for every $\epsilon > 0$,

$$L^{\frac{n}{2}+\epsilon}(\Omega) \subsetneq L^A(\Omega) \subsetneq L^{\frac{n}{2}}(\Omega).$$

For more information on Orlicz spaces, see [17].

Theorem 2 *Let $f \in L^A(\Omega)$, $A(t) = t^{\frac{n}{2}} \log(e+t)^q$, $q > \frac{n}{2}$, Q uniformly elliptic, and $v = 1$. If u is a non-negative weak solution of* (1), *then*

$$\|u\|_{L^\infty(\Omega)} \le C\|f\|_{L^A(\Omega)}.$$

Theorem 2 is a special case of our main result, Theorem 3 below. It is, however, not new: earlier, Cianchi [4] proved it using very different techniques; he also gave a better lower bound, proving it for $q > \frac{n}{2} - 1$. The above example shows that this bound is sharp.

2 Degenerate Elliptic Operators

Our main result is a generalization of Theorem 2 that holds for a large class of degenerate elliptic operators. Our approach, following our previous work in [9, 10] and the earlier work of Sawyer and Wheeden [19, 20] is to give the broadest possible hypotheses on the matrix Q and the weight v for which our results hold. We make three critical assumptions:

- $v \in L^1(\Omega)$;
- $|Q(x)|_{\mathrm{op}} = \sup\{|Q(x)\xi| : \xi \in \mathbb{R}^n, |\xi| = 1\} \le kv(x)$;
- there exists $\sigma > 1$ such that for all $\psi \in \mathrm{Lip}_0(\Omega)$ (that is, compactly supported Lipschitz functions)

$$\left(\int_\Omega |\psi(x)|^{2\sigma} v(x)\,dx\right)^{\frac{1}{2\sigma}} \le C\left(\int_\Omega |Q^{\frac{1}{2}}(x)\nabla\psi(x)|^2\,dx\right)^{\frac{1}{2}}.$$

These hypotheses hold in a number of cases: if we take $v = 1$ and let Q be uniformly elliptic, the constant k is just the upper bound on the eigenvalues of Q, and we can take $\sigma = \frac{n}{n-2}$ and use the classical Sobolev inequality. Fabes et al. [11] considered the case when the weight v satisfies the Muckenhoupt A_2 condition,

$$[v]_{A_2} = \sup_B \frac{1}{|B|}\int_B v(x)\,dx\,\frac{1}{|B|}\int_B v(x)^{-1}\,dx < \infty,$$

where B is any ball, and Q satisfies the degenerate ellipticity condition

$$\lambda v(x)|\xi|^2 \le \langle Q(x)\xi, \xi\rangle \le \Lambda v(x)|\xi|^2.$$

They proved that in this case the Sobolev inequality holds for $\sigma = \frac{n}{n-1} + \delta$, where $\delta > 0$ depends on n and $[w]_{A_2}$.

Chanillo and Wheeden [3] introduced the concept of 2-admissible pairs. They considered matrices Q that satisfy

$$w(x)|\xi|^2 \leq \langle Q(x)\xi, \xi\rangle \leq v(x)|\xi|^2,$$

where w, v are weights that satisfy $w(x) \leq v(x)$, v doubling, $w \in A_2$, and together satisfy a balance condition: there exists $\sigma > 1$ such that given $B_1 \subset B_2 \subset \Omega$,

$$\frac{r(B_1)}{r(B_2)}\left(\frac{v(B_1)}{v(B_2)}\right)^{\frac{1}{2\sigma}} \leq C\left(\frac{w(B_1)}{w(B_2)}\right)^{\frac{1}{2}}, \tag{2}$$

where $r(B)$ is the radius of the ball B. They proved that in this case a weighted Sobolev inequality holds:

$$\left(\int_\Omega |\psi(x)|^{2\sigma} v(x)\,dx\right)^{\frac{1}{2\sigma}} \leq C\left(\int_\Omega |\nabla\psi(x)|^2 w(x)\,dx\right)^{\frac{1}{2}} \leq \left(\int_\Omega |Q^{\frac{1}{2}}(x)\nabla\psi(x)|^2\,dx\right)^{\frac{1}{2}}.$$

Explicit examples of weights that satisfy the balance condition (2) can be found in [5].

The next step is to give a definition of weak solutions that is adapted to our operator. We follow the approach developed in [9]. Define the degenerate Sobolev space $QH_0^1(v;\Omega)$ to be the closure of $\mathrm{Lip}_0(\Omega)$ with respect to the norm

$$\|\psi\|_{QH_0^1(v;\Omega)} = \|\psi\|_{L^2(v;\Omega)} + \|\nabla\psi\|_{L^2_Q(\Omega)} = \left(\int_\Omega |\psi|^2 v\,dx\right)^{\frac{1}{2}} + \left(\int_\Omega |\sqrt{Q(x)}\nabla\psi|^2\,dx\right)^{\frac{1}{2}}.$$

Formally, $QH_0^1(v;\Omega)$ consists of equivalence classes of Cauchy sequences; however, we can define a unique pair of functions that is associated with each such class. Given a Cauchy sequence $\{\phi_k\}_{k=1}^\infty$, since both $L^2(v;\Omega)$ and $L^2_Q(\Omega)$ are Banach spaces, the sequence converges to some function u in $L^2(v;\Omega)$, and the sequence $\{\nabla\phi_k\}_{k=1}^\infty$ converges to some vector-valued function $\mathbf{g}$ in $L^2_Q(\Omega)$. We will write $\nabla u = \mathbf{g}$ and think of it as the weak gradient of u. However, it is important to note that in this setting, $\mathbf{g}$ may not be a weak derivative in the classical sense if the matrix Q is too degenerate. In [11], the authors give an example of a matrix Q and a pair $(u, \mathbf{g})$ such that u is non-constant, but $\mathbf{g} = 0$.

We now define weak solutions of the Dirichlet problem (1) to be any pair $\mathbf{u} = (u, \nabla u) \in QH_0^1(v; \Omega)$ that satisfies

$$\int_\Omega \nabla\psi(x) \cdot Q(x)\nabla u(x)\, dx = \int_\Omega f(x)\psi(x)v(x)\, dx$$

for every $\psi \in \mathrm{Lip}_0(\Omega)$.

We can now state our main result. Here $L^A(v; \Omega)$ is an Orlicz space defined as above but with respect to the measure $v\, dx$.

Theorem 3 *Let Q and v satisfy the above hypotheses with gain $\sigma > 1$ in the Sobolev inequality. Let $f \in L^A(v; \Omega)$, where $A(t) = t^{\sigma'} \log(e + t)^q$, $q > \sigma'$. If $\mathbf{u} = (u, \nabla u) \in QH_0^1(v; \Omega)$ is a non-negative weak solution of (1), then*

$$\|u\|_{L^\infty(v;\Omega)} \leq C\|f\|_{L^A(v;\Omega)}.$$

The proof of Theorem 3 is loosely modeled on one of the proofs of Theorem 1. As we noted above, the typical proof of this result uses Moser iteration, but we were unable to make it work in our setting. Instead, we used a proof that relied on De Giorgi iteration, adapting ideas from the recent work of Korobenko, et al. [14].

The first step in the proof is technical: as we noted above, given a pair $(u, \nabla u) \in QH_0^1(v; \Omega)$, ∇u may not be a classical weak derivative. Nevertheless, we need it to satisfy many of the same properties; in proving that they do, we make very heavy use of the hypothesis that $|Q|_{\mathrm{op}} \leq kv$. This should be contrasted with our results in [9] which did not require this assumption.

Given these properties, we can now begin the process of De Giorgi iteration. For each $r > 0$, define

$$\phi = \phi_r(u) = (u - r)_+.$$

Let $S(r) = \{x : u(x) > r\}$. Then we have that

$$(\phi, \nabla\phi) = ((u - r)_+, \chi_{S(r)}\nabla u) \in QH_0^1(\Omega).$$

By the Sobolev inequality we assume to hold, and by the definition of weak solutions (using ϕ as our test function), we have that

$$\|\phi\|_{L^{2\sigma}(v;\Omega)}^2 \leq C_0\|f\|_{L^{(2\sigma)'}(v;\Omega)}\|\phi\|_{L^{2\sigma}(v;S(r))}.$$

By Hölder's inequality in the scale of Orlicz spaces, and by the definition of the Orlicz norm, for $s > r$ we have that

$$\begin{aligned} v(S(s))^{\frac{1}{2\sigma}}(s-r) &\le \|\phi\|_{L^{2\sigma}(v;\Omega)} \\ &\le C\|f\|_{L^{(2\sigma)'}(v;S(r))} \\ &\le C\|f\|_{L^A(v;\Omega)}\|\chi_{S(r)}\|_{L^B(v;\Omega)} \\ &\le C\|f\|_{L^A(v;\Omega)}\frac{v(S(r))^{\frac{1}{2\sigma}}}{\log(e+(v(S(r)))^{-1})^{q\left(\frac{(2\sigma)'}{\sigma'}\right)}} \end{aligned} \tag{3}$$

If we now define

$$C_k = \tau_0\|f\|_{L^A(v;\Omega)}\left(1-\frac{1}{(k+1)^{\epsilon}}\right),$$

then to complete the proof we need to show that

$$v(S(\tau_0\|f\|_{L^A(v;\Omega)})) = \lim_{k\to\infty} v(S(C_k)) = 0.$$

To prove this, let $m_k = -\log(v(S(C_k))$; then the above limit is equivalent to showing that $m_k \to \infty$ as $k \to \infty$.

In inequality (3), let $s = C_{k+1}$, $r = C_k$. If we fix $\epsilon = \frac{q}{\sigma'} - 1 > 0$, then

$$m_{k+1} \ge \log\left(\frac{\epsilon\tau_0}{C}\right) + \log\left(\frac{m_k}{k+2}\right)^{\frac{2\sigma q}{\sigma'}} + m_k.$$

By induction, we can show that there exists $\tau_0 > 0$ (very large) such that $m_k \ge m_0 + k$. Therefore,

$$\lim_{k\to\infty} m_k = \infty.$$

3 Further Remarks

Theorem 3 is part of a larger project to develop a theory of existence, uniqueness, and regularity for degenerate PDEs. We want to close this note by outlining some further directions. Motivated directly by the work in [7], we see three immediate problems. First, as we noted, Theorem 2, which is just a special case of Theorem 3, is not sharp. We have not, despite repeated efforts, been able to improve our argument. Therefore, it is open whether Theorem 3 can be improved so that in the classical case we get a sharp result. Second, and perhaps related, our proof uses De Giorgi iteration. We originally attempted to use Moser iteration but were not successful. It would be interesting to make Moser iteration work in this setting. Towards this end, we have generalized the classical identiy

$$\lim_{p\to\infty} \|f\|_{L^p(\Omega)} = \|f\|_{L^\infty(\Omega)},$$

which plays a central role in Moser iteration, to the scale of Orlicz spaces: see [8]. Third, the classical result is known for equations with lower order terms, so it should be possible to formulate and prove a similar result for degenerate equations.

Looking beyond this, there are several directions we believe are worth exploring. The first is whether we can extend these results to data in other function spaces, such as Lorentz spaces, the small Lebesgue spaces (that is, the dual spaces of the grand Lebesgue spaces of Iwaniecz and Sbordone–see [2]), or the variable Lebesgue spaces. Each of these families of function spaces would provide new insight into the behavior of solutions as the data function gets close to the endpoint space.

Second, we want to examine whether the hypothesis that we have a Sobolev inequality with gain σ can be weakened. Motivated by problems in the study of hypoelliptic operators, Korobenko, et al. [14] have introduced Sobolev inequalities with gain in the scale of Orlicz spaces: roughly, replacing the $L^{2\sigma}$ norm on the lefthand side with an Orlicz norm given by $A(t) = t^2 \log(e+t)^\sigma$. We think that it is worth exploring whether Theorem 3 can be extended in this direction.

Third, having given conditions when solutions are bounded, the next step is to see whether we can prove regularity of solutions. At the heart of this problem will be to determine whether additional hypotheses are required. Preliminary results in this direction were obtained in [5, 6].

Finally, we are interested in understanding further the hypothesis that a Sobolev inequality (either in the scale of Lebesgue or Orlicz spaces) exists. Given a matrix Q and a weight v, what are necessary and/or sufficient conditions on the domain Ω for some kind of Sobolev inequality to exist? We are particularly interested in generalizing the geometric characterizations in [13] (see also [1]).

Acknowledgments The author acknowledges support from research funds from the Dean of the College of Arts & Sciences, the University of Alabama, and from a Simons Foundation Travel Support for Mathematicians Grant.

References

1. R. Alvarado, P. Górka, P. Hajłasz, Sobolev embedding for $M^{1,p}$ spaces is equivalent to a lower bound of the measure. J. Funct. Anal. **279**(7), 108628, 39 (2020)
2. C. Capone, A. Fiorenza, On small Lebesgue spaces. J. Funct. Spaces Appl. **3**(1), 73–89 (2005)
3. S. Chanillo, R.L. Wheeden, Weighted Poincaré and Sobolev inequalities and estimates for weighted Peano maximal functions. Am. J. Math. **107**(5), 1191–1226 (1985)
4. A. Cianchi, Strong and weak type inequalities for some classical operators in Orlicz spaces. J. Lond. Math. Soc. **60**(1), 187–202 (1999)

5. D. Cruz-Uribe, P. Di Gironimo, C. Sbordone, On the continuity of solutions to degenerate elliptic equations. J. Differ. Equ. **250**(6), 2671–2686 (2011)
6. D. Cruz-Uribe, K. Moen, V. Naibo, Regularity of solutions to degenerate p-Laplacian equations. J. Math. Anal. Appl. **401**(1), 458–478 (2013)
7. D. Cruz-Uribe, S. Rodney, Bounded weak solutions to elliptic PDE with data in Orlicz spaces. J. Differ. Equ. **297**, 409–432 (2021)
8. D. Cruz-Uribe, S. Rodney, A note on the limit of Orlicz norms. Real Anal. Exchange **48**(1), 77–81 (2023)
9. D. Cruz-Uribe, S. Rodney, E. Rosta, Poincaré inequalities and Neumann problems for the p-Laplacian. Can. Math. Bull. **61**(4), 738–753 (2018)
10. D. Cruz-Uribe, S. Rodney, E. Rosta, Global Sobolev inequalities and degenerate p-Laplacian equations. J. Differ. Equ. **268**(10), 6189–6210 (2020)
11. E.B. Fabes, C.E. Kenig, R.P. Serapioni, The local regularity of solutions of degenerate elliptic equations. Comm. Partial Differ. Equ. **7**(1), 77–116 (1982)
12. B. Franchi, G. Lu, R.L. Wheeden, Weighted Poincaré inequalities for Hörmander vector fields and local regularity for a class of degenerate elliptic equations, in *Potential Theory and Degenerate Partial Differential Operators* (Parma), vol. 4 (Springer, Berlin, 1995), pp. 361–375
13. P. Hajłasz, P. Koskela, H. Tuominen, Sobolev embeddings, extensions and measure density condition. J. Funct. Anal. **254**(5), 1217–1234 (2008)
14. L. Korobenko, C. Rios, E.T. Sawyer, R. Shen, Local boundedness, maximum principles, and continuity of solutions to infinitely degenerate elliptic equations with rough coefficients. Mem. Am. Math. Soc. **269**(1311), vii+130 (2021)
15. V.G. Maz'ya, Some estimates of solutions of second-order elliptic equations. Dokl. Akad. Nauk SSSR **137**, 1057–1059 (1961)
16. V.G. Maz'ya, Weak solutions of the Dirichlet and Neumann problems. Trudy Moskov. Mat. Obšč. **20**, 137–172 (1969)
17. M.M. Rao, Z.D. Ren, *Theory of Orlicz spaces*. Monographs and Textbooks in Pure and Applied Mathematics, vol. 146 (Marcel Dekker, New York, 1991)
18. E.T. Sawyer, R.L. Wheeden, Weighted inequalities for fractional integrals on Euclidean and homogeneous spaces. Am. J. Math. **114**(4), 813–874 (1992)
19. E.T. Sawyer, R.L. Wheeden, Hölder continuity of weak solutions to subelliptic equations with rough coefficients. Mem. Am. Math. Soc. **180**(847), x+157 (2006)
20. E.T. Sawyer, R.L. Wheeden, Degenerate Sobolev spaces and regularity of subelliptic equations. Trans. Am. Math. Soc. **362**(4), 1869–1906 (2010)
21. G. Stampacchia, Le problème de Dirichlet pour les équations elliptiques du second ordre à coefficients discontinus. Ann. Inst. Fourier (Grenoble) **15**(fasc. 1), 189–258 (1965)
22. N.S. Trudinger, Linear elliptic operators with measurable coefficients. Ann. Scuola Norm. Sup. Pisa Cl. Sci. **27**, 265–308 (1973)

Recent Progress on the Mathematical Theory of Wave Turbulence

Yu Deng

Abstract In this note we review some recent progress on the mathematical theory of wave turbulence.

Keywords Wave turbulence theory · Nonlinear dispersive equations · Wave equations

2020 Mathematics Subject Classification 74A25, 35Q55, 76B15

1 Introduction

In this note we will review some recent works on the mathematical theory of wave turbulence, a subject that has seen tremendous progress in the last few years.

1.1 Description of the Theory

The wave turbulence theory, or wave kinetic theory, concerns the behavior of statistical quantities for nonlinear dispersive equations in the kinetic limit. It is the wave analog of the classical kinetic theory for particles, and has played a central role in many physical and scientific applications.

To describe the idea, one starts with an arbitrary dispersive or wave equation as a first principle; here we will consider the cubic nonlinear Schrödinger equation, which is in some sense the *universal* Hamiltonian dispsesive PDE [42]. This equation is posed on a *large (periodic) box* $\mathbb{T}_L^d := [0, L]^d$, and is viewed as a system of interacting *waves*, which are represented by the Fourier modes of $\mathbb{T}_L^d$.

Y. Deng (✉)
Department of Mathematics, University of Southern California, Los Angeles, CA, USA
e-mail: yudeng@usc.edu

D. Cardona et al. (eds.), *Extended Abstracts 2021/2022*, Research Perspectives Ghent Analysis and PDE Center 3, https://doi.org/10.1007/978-3-031-48579-4_10

These waves play the role of particles in the classical kinetic theory of Boltzmann, while the nonlinear Schrödinger equation replaces the Newtonian dynamics with collision.

Thus, we consider the equation

$$(i\partial_t + \Delta)u = \alpha|u|^2 u \tag{1}$$

on $\mathbb{T}_L^d$, where α is a parameter measuring the strength of the interaction between waves. Note that the Fourier modes are $e^{ik\cdot x}$ for $k \in \mathbb{Z}_L^d := (L^{-1}\mathbb{Z})^d$, so if we restrict to unit size frequencies $|k| \sim 1$, then the number of different modes (i.e. degree of freedom) is $\sim L^d$.

The initial data of (1) is chosen to be

$$u(0,x) = \frac{1}{L^{d/2}} \sum_{k\in{}_L^d} \widehat{u}(0,k) e^{2\pi i k\cdot x}; \qquad \widehat{u}(0,k) = \sqrt{n_{\mathrm{in}}(k)} \cdot \eta_k(\omega), \tag{2}$$

where $\eta_k(\omega)$ are i.i.d. random variables with $\mathbb{E}\eta_k = 0$ and $\mathbb{E}|\eta_k|^2 = 1$. In particular the state of different waves (Fourier modes) are *independent* initially; this same independence assumption is also made in the rigorous derivation of Boltzmann equation [25, 33].

In practice, the law of $\eta_k(\omega)$ is usually chosen to be symmetric with respect to complex rotations, which is an important physical assumption known as *random phase*. Under this assumption we have that $\mathbb{E}|u(0,x)|^2$ is independent of x and has size ~ 1; this is referred to as the *(spatially) homogeneous* setting. The corresponding *inhomogeneous* setting corresponds to replacing the plane waves $e^{2\pi i k\cdot x}$ by suitable wave packets whose strength depends on x, but this only leads to technical differences for the theory.

1.1.1 Scope of the Theory

Define the *kinetic time scale*, or *Van-Hove time*, which is

$$T_{\mathrm{kin}} := \alpha^{-2}.$$

We shall consider the system (1) and (2) in the *kinetic limit*, i.e. the large box and weak nonlinearity limit, where $L \to \infty$ and $\alpha \to 0$. We shall assume a *scaling law* that quantifies the way in which these limits are taken, namely a value $\gamma \in (0, +\infty)$ such that

$$\alpha \sim L^{-\gamma}.$$

There are also endpoint cases $\gamma \in \{0, +\infty\}$, which correspond to taking iterated limits, but we will not discuss them here (they are more relevant in the *discrete* setting where $\mathbb{T}_L^d$ is replaced by a lattice, see [34]).

The main predictions of the wave kinetic theory include the followings:

(1) The wave kinetic equation: the effect of the nonlinear interaction (1) is expected to emerge precisely at the kinetic time scale T_{kin}, in terms of the variance $\mathbb{E}|\widehat{u}(t,k)|^2$. This variance, which corresponds to the density function for particle systems, satisfies the *wave kinetic equation*, which is a Boltzmann type equation, in the kinetic limit.

The wave kinetic equation is the central object of the wave kinetic theory; for (1) it has the form

$$\begin{cases} \partial_\tau n(\tau,k) = \mathcal{K}(n(\tau),n(\tau),n(\tau),k), \\ n(0,k) = n_{\mathrm{in}}(k), \end{cases} \tag{3}$$

where the collision term is

$$\begin{aligned} \mathcal{K}(\phi,\phi,\phi)(k) = \int_{(\mathbb{R}^d)^3} & \phi(k)\phi(k_1)\phi(k_2)\phi(k_3) \\ & \left[\frac{1}{\phi(k)} - \frac{1}{\phi(k_1)} + \frac{1}{\phi(k_2)} - \frac{1}{\phi(k_3)}\right] \\ & \times \delta(k - k_1 + k_2 - k_3)\delta(|k|^2 - |k_1|^2 + |k_2|^2 - |k_3|^2) \\ & \mathrm{d}k_1\mathrm{d}k_2\mathrm{d}k_3, \end{aligned} \tag{4}$$

and δ is the Dirac δ function. Then, one major prediction of the wave kinetic theory is that

$$\lim_{\substack{L\to\infty \\ \alpha\sim L^{-\gamma}}} \mathbb{E}|\widehat{u}(\tau\cdot T_{\mathrm{kin}},k)|^2 = n(\tau,k) \tag{5}$$

for any (small) value of τ and wave number $k \in \mathbb{Z}^d_L$, where $n(\tau,k)$ is the solution to (3).

(2) Propagation of chaos: note that the Fourier modes of u, which are independent at time 0 by (2), will not remain independent for times $t > 0$ due to the nonlinear interaction (1). However, the physical observation of *propagation of chaos*, which also plays a key role in the classical kinetic theory of Boltzmann, predicts that this independence is restored *in the kinetic limit*, at least up to time T_{kin}. In other words, for any different wave numbers k_j $(1 \le j \le q)$, we have

$$\lim_{\substack{L\to\infty \\ \alpha\sim L^{-\gamma}}} \mathbb{E}\left[\prod_{j=1}^{q} |\widehat{u}(\tau\cdot T_{\mathrm{kin}},k_j)|^2\right] = \prod_{j=1}^{q} n(\tau,k_j), \tag{6}$$

as well as the corresponding results for higher moments.

(3) Density evolution: note that each Fourier mode $\widehat{u}(t, k)$ is a complex random variable which has a density function $\rho_k(t, v)$ for $v \in \mathbb{R}^2$. Then, in the kinetic limit, this density function is expected to evolve under some linear equation

$$\partial_\tau \xi_k = \sigma_k(\tau)\Delta_v \xi_k - \gamma_k(\tau)\nabla_v \cdot (v\xi_k); \quad \xi_k(\tau, v) := \lim_{\substack{L\to\infty \\ \alpha\sim L^{-\gamma}}} \rho_k(\tau \cdot T_{\mathrm{kin}}, v),$$

where $\sigma_k(\tau)$ and $\gamma_k(\tau)$ are two quantities constructed from the solution $n(\tau, k)$ to (3). This equation was contained in the original work of Peierls [37], and has recently been rediscovered in the physics literature [6, 36].

1.1.2 Applications

The wave kinetic theory was first raised in the work [37] or Peierls in 1929 on anharmonic crystals, about fifty years after the works of Boltzmann on classical kinetic theory of particles. Since then, it has seen a substantial development, especially with the influential works of Hasselmann [29, 30] and Zakharov [44] in the 1960s. As of now, this theory has been formalized into a systematic approach to understand the effective long-time behavior of large systems of interacting waves undergoing weak nonlinear interactions [36, 40, 45], and has been applied in many different physical and scientific settings, including for example plasma theory [10, 24, 43, 46], water waves [2, 3, 29, 30], and oceanography [27, 32].

In particular, the works of Zakharov [44] (see also [45]) introduced the energy cascade spectra in wave kinetic theory, now called the *Zakharov spectra*, which are in parallel with the Kolmogorov spectra in hydrodynamic turbulence, and also has profound implications. It is for this reason that the wave kinetic theory are also referred to as the *wave turbulence* theory in modern literature.

1.2 Mathematical Literature

The famous theorem of Lanford [33] in 1975 (which was rigorously completed by Gallagher et al. [25] in 2014) provides the rigorous derivation of the Boltzmann equation from Newtonian dynamics. The corresponding result for wave phenomena, i.e. the rigorous derivation of the wave kinetic equations (3) from the corresponding nonlinear dispersive equations like (1), has been a major open problem of the subject. Compared to the classical particle counterpart, the mathematical development of wave turbulence theory started much later, partly due to lack of the necessary mathematical tools.

One of the earliest works in this direction was Spohn [39], which derived a *linear* kinetic equation from a linear Schrödinger equation with random potential. Later works also focus on linear problems, including Erdös and Yau [18] and Erdös et al. [19], which extended the derivation to inhomogeneous settings and to longer time scales (see also the recent work of Felipe Hernández [31], which provides an alternative proof to the result of [19]).

The next major advancement, which was the first result treating a *nonlinear* problem, was Lukkarinen and Spohn [34]. It studied a discrete nonlinear Schrödinger equation *at equilibrium*, i.e. with Gibbs measure initial data, and derived the precise asymptotics of time correlations at the kinetic time scale. A later result of Faou [23] also studied linearization of the wave kinetic equation around equilibrium (with some time-dependent multiplicative noise).

The first attempt on an off-equilibrium, full nonlinear problem was Buckmaster et al. [4], but the time scale reached is much shorter than T_{kin} (in fact shorter than $T_{\text{kin}}^{1/2}$). This time scale has been improved in subsequent works [7, 8, 11] by the author with Hani and independently by Collot-Germain, but still falls short of T_{kin} just by a little bit (i.e. $T_{\text{kin}}^{1-\varepsilon}$ for any $\varepsilon > 0$ depending on the scaling law γ). In the mean time, the works of Dymov and Kuksin [16, 17] treated formal expansions in the wave kinetic context of (1) with additive noise and dissipation.

Finally, concerning the solution theory to (3), there has been a number of important results obtained with exciting recent developments, including local well-posedness (Germain et al. [26]), existence of weak solutions with blowup (Escobaso and Velázquez [20, 21], Cai and Lu [5]) energy-cascade (Soffer and Tran [38]), and first mathematical studies on Zakharov spectra (Collot et al. [9]).

2 Main Results

In a series of recent works [13–15] (see also the expository note [12]) joint with Hani, we have finally completed the rigorous derivation of the wave kinetic equation (3) from the dispersive equation (1) at kinetic time scale, together with propagation of chaos and all the supplementary results. More precisely, we have the followings:

Theorem 1 (Deng and Hani [13–15]) *Suppose $d \geq 3$ and $0 < \gamma < 1$ (for $\gamma = 1$ we need to replace $\mathbb{T}_L^d$ by a generic irrational torus, see [13]). Then, under rotation symmetry and suitable integrability conditions for the i.i.d. random variables η_k, we have the followings:*

(1) Derivation of (3): for $0 < \tau \ll 1$ and any wave number k, we have

$$\lim_{\substack{L\to\infty \\ \alpha\sim L^{-\gamma}}} \mathbb{E}|\widehat{u}(\tau \cdot T_{\text{kin}}, k)|^2 = n(\tau, k), \tag{7}$$

where $n(\tau, k)$ solves the wave kinetic equation (3)–(4);

(2) *Propagation of chaos: for $0 < \tau \ll 1$, different wave numbers $k_1, \cdots, k_q$ and arbitrary positive integers p_j $(1 \le j \le q)$, we have*

$$\lim_{\substack{L\to\infty \\ \alpha\sim L^{-\gamma}}} \mathbb{E}\left[\prod_{j=1}^{q} |\widehat{u}(\tau \cdot T_{\mathrm{kin}}, k_j)|^{2p_j}\right] = \lim_{\substack{L\to\infty \\ \alpha\sim L^{-\gamma}}} \prod_{j=1}^{q} \mathbb{E}|\widehat{u}(\tau \cdot T_{\mathrm{kin}}, k_j)|^{2p_j}; \tag{8}$$

the limit of any other moment that does not have the form (8) is 0.

(3) *Evolution of density: let $\rho_k(t, v)$ be the probability density function of the random variable $\widehat{u}(t, k)$, then the limit*

$$\xi_k(\tau, v) := \lim_{\substack{L\to\infty \\ \alpha\sim L^{-\gamma}}} \rho_k(\tau \cdot T_{\mathrm{kin}}, v) \tag{9}$$

satisfies the linear evolution equation

$$\partial_\tau \xi_k = \sigma_k(\tau)\Delta_v \xi_k - \gamma_k(\tau)\nabla_v \cdot (v\xi_k), \tag{10}$$

where the relevant quantities are given by

$$\sigma_k(t) = \frac{1}{4}\int_{(\mathbb{R}^d)^3} n(t, k_1)n(t, k_2)n(t, k_3) \times \delta(k - k_1 + k_2 - k_3)\delta(|k|^2 - |k_1|^2 + |k_2|^2 - |k_3|^2)\, dk_1 dk_2 dk_3, \tag{11}$$

$$\gamma_k(t) = \frac{1}{2}\int_{(\mathbb{R}^d)^3} \big[n(t, k_1)n(t, k_3) - n(t, k_2)n(t, k_3) - n(t, k_1)n(t, k_2)\big] \times \delta(k - k_1 + k_2 - k_3)\delta(|k|^2 - |k_1|^2 + |k_2|^2 - |k_3|^2)\, dk_1 dk_2 dk_3. \tag{12}$$

2.1 Discussions

We make a few remarks regarding Theorem 1. First, the range of scaling laws $\gamma \in (0, 1)$ is optimal (with the genericity condition needed for $\gamma = 1$), which follows from a discussion comparing the sets of exact resonances with the set of quasi-resonances. See the discussion in Section 1.2 of [13] or Section 1.3 of [14].

Note that, the range $\gamma \in (0, 1)$ is the universal range for the Schrödinger equation (1). For other dispersive relation, the optimal range may be larger or smaller than $(0, 1)$ depending on the number theoretic properties of that specific dispersive relation; however the smaller range $\gamma \in (0, 1/2]$ seems to be admissible for a large class of regular dispersion relations.

Second, we note that there exist two important scaling laws for the Schrödinger problem (1), namely $\gamma = 1/2$ and $\gamma = 1$. The former is the naturally occurring scaling law in the inhomogeneous setting, which in some sense corresponds to the Boltzmann-Grad scaling law in the theorem of Lanford [25, 33]; the latter naturally links (1) to the unit time problem on the unit torus, which in particular plays a key role in the study of invariant Gibbs measures for nonlinear Schrödinger equations. See the discussion in Section 1.3 of [14] for more details.

Third, Theorem 1 also contains the rigorous derivation of the *wave kinetic hierarchy*, which applies to the case when the norms of the random variables η_k are not necessarily i.i.d. (the arguments are still i.i.d. uniformly distributed). This may come from the choice of *hybrid random data* or *super-statistical solutions* [22] with important physical meanings. Note that the i.i.d. case corresponds to factorized solution to the wave kinetic hierarchy, which is given by the solution to (3). We will not present the exact form of the hierarchy here, but see Section 1.4 of [15].

Finally, Theorem 1 leaves open a few important questions, most notably the justification of (3) from (1) for longer times $\tau \gg 1$, and (if applicable) pass the blowup time for (3). These are extremely challenging questions (as is the counterpart for Lanford Theorem and Boltzmann equation), and would require ideas and techniques completely different from the current and earlier works; on the other hand they would also have profound implications on the study of long-time dynamics of (1) and general nonlinear dispersive equations. See Section 1.5 of [14].

2.2 Related Works

We mention some important works in similar directions that appear around the same time or after [13–15]:

- Staffilani and Tran [41] and Hannani et al. [28]: these works concern the discrete Zakharov-Kuznetsov equation and derive the corresponding wave kinetic equation (first homogeneous and then inhomogeneous) also at the kinetic time scale. We remark that these works contain *time-dependent noise* that provides an additional randomization effect for arguments of Fourier modes. At this time, these works and [13–15] are the only results that reach the kinetic time T_{kin} in the non-equilibrium setting.
- Ampatzoglou et al. [1]: this is the first rigorous result obtained in the inhomogeneous setting. It concerns a modified quadratic nonlinear Schrödinger equation (without noise), and derives the wave kinetic equation at $T_{\text{kin}}^{1-\varepsilon}$ for arbitrary $\varepsilon > 0$.
- Ma [35]: this concerns also the Zakharov-Kuznetsov equation, but in a continuum setting. Note that the nonlinearity contains a loss of derivative, which has to be compensated by adding suitable dissipation, leading to new mathematical challenges. Here the wave kinetic equation is also derived up to almost optimal time scale $T_{\text{kin}}^{1-\varepsilon}$.

2.3 Main Ideas in the Proof

The proof of Theorem 1 is based on Feynman diagram expansions, and in particular very precise controls of terms occurring in the diagrammatic expansion. This requires deep understanding of the problem, on both combinatorial and analytical aspects. The main difficulty here, caused by the (probabilistic) criticality of the equation, comes from the factorial divergence of high order expansions, as well as certain specific divergent terms. To overcome this, the following main strategies are adapted in [13–15]:

- Classification of diagrams: this leads to the key notions of *regular couples* [13, 15], and a complete classification of the diagrams (called couples) using these objects.
- Rigidity theorem: this allows to prove that for those diagrams that deviates from the set of regular couples, the corresponding contributions are always more favorable [13, 14], with extra decay that is comparable to the distance of deviation. Here the notion of *molecules* introduced in [13] plays a fundamental role.
- Cancellation: this exhibits the miraculous cancellation structure between terms that may show divergence if examined individually [13, 14]. These terms must have very special forms that are precisely captured by molecules, and a complete classification of them naturally leads to the cancellation.

Acknowledgments The research is supported in part by NSF Grant DMS-2246908 and Sloan Fellowship.

References

1. I. Ampatzoglou, C. Collot, P. Germain, Derivation of the kinetic wave equation for quadratic dispersive problems in the inhomogeneous setting (2021). arXiv:2107.11819
2. B.J. Benney, A.C. Newell, Random wave closures. Stud. Appl. Math. **48**(1), 29 (1969)
3. B.J. Benney, P. Saffman, Nonlinear interaction of random waves in a dispersive medium. Proc. R. Soc. A **289**, 301–320 (1966)
4. T. Buckmaster, P. Germain, Z. Hani, J. Shatah, Onset of the wave turbulence description of the long-time behavior of the nonlinear Schrödinger equation. Invent. Math. **225**, 787–855 (2021)
5. S. Cai, X. Lu, The spatially Homogeneous Boltzmann equation for Bose-Einstein particles: rate of strong convergence to equilbrium. J. Stat. Phys. **175**, 289–350 (2019)
6. Y. Choi, Y. Lvov, S. Nazarenko, Joint statistics of amplitudes and phases in wave turbulence. Phys. D **201**, 121 (2005)
7. C. Collot, P. Germain. On the derivation of the homogeneous kinetic wave equation (2019). arXiv:1912.10368
8. C. Collot, P. Germain, Derivation of the homogeneous kinetic wave equation: longer time scales (2020). arXiv:2007.03508
9. C. Collot, H. Dietert, P. Germain, Stability and cascades for thr Kolmogorov-Zakharov spectrum of wave turbulence (2022). arXiv:2208.00947

10. R. Davidson, *Methods in Nonlinear Plasma Theory*, 1st edn. (Academic Press, Cambridge, 1972)
11. Y. Deng, Z. Hani, On the derivation of the wave kinetic equation for NLS. Forum Math. Pi **9**(e6), 1–37 (2021)
12. Y. Deng, Z. Hani, Rigorous justification of the wave kinetic theory (2022). arXiv:2207.08358
13. Y. Deng, Z. Hani, Full derivation of the wave kinetic equation. Invent. Math. (2023, to appear)
14. Y. Deng, Z. Hani, Derivation of the wave kinetic equation: full range of scaling laws (2023). arXiv:2301.07063
15. Y. Deng, Z. Hani, Propagation of chaos and the higher order statistics in the wave kinetic theory. J. Eur. Math. Soc. (to appear)
16. A. Dymov, S. Kuksin, Formal expansions in stochastic model for wave turbulence 1: kinetic limit. Commun. Math. Phys. **382**, 951–1014 (2021)
17. A. Dymov, S. Kuksin, Formal expansions in stochastic model for wave turbulence 2: method of diagram decomposition. J. Stat. Phys. **190**, 3 (2023)
18. L. Erdös, H.T. Yau, Linear Boltzmann equation as the weak coupling limit of a random Schrödinger equation
19. L. Erdös, M. Salmhofer, H.T. Yau, Quantum diffusion of the random Schrödinger evolution in the scaling limit. Acta Math. **200**(2), 211–277 (2008)
20. M. Escobedo, J. Velaázquez, Finite time blow-up and condensation for the bosonic Nordheim equation. Invent. Math. **200**, 761–847 (2015)
21. M. Escobedo, J. Velázquez, On the theory of Weak Turbulence for the Nonlinear Schrödinger Equation. Mem. Am. Math. Soc. **238**, 1124 (2015)
22. G.L. Eyink, Y.K. Shi, Kinetic wave turbulence. Phys. D Nonlinear Phenom. **241** (2012), pp. 1487–1511. Commun. Pure Appl. Math. **53**, 667–735 (2000)
23. E. Faou, Linearized wave turbulence convergence results for three-wave systems. Commun. Math. Phys. **378**, 807–849 (2020)
24. A.A. Galeev, R.Z. Sagdeev, *Review of Plasma Physics*, in ed. by Leontovich, vol. 7 (Consultants Bureau, New York, 1979), p. 307
25. I. Gallagher, L. Saint-Raymond, B. Texier, *From Newton to Boltzmann: The Case of Hard-Spheres and Short-Range Potentials* (ZLAM, 2014)
26. P. Germain, A. Ionescu, M.B. Tran, Optimal local well-posedness theory for the kinetic wave equation. J. Func. Anal. **279**, 108570 (2020)
27. World Meteorological Organization, *Guide to Wave Analysis and Forecasting* (Secretariat of the World Meteorological Organization, Geneva, 1998)
28. A. Hannani, M. Rosenzweig, G. Staffilani, M.B. Tran, On the wave turbulence theory for a stochastic KdV type equation - generalization for the inhomogeneous kinetic limit (2022). arXiv:2210.17445
29. K. Hasselmann, On the nonlinear energy transfer in a gravity wave spectrum part 1. J. Fluid Mech. **12**, 481–500 (1962)
30. K. Hasselmann, On the nonlinear energy transfer in a gravity wave spectrum. Part 2. J. Fluid Mech. **15**, 273–281 (1963)
31. F. Hernández, Quantum diffusion via an approximate semigroup property (2022). arXiv 2206.12998
32. P.A. Janssen, Progress in ocean wave forecasting. J. Comput. Phys. **227**(7), 3572–3594 (2008)
33. O.E. Lanford, *Time Evolution of Large Classical Systems*. Lecture Notes in Physics, ed. by J. Moser, vol. 38 (Springer, Heidelberg, 1975), pp. 1–111
34. J. Lukkarinen, H. Spohn, Weakly nonlinear Schrödinger equation with random initial data. Invent. Math. **183**, 79–188 (2011)
35. X. Ma, Almost sharp wave kinetic theory of multidimensional KdV type equations with $d \geq 3$ (2022). arXiv:2204.06148
36. S. Nazarenko, *Wave Turbulence*. Lecture Notes in Physics, vol. 825 (Springer, Heidelberg, 2011)
37. R.E. Peierls, Zur kinetischen Theorie der Wärmeleitung in Kristallen. Ann. Phys. **3**, 1055–1101 (1929)

38. A. Soffer, M.B. Tran, On the energy cascade of 3-wave kinetic equations: beyond Kolmogorov-Zakharov solutions. Commun. Math. Phys. **376**, 2229–2276 (2020)
39. H. Spohn, Derivation of the transport equation for electrons moving through random impurities. J. Stat. Phys. **17**(6), 385–412 (1977)
40. H. Spohn, On the Boltzmann equation for weakly nonlinear wave equations, in *Boltzmann's Legacy*. ESI Lectures in Mathematics and Physics (European Mathematical Society, Helsinki, 2008), pp. 145–159. ISBN: 978-3-03719-057-9
41. G. Staffilani, M.B. Tran, On the wave turbulence theory for the stochastic and random multidimensional KdV type equations (2021). arXiv:2106.09819
42. C. Sulem, P.L. Sulem, *The Nonlinear Schrödinger Equation: Self-Focusing and Wave Collapse*. Springer Applied Sciences Series, vol. 139 (Springer, Berlin, 2007). ISBN-13: 978-0387986111
43. A.A. Vedenov, Theory of weakly turbulent plasma, in *Reviews of Plasma Physics*, ed. by M.A. Leontovich, vol. 3 (Consultants Bureau, New York, 1967), p. 229
44. V.E. Zakharov, Weak turbulence in media with decay spectrum. Zh. Priklad. Tech. Fiz. **4**, 5–39 (1965) [J. Appl. Mech. Tech. Phys. 4, 22–24 (1965)]
45. V.E. Zakharov, V.S. L'vov, G. Falkovich. *Kolmogorov Spectra of Turbulence: I Wave Turbulence* (Springer, Berlin, 1992)
46. G.M. Zaslavskii, R. Z. Sagdeev, Sov. Phys. JETP **25**, 718 (1967)

Laplace-Beltrami Equation on Lipschitz Hypersurfaces in the Generic Bessel Potential Spaces

Roland Duduchava

Abstract The purpose of the present short note is to expose a new approach to the investigation of boundary value problems (BVPs) for the Laplace-Beltrami equation on a hypersurface $\mathcal{S} \subset \mathbb{R}^3$ with Lipschitz boundary $\Gamma = \partial\mathcal{S}$, containing a finite number of angular points (nodes) c_j of magnitude α_j, $j = 1, 2, \dots, n$. The Dirichlet, Neumann and mixed type BVPs are considered in a non-classical setting when solutions are sought in the generic Bessel potential spaces (GBPS) $\mathbb{GH}^s_p(\mathcal{S}, \rho)$, $s > 1/p$, $1 < p < \infty$ with weight $\rho(t) = \prod_{j=1}^{n} |t - c_j|^{\gamma_j}$ (the definition see below). By a localization procedure, the problem is reduced to the investigation of model Dirichlet, Neumann and mixed BVPs for the Laplace equation in a planar angular domain $\Omega_{\alpha_j} \subset \mathbb{R}^2$ of magnitude α_j, $j = 1, 2 \dots, n$. Further the model problem in the GBPS with weight $\mathbb{GH}^s_p(\Omega_{\alpha_j}, t^{\gamma_j})$ is investigated by means of Mellin convolution operators on the semi-axes $\mathbb{R}^+ = (0, \infty)$. Explicit criteria for the Fredholm property and the unique solvability of the initial BVPs are obtained and singularities of solutions at nodes to the mentioned BVPs are indicated. In contrast to the results on the same BVPs in the classical Bessel potential spaces $\mathbb{H}^s_p(\mathcal{S})$, the Fredholm property in the GBPS $\mathbb{GH}^s_p(\mathcal{S}, \rho)$ with weight is independent of the smoothness parameter s and Fredholm conditions as well as singularities of solutions are indicated very explicitly.

Keywords Boundary value problem · Laplace-Beltrami equation · Lipschitz hypersurface · Generic bessel potential space · Mellin convolution equation

2020 Mathematics Subject Classification 35J57, 45E10, 47B35

R. Duduchava (✉)
Institute of Mathematics, University of Georgia, Tbilisi, Georgia

A. Razmadze Mathematical Institute, I. Javakhishvili State University, Tbilisi, Georgia
e-mail: r.duduchava@ug.edu.ge

D. Cardona et al. (eds.), *Extended Abstracts 2021/2022*, Research Perspectives Ghent Analysis and PDE Center 3, https://doi.org/10.1007/978-3-031-48579-4_11

1 Introduction and Formulation of the Main Results

Let $\mathcal{C}$ be a smooth hypersurface in $\mathbb{R}^3$ with the Lipschitz (piecewise-smooth) boundary $\Gamma = \partial\mathcal{C}$ with a finite number of nodes $\mathcal{M}_\Gamma := \{c_1, \dots, c_n\} \subset \Gamma$ (see Fig. 1). The inner angle α_j at the node c_j is $\alpha_j \in (0, 2\pi)$ (no cusps!)

Let $\boldsymbol{\nu} := (\nu_1, \nu_2, \nu_3)^\top$ and $\boldsymbol{\nu}_\Gamma := (\nu_{\Gamma,1}, \nu_{\Gamma,2}, \nu_{\Gamma,3})^\top$ be the normal vector fields respectively, to the surface $\mathcal{C}$ and to the boundary Γ, ($\boldsymbol{\nu}_\Gamma$ is tangential to $\mathcal{C}$).

The boundary Γ is decomposed in two parts $\partial\mathcal{C} = \Gamma = \Gamma_D \cup \Gamma_N$ and we study the following mixed boundary value problem

$$\begin{cases} \Delta_{\mathcal{C}} u(t) = f(t), & t \in \mathcal{C}, \\ u^+(s) = g(s), & \text{on} \quad \Gamma_D, \\ (\partial_{\boldsymbol{\nu}_\Gamma} u)^+(s) = h(s), & \text{on} \quad \Gamma_N. \end{cases} \tag{1}$$

Here $\Delta_{\mathcal{C}} := \mathcal{D}_1^2 + \mathcal{D}_2^2 + \mathcal{D}_3^2$ is the Laplace-Beltrami operator and $\mathcal{D}_j := \partial_j - \nu_j \partial_{\boldsymbol{\nu}}$, j=1,2,3 are Günter's tangential derivatives on the surface (cf. [8]).

$\partial_{\boldsymbol{\nu}_\Gamma} := \nu_{\Gamma,1}\mathcal{D}_1 + \nu_{\Gamma,2}\mathcal{D}_2 + \nu_{\Gamma,3}\mathcal{D}_3$ is the normal derivative on the boundary Γ, tangential to the surface $\mathcal{C}$.

The pure Dirichlet and pure Neumann problems are particular cases of the BVP (1) when, respectively, $\Gamma_N = \emptyset$ and $\Gamma_D = \emptyset$.

In [9] was proved that the Mixed BVP (1) and the pure Dirichlet BVP have unique solutions in the classical weak setting:

$$u \in \mathbb{GH}^1(\mathcal{C}), \qquad f \in \widetilde{\mathbb{GH}}^{-1}(\mathcal{C}), \qquad g \in \mathbb{GH}^{1/2}(\Gamma_D), \qquad h \in \mathbb{GH}^{-1/2}(\Gamma_N), \tag{2}$$

while for the solvability of the pure Neumann BVP it's necessary and sufficient the following compatibility condition to hold:

$$(f, 1)_{\mathcal{C}} - (h, 1)_\Gamma = 0. \tag{3}$$

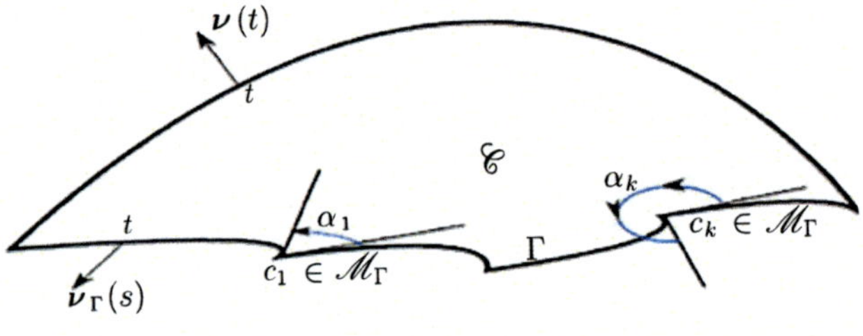

Fig. 1 Surface $\mathcal{C}$

The solvability in the classical setting (2) does not imply the continuity of a solution. But from the solvability in the non-classical setting

$$u \in \mathbb{GH}^s_p(\mathcal{C}, \rho), \ f \in \mathbb{GH}^{s-2}_p(\mathcal{C}, \rho), \ g \in \mathbb{GH}^{s-1/p}_p(\Gamma, \rho), \ h \in \mathbb{GH}^{s-1-1/p}_p(\Gamma, \rho), \tag{4}$$

$$1 < p < \infty, \ s > \frac{1}{p}, \ \rho(t) = \prod_{j=1}^{n} |t - c_j|^{\gamma_j} \ -1 < \gamma_j < p - 1, \ j = 1, \dots, n..$$

for $2 < p < \infty$, we can enjoy even a Hölder continuity of a solution. Investigation of the maximal smoothness of a solution is motivated by applications, e.g. by the numerical methods for BVPs.

To formulate the main theorems we need the following definition.

Definition 1 The BVP (1) and (4) is Fredholm if the homogeneous problem $f = g = h = 0$ has a finite number of solutions and only a finite number of orthogonality conditions on the data f, g, h ensure the solvability of the BVP.

To the set of nodes $\mathcal{M}_\Gamma$ of the surface $\mathcal{C}$ add all smoothness points on Γ where the Dirichlet and Neumann boundary conditions collide.

Let $\mathcal{M}_\Gamma = \mathcal{M}_{DD} \bigcup \mathcal{M}_{NN} \bigcup \mathcal{M}_{DN}$, where the sets $\mathcal{M}_{DD}$, $\mathcal{M}_{NN}$ and $\mathcal{M}_{DN}$ consist of nodes where, respectively, the Dirichlet, the Neumann and the DIrichlet-Neumann boundary conditions collide.

The following is the main result of the present exposition.

Theorem 2 *Let* $0 < \alpha_j < 2\pi$, $\beta_j = \dfrac{1+\gamma_j}{p}$ *(*$0 < \beta_j < 1$ *due to* (4)*),* $j = 1, 2 \dots, n$. *The mixed BVP* (1) *is Fredholm in the setting* (4) *if and only if:*

$$\beta_j = \frac{1+\gamma_1}{p} \neq \frac{\pi}{2(2\pi - \alpha_j)}, \frac{3\pi}{2(2\pi - \alpha_j)}, \frac{\pi}{2\alpha_j}, \frac{3\pi}{2\alpha_j}, \ \forall \, c_j \in \mathcal{M}_{DN}, \tag{5}$$

$$\beta_j = \frac{1+\gamma_1}{p} \neq \frac{\pi}{2\pi - \alpha_j}, \frac{\pi}{\alpha_j}, \ \forall \, c_j \in \mathcal{M}_{DD}, \tag{6}$$

$$\beta_j = \frac{1+\gamma_1}{p} \neq 1 - \frac{\pi}{2\pi - \alpha_j}, \ 1 - \frac{\pi}{\alpha_j}, \ \forall \, c_j \in \mathcal{M}_{NN}. \tag{7}$$

If the intervals $\left[\frac{1}{2}, \beta_1\right] \cup \dots \cup \left[\frac{1}{2}, \beta_n\right]$ *does not contain singular points listed (a) in* (5)*, (b) in* (6)*, (c) in* (7)*, then the corresponding (a) the Mixed BVP* (1)*, (b) the pure Dirichlet BVP (*$\Gamma_N = \emptyset$*), (c) the pure Neumann BVP (*$\Gamma_D = \emptyset$*), has a unique solutions in the setting* (4) *with the additional compatibility condition* (3) *for the pure Neumann BVP.*

Remark 3 As it follows from the foregoing theorem, the Fredholm property and solvability of BVP (1) in the setting (4) is independent of the smoothness parameter

$s > 0$. This property has the following consequence: If the mixed BVP (1) in the setting (4) is Fredholm for some $s \in \mathbb{R}$, $1 < p < \infty$, the data of BVP are infinitely smooth and the BVP has a solution $u(x)$, this solution is "infinitely smooth" in the sense that the "weighted derivatives" $\mathcal{D}^{m_1}_{1,0}\mathcal{D}^{m_2}_{2,0}\mathcal{D}^{m_3}_{3,0}u(x)$, where $\mathcal{D}_{k,0} := \prod_{j=1}^{n}(t - c_j)\mathcal{D}_k$, belong to the space $\mathbb{L}_p(\mathcal{C}, \rho)$ for all $m_1, m_2, m_3 = 1, 2, \ldots$.

Remark 4 It is well known that solutions to the BVP (1) and (4) have siongularities at the nodes $c_1, \ldots, c_n$ and these singularities depend on which boundary conditions collide there. If s_j denotes the singularity of a solution at c_j (i.e. $(x - c_j)^{s_j}u \in \bigcap_{p>1} L_p(U_{c_j})$ for some neighbourhood $U_{c_j} \subset \mathcal{C}$ of c_j), than $s_j p + \gamma_j = -1$, which gives $s_j = -(\gamma_j + 1)/p = -\beta_j$. Therefore, due to (5) and (7), a solution to the BVP (1) and (4) has the following singularities (we did not take into account logarithmic factors):

$$
\begin{aligned}
s_j &= \frac{\pi}{2(2\pi - \alpha_j)}, \frac{3\pi}{2(2\pi - \alpha_j)}, \frac{\pi}{2\alpha_j}, \frac{3\pi}{2\alpha_j}, \quad \text{for } c_j \in \mathcal{M}_{DN}, \\
s_j &= \frac{\pi}{2\pi - \alpha_j}, \frac{\pi}{\alpha_j}, \quad \text{for } \ c_j \in \mathcal{M}_{DD}, \\
s_j &= \frac{1 + \gamma_1}{p} \neq 1 - \frac{\pi}{2\pi - \alpha_j}, 1 - \frac{\pi}{\alpha_j}, \text{ for } \ c_j \in \mathcal{M}_{NN}.
\end{aligned} \tag{8}
$$

2 Generic Bessel Potential Spaces and Mellin Convolutions

It is well known that the half axes $\mathbb{R}^+ = (0, \infty)$ is a Lie group and Haar measure on $\mathbb{R}^+$ is $\frac{dt}{t}$. The group Fourier transformation on $\mathbb{R}^+$ coincides with the Mellin transformation $\mathcal{M}$ and $\mathcal{M}^{-1}$ is its inverse:

$$
\mathcal{M}_\beta\psi(\xi) := \int_0^\infty t^{\beta - i\xi}\psi(t)\,\frac{dt}{t}, \quad \xi \in \mathbb{R},
$$

$$
\mathcal{M}_\beta^{-1}\varphi(t) := \frac{1}{2\pi}\int_{-\infty}^{\infty} t^{i\xi - \beta}\varphi(\xi)\,d\xi, \quad t \in \mathbb{R}^+.
$$

Generator of the Lie algebra-the generic differenrial operator is $\mathfrak{D}\varphi(t) := t\dfrac{d\varphi(t)}{dt}$.

If $g \in \mathbb{L}_\infty(\mathbb{R})$ is an essentially bounded measurable $N \times N$ matrix function, the Mellin convolution operator $\mathfrak{M}^0_g$ is defined as follows (cf. [3])

$$\mathfrak{M}^0_g\varphi(t) := \mathcal{M}_\beta^{-1} g \mathcal{M}_\beta \varphi(t) = \frac{1}{2\pi}\int_{-\infty}^{\infty} g(\xi) \int_0^{\infty} \left(\frac{t}{\tau}\right)^{i\xi-\beta} \varphi(\tau)\,\frac{d\tau}{\tau}\,d\xi, \quad \varphi \in \mathbb{S}(\mathbb{R}^+),$$

where $\mathbb{S}(\mathbb{R}^+)$ is the Schwartz space of fast decaying smooth functions on $\mathbb{R}^+$. The function $g(\xi)$ is usually referred to as a symbol of the Mellin convolution operator $\mathfrak{M}^0_g$ and this operator is represented also in the form (cf. [3])

$$\mathfrak{M}^0_g\varphi(t) = \int_0^{\infty} k\left(\frac{t}{\tau}\right)\varphi(\tau)\,\frac{d\tau}{\tau}, \quad g(\xi) := (\mathcal{M}_\beta k)(\xi).$$

Here $k(x)$ is a generalized Hörnmander's kernel. In particular, if $k \in \mathbb{L}_1(\mathbb{R}^+, t^\gamma)$, $\beta := \dfrac{\gamma+1}{p}$, than the symbol is a regular (even continuous) function and belongs to the Wiener algebra $g \in \mathbb{W}(\mathbb{R})$. The latter is defined as follows

$$\mathbb{W}(\mathbb{R}) := \left\{g(\xi) = c + (\mathcal{M}_\beta k)(\xi), \quad c = \text{const}, \; k \in \mathbb{L}_1(\mathbb{R}^+, t^\gamma)\right\}, \quad \beta := \frac{\gamma+1}{p}$$

and is endowed with the reduced norm $\|g|W(\mathbb{R})\| = |c| + \|k|\mathbb{L}_1(\mathbb{R}^+, t^\gamma)\|$.

Let $1 \leqslant p \leqslant \infty,\; s \in \mathbb{R},\; 0 < \beta < 2$. By analogy with the Bessel potential space $\mathbb{H}^s_p(\mathbb{R})$ on the real axes (Lie group) $\mathbb{R}$, endowed with the norm

$$\begin{aligned}\|\psi \mid \mathbb{H}^s_p(\mathbb{R})\| &:= \|\mathcal{F}^{-1}\langle\xi\rangle^s \mathcal{F}\psi \mid \mathbb{L}_p(\mathbb{R})\| = \|W^0_{\langle\cdot\rangle^s}\psi \mid \mathbb{L}_p(\mathbb{R})\|,\\ \langle\xi\rangle^s &:= (1+|\xi|^2)^{s/2},\end{aligned}$$

we define the generic Bessel potential space $\mathbb{GH}^s_p(\mathbb{R}^+, t^\gamma dt/t) = \mathbb{GH}^s_p(\mathbb{R}^+, t^\gamma)$ on the half axes (Lie group) $\mathbb{R}^+$, endowed with the norm

$$\begin{aligned}\|\psi \mid \mathbb{GH}^s_p(\mathbb{R}^+, t^\gamma)\| &:= \|\mathfrak{M}^0_{\langle\cdot\rangle^s}\psi \mid \mathbb{L}_p(\mathbb{R}^+, t^\gamma)\|,\\ \|\varphi \mid \mathbb{L}_p(\mathbb{R}^+, t^\gamma)\| &:= \left[\int_0^\infty |\varphi(t)|^p t^\gamma\,dt\right]^{1/p}.\end{aligned}$$

For an integer $s = m = 1, 2, \dots$ the space $\mathbb{GH}^m_p(\mathbb{R}^+, t^\gamma)$ is isomorphic to the generic Sobolev space $\mathbb{GW}^m_p(\mathbb{R}^+, t^\gamma)$, where functions have the finite norm

$$\|\varphi \,|\, \mathbb{GW}^m_p(\mathbb{R}^+, t^\gamma)\| := \left[\sum_{k=0}^{m} \|\boldsymbol{D}^k\varphi \,|\, \mathbb{L}_p(\mathbb{R}^+, t^\gamma)\|^p\right]^{1/p}, \qquad \boldsymbol{D}\varphi(t) := t\frac{d\varphi(t)}{dt}.$$

The generic Sobolev-Slobodecki space $\mathbb{GW}^s_p(\mathbb{R}^+, t^\gamma)$ can be defined either as the trace space of $\mathbb{GH}^{s+1/p}_p(\Omega_\alpha, t^\gamma)$ on $\mathbb{R}^+$ or as the pull-back space of corresponding Sobolev-Slobodecki space on the real axes $\mathbb{W}^s_p(\mathbb{R})$ under the transformation $-\ln t : \mathbb{R}^+ \to \mathbb{R}$.

The spaces $\mathbb{GH}^s_p(\mathcal{C}, \rho)$, $\mathbb{GH}^s_p(\Gamma, \rho)$, $\mathbb{GW}^s_p(\mathcal{C}, \rho)$, $\mathbb{GH}^s_p(\Gamma, \rho)$ are defined by the standard local diffeomorphisms of $\mathcal{C} \to \Omega_{\alpha_j}$ and of $\Gamma \to \Gamma_{\alpha_j}$ and a decomposition of identity.

Theorem 5 (cf. [3]) *Convolution operator $\mathfrak{M}^0_{a_\beta}$ extends to a bounded operator*

$$\begin{aligned}\mathfrak{M}^0_{a_\beta} &: \mathbb{GH}^s_p(\mathbb{R}^+, t^\gamma) \to \mathbb{GH}^{s-r}_p(\mathbb{R}^+, t^\gamma), \\ &: \mathbb{GW}^s_p(\mathbb{R}^+, t^\gamma) \to \mathbb{GW}^{s-r}_p(\mathbb{R}^+, t^\gamma)\end{aligned} \tag{9}$$

for arbitrary $s \in \mathbb{R}$, $1 < p < \infty$, $-1 < \gamma < p - 1$, if and only if the symbol

$$a_\beta(\xi) = \int_0^\infty t^{\beta - i\xi} k(t)\frac{dt}{t}, \qquad \xi \in \mathbb{R}, \quad 0 < \beta := \frac{\gamma + 1}{p} < 1, \tag{10}$$

where $k(t)$ is the Hörmander's kernel of the operator $\mathfrak{M}^0_{a_\beta}$, belongs to the $\mathbb{L}_p$-multiplier class $a_\beta \in M_p(\mathbb{R})$. The boundedness in (9) is independent of $s \in \mathbb{R}$.

The $\mathbb{L}_p$-multiplier class $M_p(\mathbb{R})$ contains all functions of bounded variation from $V_1(\mathbb{R})$ and the Wiener functions from $W(\mathbb{R})$.

Theorem 6 (cf. [3]) *Let $1 < p < \infty$, $s, r \in \mathbb{R}$. The convolution operator $\mathfrak{M}^0_{a_\beta}$ in (9) is Fredholm if only the shifted symbol is elliptic*

$$\inf_{\xi \in \mathbb{R}} \left|\det a^{(-r)}_\beta(\xi)\right| > 0, \qquad a^{(-r)}_\beta(\xi) := \langle\xi\rangle^{-r} a_\beta(\xi). \tag{11}$$

*If the symbol has bounded variation $a^{(-r)}_\beta \in \mathbb{V}_1(\mathbb{R})$ or belongs to the Wiener class, $a^{(-r)}_\beta \in \mathbb{W}(\mathbb{R})$, the ellipticity of the symbol (11) **is sufficient for** $\mathfrak{M}^0_a$ to be invertible in the setting (9) and the inverse operator is $\mathfrak{M}^0_{a^{-1}_\beta}$.*

3 Localization and the Model Problems, Proof of the Main Result

Ω_α be a model domain, an angle of magnitude α with the vertex at 0 (see Fig. 2) and consider the following model Mixed, Dirichlet and Neumann BVPs:

$$\begin{cases} \Delta u(x) = f(x), & x \in \Omega_{\alpha_j}, \\ u^+(t) = g(t), & \text{on } \mathbb{R}^+, \\ (\partial_{x_2} u)^+(t) = h(t), & \text{on } \mathbb{R}_{\alpha_j}. \end{cases} \quad \begin{array}{l} f \in \mathbb{GH}^{s-2}_p(\Omega_{\alpha_j}, t^{\gamma_j}), \\ g \in \mathbb{GH}^{s-1/p}_p(\mathbb{R}^+, t^{\gamma_j}), \\ h \in \mathbb{GH}^{s-1-1/p}_p(\mathbb{R}_{\alpha_j}, t^{\gamma_j}) \end{array} \tag{12}$$

for a node $c_j \in \mathcal{M}_{DN}$.

$$\begin{cases} \Delta u(x) = f(x), \ x \in \Omega_{\alpha_j}, \\ u^+(t) = g(t), \quad \text{on } \Gamma_{\alpha_j}, \end{cases} \quad \begin{array}{l} f \in \mathbb{GH}^{s-2}_p(\Omega_{\alpha_j}, t^{\gamma_j}), \\ g \in \mathbb{GH}^{s-1/p}_p(\Gamma_{\alpha_j}, t^{\gamma_j}) \end{array} \tag{13}$$

for a node $c_j \in \mathcal{M}_{DD}$,

$$\begin{cases} \Delta u(x) = f(x), & x \in \Omega_{\alpha_j}, \\ (\partial_{x_2} u)^+(t) = h(t), & \text{on } \Gamma_{\alpha_j}. \end{cases} \quad \begin{array}{l} f \in \mathbb{GH}^{s-2}_p(\Omega_{\alpha_j}, t^{\gamma_j}), \\ h \in \mathbb{GH}^{s-1-1/p}_p(\Gamma_{\alpha_j}, t^{\gamma_j}) \end{array} \tag{14}$$

for a node $c_j \in \mathcal{M}_{NN}$,

Theorem 7 (Local Principle; cf. [1]) *The initial mixed boundary value problem* (1) *in the setting* (5) *is Fredholm if and only if*

- *the BVP* (12) *are Fredholm at all nodes* $c_j \in \mathcal{M}_{DN}$*;*
- *the BVP* (13) *are Fredholm at all nodes* $c_j \in \mathcal{M}_{DD}$*;*
- *the BVP* (14) *are Fredholm at all nodes* $c_j \in \mathcal{M}_{NN}$*.*

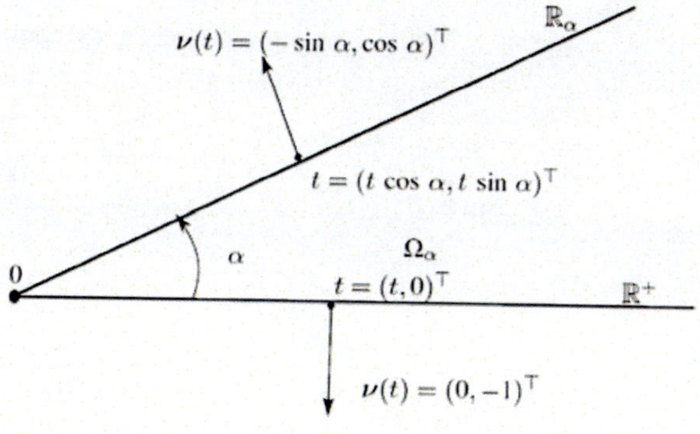

Fig. 2 Model domain Ω_α

Theorem 8 (cf. [5, 7, 9]) *The mixed, Dirichlet and Neumann boundary value problems* (12), (14) *and* (14) *respectively, have at most one solution in the space* $\mathbb{GH}^1(\Omega_{\alpha_j})$ *(the classical setting).*

Proof of Theorem 2 Due to Theorem 7 Fredholmness of the BVP (1) follows from the Fredholmness of all local representatives (12)–(14) at all nodes $c_j \in \mathcal{M}_\gamma$.

Any solution to the mixed BVPs (12)–(14) is represented by the formula

$$u(x) = \boldsymbol{N}_{\mathcal{C}} f(x) + \boldsymbol{W}_\Gamma u^+(x) - \boldsymbol{V}_\Gamma[\partial_{\boldsymbol{\nu}} u]^+(x),$$

where u^+ is the Dirichlet and $[\partial_{\boldsymbol{\nu}} u]^+$ is the Neumann trace of the solution u on the boundary and $\boldsymbol{N}_{\mathcal{C}}$ is the Newtons, $\boldsymbol{V}_\Gamma$ is the Single layer and $\boldsymbol{W}_\Gamma$ is the Double layer potentials. To find the missing Dirichlet data $\varphi = u^+ \in \mathbb{GW}^{s-1/p}_p(\mathbb{R}_\alpha, t^{\gamma_j})$ for the Neumann BVP on $\mathbb{R}_\alpha$ or find the missing Neumann data $\psi = (\partial_{\boldsymbol{\nu}} u)^+ \in \mathbb{GW}^{s-1/p-1}_p(\mathbb{R}^+, t^{\gamma_j})$ for the Dirichlet BVP on $\mathbb{R}^+$, we apply the Plemelji formulae and derive the corresponding system of boundary pseudodifferential equations (the potential method).

Now let us derive solvability conditions of BVPs (12)–(14). Let us start with the model Mixed BVP (12).

After some simplifications the obtained system is reduced to the following equivalent system of boundary integral equations (see [7, § 3] for details)

$$A_{DN}\Psi := \begin{bmatrix} I & \dfrac{1}{2}\left[\boldsymbol{K}^1_{e^{i\alpha_j}} + \boldsymbol{K}^1_{e^{i(2\pi-\alpha_j)}}\right] \\ \dfrac{1}{2}\left[\boldsymbol{K}^1_{e^{i\alpha_j}} + \boldsymbol{K}^1_{e^{i(2\pi-\alpha_j)}}\right] & I \end{bmatrix} \Psi = G,$$

$\Psi = (\Psi_1, \Psi_2)^\top,\ G = (G_1, G_2)^\top, \in \mathbb{GW}^{s-1/p}_p(\mathbb{R}^+, ^{\beta_j})$ and

$$\boldsymbol{K}^1_{e^{i\omega}}\psi(t) := \frac{1}{\pi}\int_0^\infty \frac{\psi(\tau)d\tau}{t - e^{i\omega}\tau}, \qquad 0 < \omega < \pi.$$

The derived system is a Mellin convolution and the symbol of $\boldsymbol{K}^1_{e^{i\omega}}$ (the Mellin transform $\mathcal{M}_{\beta_j}$ of the kernel) is (cf. [2, (9)])

$$\mathcal{K}^1_{e^{i\omega}}(\xi) = \frac{e^{i(\omega-\pi)(\beta_j - i\xi)}}{\sin\pi(\beta_j - i\xi)}, \qquad \beta_j = \frac{\gamma_j + 1}{p}, \qquad 0 < \omega < \pi,$$

the symbol of $\boldsymbol{A}_{DN}$ is

$$\mathcal{A}_{DN}(\xi) = \begin{bmatrix} 1 & \dfrac{\cos(\pi - \alpha_j)(\beta_j - i\xi)}{\sin\pi(\beta_j - i\xi)} \\ \dfrac{\cos(\pi - \alpha_j)(\beta_j - i\xi)}{\sin\pi(\beta_j - i\xi)} & 1 \end{bmatrix}$$

and

$$\det \mathcal{A}_{DN}(\xi) = 1 - \frac{\cos^2(\pi - \alpha_j)(\beta_j - i\xi)}{\sin^2 \pi(\beta_j - i\xi)}$$

$$= \frac{\cos^2 \pi(\beta_j - i\xi - 1/2) - \cos^2(\pi - \alpha_j)(\beta_j - i\xi)}{\sin^2 \pi(\beta_j - i\xi)}$$

$$= \frac{\cos(2\pi - \alpha_j)(\beta_j - i\xi)\cos\alpha_j(\beta_j - i\xi)}{\sin^2 \pi(\beta_j - i\xi)}, \quad c_j \in \mathcal{M}_{DN}. \quad (15)$$

For the node $c_j \in \mathcal{M}_{DD}$ the equivalent system of boundary integral equation to BVP (13) is (see [5, § 3] for details)

$$\boldsymbol{A}_{DD}\Psi := \begin{bmatrix} I & \frac{1}{2i}\left[\boldsymbol{K}^1_{e^{i\alpha_j}} - \boldsymbol{K}^1_{e^{i(2\pi-\alpha_j)}}\right] \\ \frac{1}{2i}\left[\boldsymbol{K}^1_{e^{i\alpha_j}} - \boldsymbol{K}^1_{e^{i(2\pi-\alpha_j)}}\right] & I \end{bmatrix} \Psi = F,$$

where $\Psi = (\Psi_1, \Psi_2)^\top, \ F = (F_1, F_2)^\top, \in \mathbb{GW}^{s-1/p}_p(\mathbb{R}^+, \gamma_j)$. The symbol of $\boldsymbol{A}_{DD}$ is

$$\mathcal{A}_{DD}(\xi) = \begin{bmatrix} 1 & -\frac{\sin(\pi - \alpha_j)(\beta_j - i\xi)}{\sin \pi(\beta_j - i\xi)} \\ -\frac{\sin(\pi - \alpha_j)(\beta_j - i\xi)}{\sin \pi(\beta_j - i\xi)} & 1 \end{bmatrix},$$

$$\det \mathcal{A}_{DD}(\xi) = 1 - \frac{\sin^2(\pi - \alpha_j)(\beta_j - i\xi)}{\sin^2 \pi(\beta_j - i\xi)}$$

$$= \frac{\sin^2 \pi(\beta_j - i\xi) - \sin^2(\pi - \alpha_j)(\beta_j - i\xi)}{\sin^2 \pi(\beta_j - i\xi)}$$

$$= \frac{\sin(2\pi - \alpha_j)(\beta_j - i\xi)\sin\alpha_j(\beta_j - i\xi)}{\sin^2 \pi(\beta_j - i\xi)}, \quad c_j \in \mathcal{M}_{DD}. \quad (16)$$

For the node $c_j \in \mathcal{M}_{NN}$ the equivalent system of boundary integral equations to BVP (14) is (see [5, § 3] for details)

$$\boldsymbol{A}_{NN}\Psi := \begin{bmatrix} I & \frac{1}{2i}\left[e^{i\alpha_j}\boldsymbol{K}^1_{e^{i\alpha_j}} - e^{-i\alpha_j}\boldsymbol{K}^1_{e^{i(2\pi-\alpha_j)}}\right] \\ \frac{1}{2i}\left[e^{i\alpha_j}\boldsymbol{K}^1_{e^{i\alpha_j}} - e^{-i\alpha_j}\boldsymbol{K}^1_{e^{i(2\pi-\alpha_j)}}\right] & I \end{bmatrix} \Psi = H,$$

where $\Psi = (\Psi_1, \Psi_2)^\top$, $H = (H_1, H_2)^\top, \in \mathbb{GW}^{s-1/p}_p(\mathbb{R}^+, {}^{\beta_j})$. If we replace $e^{-i\alpha_j} = -e^{-i(\alpha_j-\pi)}$ and $e^{i\alpha_j} = -e^{i(\alpha_j-\pi)}$, the symbol of $\boldsymbol{A}_{NN}$ acquires the following form:

$$\mathcal{A}_{NN}(\xi) = \begin{bmatrix} 1 & -\dfrac{\sin(\pi-\alpha_j)(\beta_j - i\xi - 1)}{\sin\pi(\beta_j - i\xi)} \\ -\dfrac{\sin(\pi-\alpha_j)(\beta_j - i\xi - 1)}{\sin\pi(\beta_j - i\xi)} & 1 \end{bmatrix},$$

$$\det \mathcal{A}_{NN}(\xi) = 1 - \frac{\sin^2(\pi-\alpha_j)(\beta_j - i\xi - 1)}{\sin^2\pi(\beta_j - i\xi)}$$
$$= \frac{\sin(2\pi-\alpha_j)(\beta_j - i\xi - 1)\sin\alpha_j(\beta_j - i\xi - 1)}{\sin^2\pi(\beta_j - i\xi - 1)}. \quad (17)$$

From Theorem 6 follows:

- BVP (12) is Fredholm iff (cf. (16)) $(2\pi - \alpha_j)\beta_j,\ \alpha_j\beta_j \neq (k + 1/2)\pi,\ k = 0, \pm 1, \ldots\ \forall\, c_j \in \mathcal{M}_{DN}$;
- BVP (13) is Fredholm iff (cf. (16)) $(2\pi - \alpha_j)\beta_j,\ \alpha_j\beta_j \neq k\pi,\ k = 0, \pm 1, \ldots\ \forall\, c_j \in \mathcal{M}_{DD}$;
- BVP (14) is Fredholm iff (cf. (17)) $(2\pi - \alpha_j)(\beta_j - 1),\ \alpha_j(\beta_j - 1) \neq k\pi$, $k = 0, \pm 1, \ldots\ \forall\, c_j \in \mathcal{M}_{NN}$.

To check the unique solvability property of the Dirichlet, Neumann and Mixed type BVPs, stated in Theorem 2, we pick up the intervals I^0_{DN}, I^0_D and I^0_N, described in Theorem 2. The corresponding BVPs are Fredholm for the parameter $(\gamma + 1)/p$ from these intervals and for $\gamma = 0$, $p = 2$, $s = 1$ (s can be arbitrary), which corresponds to the classical setting of the BVPs, are uniquelly solvable (with the compatibility condition (3) in case of the Neumann BVP). Then, due to [4, Corollary 6.3] (also see [2, Corollary 5.6]), the DIrivhlet, Neumann and Mixed type BVPs have unique solutions for all values of the parameter $(1 + \gamma)/p$ from the intervals I^0_{DN}, I^0_D and I^0_N, respectively. ■

Acknowledgments The author was supported by the Grant "EU fellowships for Georgian researchers, 2023 (57655523)".

References

1. T. Buchukuri, R. Duduchava, D. Kapanadze, M. Tsaava, Localization of a Helmholtz boundary value problem in a domain with piecewise-smooth boundary. Proc. A. Razmadze Math. Inst. **162**, 37–44 (2013)
2. V. Didenko, R. Duduchava, Mellin convolution operators in the Bessel potential spaces. J. Anal. Appl. **443**, 707–731 (2016)
3. R. Duduchava, *Integral Equations with Fixed Singularities* (Teubner, Leipzig, 1979)

4. R. Duduchava, Mellin convolution operators in Bessel potential spaces with admissible meromorphic kernels. Mem. Differ. Equ. Math. Phys. **60**, 135–177 (2013). Corrected and extended version: http://arxiv.org/pdf/1502.02756.pdf
5. R. Duduchava, Mixed type boundary value problems for Laplace-Beltrami equation on a surface with the Lipschitz boundary. Georgian Math. J. **28**(2), 219–232 (2021)
6. R. Duduchava, M. Tsaava, Mixed boundary value problems for the Laplace-Beltrami equation. Complex Variables Elliptic Equ. **63**(10), 1468–1496 (2018)
7. R. Duduchava, M. Tsaava, Mixed boundary value problems for the Helmholtz equation in a model 2D angular domain. Georgian Math. J. **27**(2), 211–231 (2020)
8. R. Duduchava, D. Mitrea, M. Mitrea, Differential operators and boundary value problems on surfaces. Math. Nachrichten **279**(9–10), 996–1023 (2006)
9. R. Duduchava, M. Tsaava, T. Tsutsunava, Mixed boundary value problem on hypersurfaces. Int. J. Differ. Equ. **2014**, 245350, 8pp. (2014)

On the Convergence Fourier Series and Greedy Algorithm by Multiplicative System

M. G. Grigoryan, T. M. Grigoryan, and L. S. Simonyan

Abstract In this work we discuss the behavior of Fourier coefficients with respect to the multiplicative system (Vilenkin system), as well as convergence of the Fourier series and gredy algorithm with respect to the multiplicative system after modification of functions.

Keywords Fourier series · Multiplicative Systems · Greedy algorithm

2020 Mathematics Subject Classification 42A16, 42A10, 42A15

1 Introduction

Let $|E|$ be the Lebesgue measure of a measurable set $E \subseteq [0, 1)$ (or $E \subseteq [0, 1) \times [0, 1) = [0, 1)^2$), and let $L^r[0, 1)$, $r \geq 1$, be the class of all those measurable functions $f(x)$ on $[0, 1)$ such that

$$\int_0^1 |f(x)|^r dx < \infty .$$

Let $\mu(x)$ be a positive Lebesgue-measurable function (weight function) defined on $[0, 1)$. By $L^r_\mu[0, 1)$ we denote the space of all measurable functions on $[0, 1)$ with the norm

$$\|\cdot\|_{L^p_\mu} = \left(\int_0^1 |.|^r \ \mu(x)dx\right)^{\frac{1}{r}} < \infty : \ r \in [1, \infty).$$

This work was supported by the RAMES State Committee of Science.

M. G. Grigoryan (✉) · T. M. Grigoryan · L. S. Simonyan
Yerevan State University, Yerevan, Armenia
e-mail: gmarting@ysu.am; t.grigoryan@ysu.am; lussimonyan@mail.ru; lysine.simonyan@ysu.am

D. Cardona et al. (eds.), *Extended Abstracts 2021/2022*, Research Perspectives Ghent Analysis and PDE Center 3, https://doi.org/10.1007/978-3-031-48579-4_12

In the sequel, we will accept the terms "measure" and "measurable" in sense of Lebesgue.

Recall the definition of multiplicative systems (see [6]) (systems of Vilenkin) of functionsS. Consider the arbitrary sequence of natural numbers $P \equiv \{p_1, p_2, \ldots, p_k, \ldots\}$ where $p_j \geq 2$ for all $j \in \mathbb{N}$. We set

$$m_0 = 1,\ m_k = \prod_{j=1}^{k} p_j\,, \quad k \in \mathbb{N}$$

It is not difficult to notice that for each point $x \in [0, 1)$ and for any $n \in [m_k, m_{k+1}) \cap \mathbb{N},\ k \in \mathbb{N}$, there exist numbers $x_j, \alpha_j \in \{0, 1, \ldots p_j - 1\}$ such that

$$n = \sum_{j=1}^{k} \alpha_j m_{j-1} \text{ and } x = \sum_{j=1}^{\infty} \frac{x_j}{m_j}, \ (P\text{-order expansions}).$$

Note that all points of type $\frac{l}{m_k}$ with $l, k \in \mathbb{N};\ 0 \leq l \leq m_k - 1$, have two different expansions: finite and infinite, and to have only unique expansions we take only finite expansions for such points. As a result we get the correspondences

$$n \to \{\alpha_1, \alpha_2, \ldots, \alpha_k\},\ x \to \{x_1, x_2, \ldots, x_k, \ldots\}.$$

The multiplicative system (Vilenkin system) for sequence P is defined as follows:

$$V_0(x) \equiv 1;\ V_n(x) = \exp\left(2\pi i \sum_{j=1}^{k} \alpha_j \frac{x_j}{p_j}\right).$$

The expression we can change to the form

$$V_n(x) = \exp\left(2\pi i \sum_{j=1}^{k} \alpha_j \frac{x_j}{p_j}\right) = \prod_{j=1}^{k} \left(\exp\left(2\pi i \frac{x_j}{p_j}\right)\right)^{\alpha_j}.$$

From this it follows

$$V_{m_{j-1}}(x) = \exp\left(2\pi i \frac{x_j}{p_j}\right),.$$

and for the n-th function we obtain the expression

$$V_n(x) = \prod_{j=1}^{k} (V_{m_{j-1}}(x))^{\alpha_j}.$$

It is not difficult to notice that

$$\int_0^1 V_n(t)\overline{V_k(t)}dt = \begin{cases} 1, & \text{if } k = n; \\ 0, & \text{if } k \neq n, \end{cases}$$ where $\overline{z}$ denotes the complex conjugate of z.

Let $V = \{V_k(x)\}$—be eiter unboundet or boundet Vilenkin system and let $f(x)$ be a real valued function from $L^r[0, 1)$, $r \geq 1$ and let $c_n(f)$ be the Fourier-Vilenkin coefficients of function f, that is

$$c_k(f) = \int_0^1 f(x)\overline{V_k(t)}dx .$$

We denote by $S_N(x, f)$ – N-th partial summ of Fourier-Vilenkin series of function $f(x)$, that is

$$S_N(x, f) = \sum_{k=0}^{N} c_k(f)V_k(x) .$$

The spectrum of $f(x)$ (denoted by $spec(f)$) is the support of $c_k(f)$, i.e. the set of integers where $c_k(f)$ is non-zero, that is

$$spec(f) = \{k \in N, c_k(f) \neq 0\}.$$

It's obvious that systems corresponding to different sequences $\{p_k\}$, differ from each other (in case $P \equiv \{2, 2, \ldots, 2, \ldots\}$ Vilenkin system coincides with the Walsh system (see [25]). In case $\sup\{p_k\} = \infty$ ($\sup\{p_k\} < \infty$) the system $\{V_n(x)\}$ is said to be unbounded (accordingly bounded).

The theory of such systems have been introduced by N. Ya. Vilenkin [24] in 1946. Then there are interesting results for Vilenkin system (see [24, 26, 28]).

In case of bounded Vilenkin systems it is known that if f is a function of bounded variation then $c_n(f) = O(n^{-1})$, and if $f \in \text{Lip}\,\alpha$ it is true $c_n(f) = O(n^{-\alpha})$. In contrast with this note that in case of unbounded Vilenkin systems certain basic properties of classical orthonormal systems no longer hold, for instance if $\limsup p_n = \infty$, then $\limsup n|c_n(\psi)| = \infty$ where $\psi(x) = x - [x]$. Moreover, there exists $f_0 \in \text{Lip}\,\alpha$ such that $\limsup n|c_n(f_0)| = \infty$.

In 1957 C. Watari [26] proved that the bounded Vilenkin system is basis in the spaces $L^r[0, 1)$ for all $r > 1$, that is, any function $f(x) \in L^r[0, 1)$ can be uniquely represented by the series $\sum_{k=0}^{\infty} c_k(f)V_k(x)$ which converges to f in the $L^r[0, 1)$-norm.

Then, in 1976, W.S. Young [28] for arbitrary sequence $\{p_k\}$ (i.e. both for bounded and unbounded Vilenkin systems) established the basicity of Vilenkin system in L^r when $r > 1$.

Note that the following problem remains open: Is the Fourier series of function from $L^2[0, 1)$ with respect to the unbounded Vilenkin systems convergent almost everywhere or no (in [1] P. Billard established that this problem has a positive answer for the Walsh system) ?

Note also that in for the unbounded Vilenkin systems [16] is proved

Theorem 1 *Let $V = \{V_k(x)\}$ -be eiter unboundet or boundet Vilenkin system. Then for any $\epsilon > 0$ there exists a measurable set $E \subset [0, 1]$ of measure $|E| > 1 - \epsilon$ such that for each function $f \in L^1[0, 1]$ one can find a function $g \in L^1[0, 1)$ equal to $f(x)$ on E and such that the sequence $\{|c_k(g)|\ ,\ k \in spec(g)\}$ is monotonically decreasing.*

Now we recall the definition of the greedy algorithm.

Let $\Psi = \{\psi_n\}_{n=1}^{\infty}$ be a normalized basis in Banach space X. Then for each element $f \in X$ there exists a unique series by system $\{\psi_n\}_{n=1}^{\infty}$ converging to f in the norm of space X:

$$f = \sum_{n=0}^{\infty} c_n(f)\psi_n \ ,$$

Let an element $f \in X$ be given. We call a permutation $\sigma = \{\sigma(n)\}_{n=1}^{\infty}$ of nonnegative integers decreasing and write $\sigma \in D(f, \Psi)$, if

$$|c_{\sigma(n)}(f)| \geq |c_{\sigma(n+1)}(f)|, \quad n = 1, 2, \dots .$$

In the case of strict inequalities here $D(f, \Psi)$ consists of only one permutation.

We define the m-th greedy approximant of f with regard to the basis Ψ corresponding to a permutation $\sigma \in D(f, \Psi)$ by formula

$$G_m(f) := G_m(f, \Psi, \sigma) := \sum_{n=1}^{m} c_{\sigma(n)}(f)\psi_{\sigma(n)} \ .$$

We say that the greedy approximant of element f by system Ψ converges, if for some $\sigma \in D(f, \Psi)$ we have

$$\lim_{m\to\infty} \|G_m(f, \Psi, \sigma) - f\|_X = 0 \ .$$

Greedy algorithms in Banach spaces with respect to normalized bases have been investigated by Temlyakov, DeVore, Konyagin, Wojtaszczyk, Korner and other authors .

Greedy algorithms in Banach spaces with respect to normalized bases have been considered in [2–4, 7, 8, 10, 12, 13, 17, 18, 20, 21] and [27]

Here we present results having a direct bearing on the present work.

In [21] T.W. Korner answering a question raised by Carleson and Coifman constructed a continuous function, whose greedy algorithm with respect to the trigonometric systems diverges almost everywhere.

In [23] V.N. Temlyakov constructed a function f that belongs to any L^r , $1 \leq r < 2$ (respectively $r > 2$), whose greedy algorithm with respect to the trigonometric system diverges in measure (respectively in $L^r, r > 2$).

In [7] R. Gribonval and M. Nielsen proved that for any $r \neq 2$ there exists a function from $L^r[0, 1]$, whose greedy algorithm with respect to the Walsh system diverges in $L^r[0, 1]$.

In [10] M. G. Grigorian and A. A. Sargsyan, constructed a continuous function $f \in C[0, 1]$, whose greedy algorithm with respect to the Faber-Shauder system diverges in measure.

The following question arises here.

Question 1 Does there exist a set e of arbitrarily small measure such that, after modifying the values of an arbitrar function in $L^r[0, 1]$, $r > 1$ on e, the greedy algorithms for the modified function with respect to the classica (trigonometric,Walsh and Haar, Faber-Shauder,...) systems converge to it (almost everywhere, in the norm of $L^r[0, 1]$, uniformly)?

Note that the results obtained in the framework of this direction we have been published in papers [5, 8, 9, 12–14, 16, 17] and [19].

It is important to note that, as was shown in the article [17] by M. G. Grigoryan, K. S. Kazaryan and F. Soria, there exist a complete orthonormal system $\{\varphi_k(x)\}$ and a function $f(x) \in L^r[0, 1]$, $r > 2$, such that if g(x) is any function in $L^r[0, 1]$ with measure

$$mes\{x \in [0, 1] ;\ f(x) = g(x)\} > 0 ,$$

then it's greedy algorithm with respect to the system $\{\varphi_k(x)\}$ diverges in $L^r[0, 1]$.

2 New Results: (Theorems)

The following theorems are true

Theorem 2 *Let* $V = \{V_k(x)\}$*—be boundet multiplicative system. Then for each* $0 < \epsilon < 1$ *there exist a measurable set* $E \subset [0, 1]$, *with measure* $|E| > 1 - \epsilon$ *and a weight function* $\mu(x)$; $0 < \mu(x) \leq 1$; $\mu(x) = 1$ *on* E *such that for every* $r \in [1, \infty)$ *and for each* $f(x) \in L^r_\mu[0, 1]$ *there is a function* $g(x) \in L^1[0, 1] \cap L^r_\mu[0, 1]$ *coinciding with* $f(x)$ *on* E *the the Fourier series and greedy algorithm of* $g(x)$ *with respect to the multiplicative system and the sequence* $\{|c_k(g)| \ , \ k \in spec(g)\}$ *is monotonically decreasing.*

Theorem 3 *Let* $V = \{V_k(x)\}$*—be boundet multiplicative system. Then for any* $\epsilon > 0$ *there exists a measurable set* $E \subset [0, 1]$, *with measure* $|E| > 1 - \epsilon$ *such that* *for any function* $f(x) \in L^1[0, 1)$ *one can find a function* $g(x) \in L^1[0, 1)$ *equal to* $f(x)$ *on E such that its the Fourier series and greedy algorithm with respect to the* $\{V_k\}_{k=0}^{\infty}$ *system converges to it both in the* L^1 *norm and almost everywhere on [0, 1] and the sequence* $\{|c_k(g)|\ ,\ k \in spec(g)\}$ *is monotonically decreasing.*

Theorem 4 *Let* $V = \{V_k(x)\}$*—be eiter unboundet or boundet Vilenkin system. Then there exists such a (universal) function* $U \in L^1[0, 1)$ *with strictly decreasing Fourier —Vilenkin coefficients, with the follwing property: for any* $\epsilon > 0$ *there exists a measurable set* $E \subset [0, 1]$ *of measure* $|E| > 1 - \epsilon$ *such that for each function* $f \in L^1[0, 1]$ *one can find a function* $g \in L^1[0, 1)$ *equal to* $f(x)$ *on E and* $\{|c_k(g)| = c_k(U),\ k \in spec(g)\}$ *.*

Note that these theorems for the Walsh system were proved in [5] and [9].

Note also that D. E. Menshov [21] obtained the following result:

Let $f(x)$ be an almost everywhere finite measurable function on $[0, 2\pi]$. Then for each $\epsilon > 0$ one can define a continuous function $g(x)$ coinciding with $f(x)$ on a subset E of measure $|E| > 2\pi - \epsilon$ such that its Fourier series with respect to the trigonometric system converges uniformly on $[0, 2\pi]$.

The following problem remains open:

Question 2 Is the theorem 2 true for the trigonometric system?

Question 3 Is the theorem 3 true for the unboundet Vilenkin system?

References

1. P. Billard, Sur la convergence presque partout des series de Fourier-Walsh des fonctions de l'espace $L^2(0, 1)$. Stud. Math. **28**(3), 363–388 (1967)
2. R.A. DeVore, Some remarks on greedy algorithms. Adv. Comput. Math. **5**, 173–187 (1996)
3. S.A. Episkoposian, M.G. Grigoryan, Convergence of greedy algorithm by generalized Walsh system. J. Math. Anal. Appl. **389**, 1374–1379 (2012)
4. S.A. Episkoposian, T.G. Grigorian, L.S. Simonyan, On L^p−greedy universal functions with respect to the Walsh generalized system. J. Contemp. Math. Anal. **57**(2), 17–32 (2022)
5. M.G. Grigoryan, Functions with universal Fourier–Walsh series. Mat. Sb. **211**(6), 107–131 (2020)
6. B.I. Golubov, Multiple Furier series and integrals. Itogi Nauki i Tekhniki Ser. Mat. Anal. **47**(5), 97–162 (1992). (English transl. in J. Soviet Math. 24 (1984))
7. R. Gribonval, M. Nielsen, On the quasi-greedy property and uniformly bounded orthonormal systems. http://www.math.auc.dk/research/reports/R-2003-09.pdf
8. M.G. Grigoryan Uniform convergence of the greedy algorithm with respect to the Walsh system. Stud. Math. **198**(2), 197–206 (2010)
9. M.G. Grigoryan, Modifications of functions, Fourier coefficients and nonlinear approximation. Matem. Sb. **203**(3), 49–78 (2012)
10. M.G. Grigorian, A.A. Sargsyan, Non-linear approximation of continuou functions by the Faber–Schauder system Sbornik. Mathematics **199**(5), 629–653 (2008)
11. M.G. Grigorian, S.A. Sargsyan, On the L1-convergence and behavior of coefficients of Fourier–Vilenkin series. Positivity **22**(3), 897–918 (2018)

12. M.G. Grigorian, R.E. Zink, Greedy approximation with respect to certain subsystems of the Walsh orthonormal system. Proc. Am. Math. Soc. **134**(12), 3495–3505 (2006)
13. M.G. Grigoryan, S.L. Gogyan, On nonlinear approximation with respect to the Haar system and modifications of functions. An. Math. **32**, 49–80 (2006)
14. M.G. Grigoryan, L.N. Galoyan, On the uniform convergence of negative order Cesaro means of Fourier series. J. Math. Anal. Appl. **4341**, 554–567 (2016)
15. M.G. Grigoryan, A.A. Sargsyan, On the universal functions fr the class Lp[0,1]. J. Funct. Anal. **270**, 3111–3133 (2016)
16. M.G. Grigoryan, S.A. Sargsyan, On the Fourier-Vilenkin coefficients. Acta Math. Sci. **37**(2), 293–300 (2017)
17. M.G. Grigoryan, K.S. Kazarian, F. Soria, Mean convergence of orthonormal Fourier series of modified funcions. Trans. Am. Math. Soc. **352**(8), 3777–3799 (2000)
18. M.G. Grigoryan, T.M. Grigoryan, L.S. Simonyan, *Convergence of Fourier-Walsh Double Series in Weighted* $L^p_\mu[0;1)^2$. Springer Proceedings in Mathematics and Statistics 2, vol. 275 (Springer, Cham, 2019), pp. 109–137
19. S.V. Konyagin, V.N. Temlyakov, A remark on Greedy approximation in Banach spaces. East J. Approx. **5**, 1–15 (1999)
20. T.W. Körner, Decreasing rearranged Fourier series. J. Fourier Anal. Appl. **5**, 1–19 (1999)
21. D.E. Menchoff [D. E. Men'shov], Sur la convergence uniforme des series de Fourier (French; Russian summary). Mat. Sb. **11**(53), 67–96 (1942)
22. R.E.A.C. Paley, A remarkable set of orthogonal functions. Proc. Lond. Math. Soc. **34**, 241–279 (1932)
23. V.N. Temlyakov, Greedy Algoritm and m-term Trigonometric approximation. Constructive Approx. **14**, 569–587 (1998)
24. N.J. Vilenkin, On a class of complete orthonormal systems. Am. Math. Soc. Transl. **28**(2), 1–35 (1963)
25. J.L. Walsh, A closed set of normal orthogonal functions. Am. J. Math. **45**, 5–24 (1923)
26. C. Watari, On generalizes Walsh-Fourier series. Proc. Japan Acad. **73**(8), 435–438 (1957)
27. P. Wojtaszczyk, Greedy algoritm for general biorthogonal systems. J. Approx. Theory **107**, 293–314 (2000)
28. W.S. Young, Mean convergence of generalized Walsh-Fourier series. Trans Am. Math. Soc. **218**, 311–320 (1976)

Asymptotics of Harmonic Functions in the Absence of Monotonicity Formulas

Zongyuan Li

Abstract In this article, we study the asymptotics of harmonic functions. In literature, one typical method is by proving the monotonicity of a version of rescaled Dirichlet energies, and use it to study the renormalized solution—the Almgren's blowup. However, such monotonicity formulas require strong smoothness assumptions on domains and operators. We are interested in the cases when monotonicity formulas are not available, including variable coefficient equations with unbounded lower order terms, Dirichlet problems on rough (non-C^1) domains, and Robin problems with rough Robin potentials.

Keywords Unique continuation · Asymptotic expansion · Doubling index · Almgren's monotonicity formula

2010 Mathematics Subject Classification 35J15, 35J25, 35B40

1 Introduction

We discuss asymptotics of solutions to elliptic equations near both interior and boundary points. Let us start from a simple case. Consider a harmonic function u in a bounded domain $\Omega \subset \mathbb{R}^d$. Near an interior point $x_0 \in \Omega$, we know that u is analytic:

$$\begin{aligned} u = \sum_{\alpha} \frac{D^\alpha u(x_0)}{\alpha!}(x - x_0)^\alpha &= \sum_k P_k(x - x_0) \\ &= P_N(x - x_0) + O(|x - x_0|^{N+1}). \end{aligned} \tag{1}$$

Z. Li (✉)
Department of Mathematics, Rutgers University, Piscataway, NJ, USA
e-mail: zongyuan.li@rutgers.edu

D. Cardona et al. (eds.), *Extended Abstracts 2021/2022*, Research Perspectives Ghent Analysis and PDE Center 3, https://doi.org/10.1007/978-3-031-48579-4_13

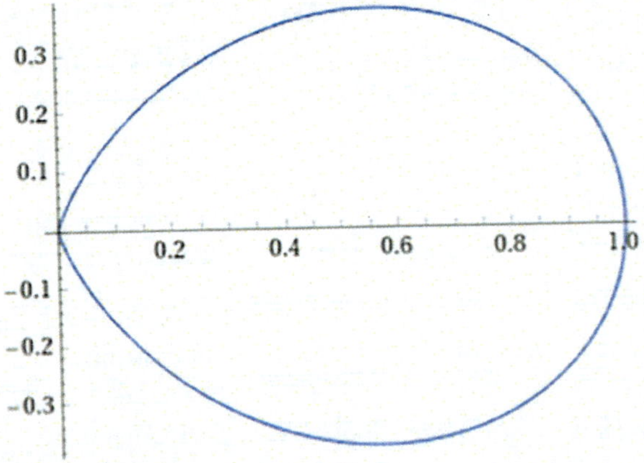

Fig. 1 Nodal set of $\mathrm{Re}(z/\log z)$

Here P_k is a homogeneous harmonic polynomial of degree k and P_N represents the leading term. As is commonly known, expansion formulas like (1) can be useful, which are, however, not always available in the presence of variable coefficient operators or rough domains. For instance, under polar coordinates (r, θ) of $\mathbb{R}^2$, consider

$$u = \mathrm{Re}\frac{re^{i\theta}}{\log(re^{i\theta})}, \quad r > 0, \theta \neq \pi. \tag{2}$$

One can see that u is harmonic in the enclosed region in Fig. 1, and equals to zero on the boundary given by $r = e^{\theta \tan\theta}$, except at $(r, \theta) = (1, 0)$ where u has a pole. Clearly it is impossible to write down an expansion like (1), due to the presence of the log drift.

To capture the asymptotics of functions like (2), one typically uses the "Almgren's blowup"—the rescaled limit as $\lambda \to 0$ of

$$u_\lambda(\cdot) = u(\lambda\cdot)/(\fint_{\partial B_\lambda} |u|^2)^{1/2}. \tag{3}$$

For u in (2), one can simply see that $(\fint_{\partial B_\lambda} |u|^2)^{1/2} \approx \lambda \log(\lambda)$ and $u_\lambda \to Cr\cos(\theta)$ as $\lambda \to 0$, where C is a normalizing factor.

Actually the convergence in the above example is guaranteed by a general theorem—Almgren's monotonicity formulas on convex domains. Let us describe the motivation and method. In general, one hopes to prove that the family $\{u_\lambda(\cdot)\}_{\lambda\in(0,1)}$ is compact in certain space. For this, we bound a rescaled Dirichlet energy

$$F(r) = \frac{rD(r)}{H(r)} = \frac{r\int_{B_r} |\nabla u|^2}{\int_{\partial B_r} |u|^2}. \tag{4}$$

In [3], Almgren observed that if $\Delta u = 0$ in B_1, $F(r)$ is monotonically increasing for $r \in (0, 1)$. From this, $\{u_\lambda\}_{\lambda \in (0,1)}$ is uniformly bounded in H^1, and hence is compact in L_2. In literature, a quantity like (4) is usually called a (generalized) Almgren's frequency. Its monotonicity property play an important role in blowup analysis. In this work, we are interested in three more general problems.

Variable Coefficient Equations, Interior

$$D_i(a_{ij}D_j u + b_i u) + W_i D_i u + Vu = 0 \quad \text{in } B_1,$$

where a_{ij} are symmetric, bounded, and uniformly elliptic. In [7], Garofalo-Lin proved that if $a_{ij} \in C^{0,1}$, $b_i = 0$, $W_i, V \in L_\infty$, a modified version of F in (4) is almost monotone, and hence, is bounded. The condition $a_{ij} \in C^{0,1}$ cannot be improved, due to the classical counterexample in unique continuation. Later, we will discuss the cases with unbounded b_i, W_i, V.

Dirichlet Problem, Boundary Suppose $\Omega \subset \mathbb{R}^d$ and $0 \in \partial\Omega$. Consider

$$\begin{cases} \Delta u = 0 & \text{in } \Omega \cap B_1, \\ u = 0 & \text{on } \partial\Omega \cap B_1. \end{cases} \tag{5}$$

When Ω is half space or a cone, the monotonicity formula holds. For curved domains, in [1, 2, 12], certain variations of F in (4) were proved to be almost monotone on $C^{1,1}$, convex, and $C^{1,Dini}$ domains, respectively. Some discussions on C^1 domains were also made in [14]. It is worth mentioning that, the continuity of the normal direction $\boldsymbol{n}|_{\partial\Omega}$ is essential in deriving the monotonicity formula, which is not available for rough domains, for instance general Lipschitz domains.

Neumann and Robin Problem, Boundary Suppose $\Omega \subset \mathbb{R}^d$ and $0 \in \partial\Omega$. Consider

$$\begin{cases} \Delta u = 0 & \text{in } \Omega \cap B_1, \\ \frac{\partial u}{\partial n} = \eta u & \text{on } \partial\Omega \cap B_1. \end{cases} \tag{6}$$

Again, when Ω is half space or a cone and when $\eta = 0$ (Neumann), the monotonicity formula holds. In [1, 6], this was further generalized to the case when $\partial\Omega \in C^{1,1}$ and $\eta \in C^{0,1}$ (or $\eta \in W^{1,1}$ with some pointwise control on $\nabla\eta$). See also a sharp quantitative version in [11]. In all these works, the differentiability of η cannot be dropped, which leaves the asymptotic analysis of (6) with rough η widely open, even in the case when η is non-negative and bounded. For instance, see the open question in [4].

2 Alternative for Motonicity Formula: Convergence of Doubling Index

Robin Problems and Variable Coefficient Equations In a recent work, we prove the following.

Theorem 1 ([10], Theorem 1.1 (b)) *Let $\Omega(\subset \mathbb{R}^d) \in C^{1,1}$, $d \geq 2$, and $\eta \in L_p(\partial\Omega)$ for some $p > d-1$. Then for any nontrivial solution $u \in H^1$ to (6), we have*

$$\dim(\{u = 0\} \cap \partial\Omega) \leq d - 2.$$

Such estimate relies on blowup analysis near boundary points. However, as mentioned before, monotonicity formulas (the usual tool) are only proved when η is differentiable and $\partial\Omega \in C^{1,1}$. This requires us to design more robust methods for blowup analysis. It turns out the Federer's dimension reduction argument, which we used to prove Theorem 1, only needs the following:

(a) a uniform C^0 estimate for the "rescaled" boundary value problems;
(b) compactness of the blowup sequence (3), as $\lambda \to 0$;
(c) the homogeneity of the blowup limit of (3), along subsequences.

In [10], (a) was achieved by a De Giorgi-type estimate. For (b) and (c), which are typically proved via monotonicity formula, we prove the following alternative.

Lemma 2 *Let $u \in H^1$ be a weak solution to (6) with*

$$\partial\Omega \in C^1 \quad \text{and} \quad \eta \in L_{d-1+\varepsilon}(\partial\Omega). \tag{7}$$

Then,

$$\log_2 \left(\frac{\fint_{B_{2r}\cap\Omega} |u|^2}{\fint_{B_r\cap\Omega} |u|^2} \right)^{1/2} \to \mathbb{N} \cup \{+\infty\}, \quad \text{as } r \to 0.$$

Remark 3 The condition (7) appears naturally after scaling: if u solves (6), then u_λ solves

$$\Delta u_\lambda = 0 \text{ in } \Omega_\lambda = (\Omega \cap B_\lambda)/\lambda, \quad \partial u_\lambda/\partial \boldsymbol{n} = \lambda\eta(\lambda\cdot)u_\lambda \text{ on } (\partial\Omega \cap B_\lambda)/\lambda.$$

From (7), $\lambda\eta(\lambda\cdot)$ converges to 0 in a proper space and Ω_λ converges to $\mathbb{R}^d_+ \cap B_1$.

Here in Lemma 2, we study an averaged version of the doubling index

$$N(r) := \log_2 \frac{(\fint_{\partial B_{2r}} |u|^2)^{1/2}}{(\fint_{\partial B_r} |u|^2)^{1/2}}$$

instead of the frequency $F(r)$. Note that when u is exactly a homogeneous polynomial of degree k, $N(r) \equiv k$. Hence, Lemma 2 can be interpreted as "the existence of the limiting homogeneity". Simple computation shows for harmonic functions, near an interior point

$$N(r) = \int_r^{2r} \frac{F(s)}{s} ds.$$

Hence, the monotonicity of F implies the convergence of N, as $r \to 0$. However, the condition in Lemma 2 is much weaker than that of a monotonicity formula—recall in [1, 6, 11], it was required that $\partial\Omega \in C^{1,1}$ and η is differentiable. Hence, we expect the conclusion of Lemma 2 can serve as a more robust tool in blowup analysis. Indeed, in [10], we obtain the asymptotic behavior of solutions to Robin/Neumann problems on C^1 domains.

The proof of Lemma 2 borrows ideas of Lin-Shen [13] when studying homogenization. Essentially, it is relies on fact that the monotonicity formula of harmonic functions has a rigidity property.

Lemma 4 *Suppose u is a harmonic function in B_1^+ satisfying $\partial u/\partial \boldsymbol{n} = 0$ on the flat portion of the boundary. Then its Almgren's frequency F ((4)) is either strictly increasing for $r \in (0, 1)$, or for some $k \in \mathbb{N}$, $F \equiv k/\log 2$ and u is a homogeneous harmonic polynomial of degree k.*

From Lemma 4, it can be shown that for any non-integer real number μ, as r decreases, after certain small scale, the doubling index $N(r)$ of u can no longer jump from below μ to above. Hence, $N(r)$ is trapped near an integer, from which Lemma 2 follows.

Dirichlet Problem Near a Conical Point

In a recent joint work with Dennis Kriventsov, we also study the boundary asymptotics of harmonic functions under Dirichlet boundary conditions. Our study is closely related to a long-standing conjecture.

Conjecture 5 Suppose $u \in H^1$ is a weak solution to (5) on a Lipschitz domain Ω. Then, if $\{\partial u/\partial \boldsymbol{n} = 0\} \cap \partial\Omega$ has a postive surface measure, we must have $u \equiv 0$.

The conjectured was proved in the case when $\Omega \in C^{1,1}, C^{1,Dini}$, and C^1 in [2, 12], and [1], via some variants of Almgren's monotonicity formulas. For such formulas, the continuity of $\boldsymbol{n}|_{\partial\Omega}$ seems inevitable, which is typically not true on general Lipschitz domains. We aim to discover the case when $\boldsymbol{n}$ is not continuous. A point $x_0 \in \partial\Omega$ is called conical, if

$$\frac{\Omega - x_0}{r} \to \Gamma_{x_0} = \text{cone}.$$

Clearly, all differentiable boundary points are conical with Γ being the tangent plane. Moreover, any boundary point of a convex domain is conical, due to the monotonicity. In [9], we prove the following.

Theorem 6 ([9]) *Suppose $0 \in \partial\Omega$ is a conical point and $u \in H^1$ is a nontrivial solution to (5). Then the limiting homogeneity of u exists. That is,*

$$\log_2 \frac{(\fint_{\partial B_{2r}\cap\Omega} |u|^2)^{1/2}}{(\fint_{\partial B_r\cap\Omega} |u|^2)^{1/2}} \to \{\mu_j\}_j \cup \{+\infty\} \quad \text{as } r \to 0,$$

where μ_j are numbers determined by the spectrum of Δ on the limit cone Γ.

It is worth mentioning that, in our theorem only an one-point condition at $0 \in \partial\Omega$ is assumed—no smoothness of Ω is needed.

3 Uniqueness of Blowup and Expansion Formula

It is not difficult to check that the "asymptotic homogeneity" in Lemma 2 and Theorem 6 combined with strong unique continuation properties implies the existence of (subsequence) limit in (3). It is naturally to ask

Problem 1 When is the subsequence limit in (3) unique?

One the one hand, naturally one may further ask.

Problem 2 Does a monotonicity formula, which guarantees the existence of blowup limits, also guarantees the uniqueness of such limit?

The answer is yes when the dimension is two. This is simply due to the fact that in 2D, all eigenspaces of the Laplace operator are one-dimensional. In higher dimension, the answer is no in general. In [9], we constructed a convex domain Ω and a harmonic function u vanishing locally on the boundary. From [2], the Almgren's monotonicity formula holds due to the convexity of the domain. However, along different subsequences, the blowup limits can be different. Actually, our $\{u_\lambda\}_{\lambda\in(0,1)}$ rotates within a two-dimensional eigenspace.

On the other hand, clearly an expansion formula like (1) leads to the uniqueness of blowup limit. One can simply see that the limit has to be exactly the leading term P_N up to normalization. For Dirichlet problems, in [9] we prove that a slightly stronger condition than "conical"—the Hölder conical, will lead to an expansion formula. A point $x_0 \in \partial\Omega$ is called Hölder conical, if for a cone Γ_{x_0},

$$\frac{\mathrm{dist}((\Omega - x_0) \cap B_r, \Gamma_{x_0})}{r} \le Cr^\alpha, \quad \text{for some } \alpha > 0.$$

Theorem 7 ([9]) *Suppose $0 \in \partial\Omega$ is Hölder conical. Then for any non-trivial solution u to (5), either $u = O(|x|^N)$ for all $N > 0$, or there exists a homogeneous harmonic polynomial P on the cone Γ, such that*

$$u(x) = P(x)(1 + O(|x|^\varepsilon)).$$

An interior version was proved in [5] for higher order elliptic equations with scaling subcritical lower order terms. See also [8]. For Neumann and Robin problems, similar results are expected. For instance, (6) near an α-conical boundary point, with $\eta \in L_{d-1+\varepsilon}(\partial\Omega)$.

Acknowledgments Z. Li was partially supported by an AMS-Simons travel grant.

References

1. V. Adolfsson, L. Escauriaza, $C^{1,\alpha}$ domains and unique continuation at the boundary. Commun. Pure Appl. Math. **50**(10), 935–969 (1997)
2. V. Adolfsson, L. Escauriaza, C. Kenig, Convex domains and unique continuation at the boundary. Rev. Mat. Iberoamericana **11**(3), 513–525 (1995)
3. F.J. Almgren, Jr., Dirichlet's problem for multiple valued functions and the regularity of mass minimizing integral currents, in *Minimal Submanifolds and Geodesics (Proceedings of Japan-United States Seminar, Tokyo, 1977)* (North-Holland, Amsterdam, 1979), pp. 1–6
4. L. Baratchart, L. Bourgeois, J. Leblond, Uniqueness results for inverse Robin problems with bounded coefficient. J. Funct. Anal. **270**(7), 2508–2542 (2016)
5. X.Y. Chen, Polynomial asymptotics near zero points of solutions of general elliptic equations. Commun. Partial Differ. Equ. **22**(7–8), 1227–1250 (1997)
6. S. Dipierro, V. Felli, E. Valdinoci, Unique continuation principles in cones under nonzero Neumann boundary conditions. Ann. Inst. H. Poincaré C Anal. Non Linéaire **37**(4), 785–815 (2020)
7. N. Garofalo, F.H. Lin, Unique continuation for elliptic operators: a geometric-variational approach. Commun. Pure Appl. Math. **40**(3), 347–366 (1987)
8. Q. Han, Singular sets of solutions to elliptic equations. Ind. Univ. Math. J. **43**(3), 983–1002 (1994)
9. D. Kriventsov, Z. Li, Asymptotic expansions for harmonic functions at conical boundary points (2023). arXiv:2307.10517
10. Z. Li, Unique continuation on Robin problems with measurable potentials (2023). arXiv:2304.04342
11. Z. Li, W. Wang, Quantitative unique continuation for Robin boundary value problems on $C^{1,1}$ domains (2021). arXiv:2111.01766
12. F.H. Lin, Nodal sets of solutions of elliptic and parabolic equations. Commun. Pure Appl. Math. **44**(3), 287–308 (1991)
13. F. Lin, Z. Shen, Nodal sets and doubling conditions in elliptic homogenization. Acta Math. Sin. **35**(6), 815–831 (2019)
14. X. Tolsa, Unique continuation at the boundary for harmonic functions in C^1 domains and Lipschitz domains with small constant. Commun. Pure Appl. Math. **76**(2), 305–336 (2023)

Semiregular Non-commutative Harmonic Oscillators: Some Spectral Asymptotic Properties

Marcello Malagutti and Alberto Parmeggiani

Abstract The study is devoted to spectral analysis of systems of PDEs, namely, a class of systems containing certain quantum optics models such as the Jaynes-Cummings model. More in detail, the research deals with spectral Weyl asymptotics for a semiregular system, extending to the vector-valued case results of Helffer and Robert, and more recently of Doll, Gannot and Wunsch.

Keywords Spectral theory and eigenvalue problems for PDEs · Non-commutative harmonic oscillators · Weyl-calculus

2000 Mathematics Subject Classification Primary 35P20; Secondary 35S05 81Q10

1 Introduction

In this paper we deal with the determination of asymptotic spectral properties of a class of pseudodifferential systems containing model relevant in Quantum Optics like the Jaynes-Cummings model (see Sect. 3.1). The main class analyzed is the one of Semiregular Metric Global Elliptic Systems (SMGES). Namely, we are considering those global semiregular (see Sect. 4.1) systems with scalar elliptic principal symbol such that there is not only a positively homogeneous of order 0 unitary matrix-valued function whose conjugation diagonalizes both the principal and semiprincipal symbol but also separation of the eigenvalues for the semiprincipal symbol. Namely, the object of the study is the spectral Weyl asymptotics for a semiregular global system, extending to the vector-valued case results of Helffer and Robert [3], and more recently of Doll, Gannot and Wunsch [2]. The investigation starts by defining the class of systems we will be concerned

M. Malagutti (✉) · A. Parmeggiani
Department of Mathematics, University of Bologna, Bologna, Italy
e-mail: marcello.malagutti2@unibo.it; alberto.parmeggiani@unibo.it

D. Cardona et al. (eds.), *Extended Abstracts 2021/2022*, Research Perspectives Ghent Analysis and PDE Center 3, https://doi.org/10.1007/978-3-031-48579-4_14

with here. Then, we recall the Jaynes-Cummings model and its variations to include the cases of atoms with several energetic levels. Actually, it is possible to associate such systems describing quantum Physics phenomena with models concerning geometrical complexes of vector-valued differential forms. Hence, likely higher Lie groups of symmetries could be introduced in the theory. After that, we state a decoupling theorem in the semiregular case, that leads to a method for a pseudodifferential block-reduction of the systems in the class under study. This reduction is the primary object to obtain a parametrix of the Schrödinger flow associated with the considered system, which, really, is the fundamental tool to investigate for achieving the Weyl asymptotics we are looking for. Finally, the Weyl asymptotic results are stated: with the first one we generalize the scalar asymptotics due to Helffer and Robert to our class of systems, while with the second we achive a smaller error term under the hypothesis that the angular gradients of the X-ray transform of the eigenvalues of the semiprincipal symbol vanish to infinite order exactly on a subset of measure zero of $\mathbb{S}^{2n-1}$. Similarly to the approach of Doll, Gannot and Wunsch [2] these asymptotics are obtained by a reduced propagator computation: here the use of the decoupling theorem is essential. In truth to generalize to systems is not trivial, since we need to conjugate the Fourier integral operators (FIOs) with quadratic phases by the pseudodifferential decoupler, with no property loss of the symbol-calculus. To overcome this crucial step we consider the approach of Doll and Zelditch [1] which, however, needs to be revised to fit our case.

2 Non-commutative Harmonic Oscillators (NCHOs)

Let us introduce the Non-Commutative Harmonic Oscillators (NCHOs).

Definitions 2.1 A **Non-Commutative Harmonic Oscillator (NCHO)** is the Weyl-quantization $a^{\mathrm{w}}(x, D)$ of an $N \times N$ system of the form

$$a(x, \xi) = a_2(x, \xi) + a_0, \quad (x, \xi) \in \mathbb{R}^n \times \mathbb{R}^n = T^*\mathbb{R}^n,$$

where $a_2(x, \xi)$is an $N \times N$ matrix whose entries are *homogeneous polynomials of degree* 2 *in the* (x, ξ) *variables*, and a_0 is a constant $N \times N$ matrix. (Introduced by A. Parmeggiani and M. Wakayama in [11, 12].)

Therefore it can also be said that an NCHO comes from the Weyl-quantization of a matrix-valued quadratic form in (x, ξ) adding a constant matrix term.

Remark A. Parmeggiani and M. Wakayama choose the name NCHO for two main reasons:

- the fact that a scalar harmonic oscillator is a single quadratic form in (x, ξ);
- the two levels of non-commutativity that we have to deal with when studying these systems: one due to the matrix-valued nature of the symbol of the system,

and the other to the Weyl-quantization rule

$$x_k \xi_j \leftrightarrow (x_k D_{x_j} + D_{x_j} x_k)/2, \quad \text{(where } D = -i\partial\text{)},$$

reflected through symplectic geometry by the Poisson-bracket relations

$$\{\xi_j, x_k\} = \delta_{jk}, \quad 1 \leq j, k \leq n.$$

Definitions 2.2 A NCHO $a^{\mathrm{w}}(x, D)$ is said to be *elliptic* when

$$a_2 \text{ is a } N \times N \text{ matrix and } \det a_2(x, \xi) \text{ behaves exactly like } (|x|^2 + |\xi|^2)^N$$

for $|(x, \xi)|$ large.

When a_2 and a_0 are *Hermitian matrices*, the operator $a^{\mathrm{w}}(x, D)$ is "formally self-adjoint" (i.e. symmetric on $\mathscr{S}(\mathbb{R}^n; \mathbb{C}^N)$). Moreover if in addition $a^{\mathrm{w}}(x, D)$ is positive elliptic (i.e. the matrix $a_2(x, \xi)$ is positive definite for $(x, \xi) \neq (0, 0)$), then it is *self-adjoint as an unbounded operator in* $L^2(\mathbb{R}^n; \mathbb{C}^N)$ *with a discrete real spectrum*.

Remark We note that while scalar harmonic oscillators have been deeply studied, very little has been investigated about the spectral properties of selfadjoint elliptic systems, even in the basic case of NCHOs.

The system written below is a particularly important example of NCHO

$$Q^{\mathrm{w}}_{(\alpha,\beta)}(x, D) = \begin{bmatrix} \alpha\left(-\frac{\partial_x^2}{2} + \frac{x^2}{2}\right) & -\left(x\partial_x + \frac{1}{2}\right) \\ x\partial_x + \frac{1}{2} & \beta\left(-\frac{\partial_x^2}{2} + \frac{x^2}{2}\right) \end{bmatrix}, \qquad x \in \mathbb{R},\ \alpha, \beta \in \mathbb{C}.$$

This is the Weyl-quantization of the matrix

$$Q_{(\alpha,\beta)}(x, \xi) = \begin{bmatrix} \alpha\left(\frac{\xi^2+x^2}{2}\right) & -ix\xi \\ ix\xi & \beta\left(\frac{\xi^2+x^2}{2}\right) \end{bmatrix}, \qquad (x, \xi) \in \mathbb{R} \times \mathbb{R},$$

introduced by Parmeggiani and Wakayama [11, 12]. When $\alpha, \beta > 0$ with $\alpha\beta > 1$, the system *is positive elliptic*, self-adjoint, and so it has a discrete spectrum in $L^2(\mathbb{R}; \mathbb{C}^2)$, and a very rich and remarkable structure.

It is worth remarking that in [12] the eigenvalues are described in terms of a *scalar three-term recurrence*, that means in terms of a continued fraction (nevertheless, it is very difficult to get a direct and explicit expression of them).

In addition we mention that it can be worth to underline that when $\alpha = \beta > 1$, $Q^{\mathrm{w}}_{(\alpha,\alpha)}(x, D)$ is *unitarily* equivalent [11, 12] to a scalar harmonic oscillator times the identity 2×2 matrix, hence its spectral properties are governed by the tensor product of the oscillator representation and the 2-dimensional trivial representation

of $\mathfrak{sl}_2(\mathbb{R})$ [5], i.e. one has matrix-valued creation/annihilation operators that can be used to "construct" the spectrum.
Therefore, when $\boldsymbol{\alpha}, \boldsymbol{\beta} > 0$ and $\boldsymbol{\alpha\beta} > 1$, we have that $Q^{\mathrm{w}}_{(\alpha,\beta)}(x, D)$ can be seen as *a matrix-valued deformation of the scalar harmonic oscillator*. In the case of $\boldsymbol{\alpha} \neq \boldsymbol{\beta}$ and $\boldsymbol{\alpha}, \boldsymbol{\beta} > 0$ it was proved by Parmeggiani in [9] (Theorem 4.4, pp. 351–353) that $Q^{\mathrm{w}}_{(\alpha,\beta)}(x, D)$ does not admit creation/annihilation operators.

Finally it can be worth underlined that a motivation for investigating systems like $Q^{\mathrm{w}}_{(\alpha,\beta)}(x, D)$ originates from PDEs, that is from the study of a-priori lower bound estimates, such as Melin's or Hörmander's or Fefferman-Phong's, for pseudodifferential systems (see [6], and also [7, 8] and references therein).

3 Jaynes-Cumming Model Hamiltonian Analytical Study

We give here a few examples of semiregular (see Sect. 4.1) NCHOs in the class SMGES, relevant to Quantum Optics (see [13]), that serve as a model of the class we consider in this work.

It will be convenient to use the following notation. We denote by $\boldsymbol{\sigma}_j$, $j = 0, \ldots, 3$, the Pauli-matrices, i.e.

$$\boldsymbol{\sigma}_0 = I_2, \quad \boldsymbol{\sigma}_1 = \begin{bmatrix} 0 & 1 \\ 1 & 0 \end{bmatrix}, \quad \boldsymbol{\sigma}_2 = \begin{bmatrix} 0 & -i \\ i & 0 \end{bmatrix}, \quad \boldsymbol{\sigma}_3 = \begin{bmatrix} 1 & 0 \\ 0 & -1 \end{bmatrix},$$

and

$$\boldsymbol{\sigma}_\pm = \frac{1}{2}(\boldsymbol{\sigma}_1 \pm i\boldsymbol{\sigma}_2).$$

Let $\langle \cdot, \cdot \rangle$ be the canonical Hermitian product in $\mathbb{C}^N$, and $e_1, \ldots, e_N$ be the canonical basis of $\mathbb{C}^N$. Let

$$E_{jk} := e_k^* \otimes e_j, \quad 1 \leq j, k \leq N,$$

be the basis of $\mathsf{M}_N(\mathbb{C}) = \mathfrak{gl}(N, \mathbb{C})$, where E_{jk} acts on $\mathbb{C}^N$ as

$$E_{jk} w = \langle w, e_k \rangle e_j, \quad w \in \mathbb{C}^N.$$

Hence we have the relation

$$E_{jk} E_{hm} = (e_k^* \otimes e_j)(e_m^* \otimes e_h) = e_k^*(e_h)(e_m^* \otimes e_j) = \langle e_h, e_k \rangle (e_m^* \otimes e_j) = \delta_{hk} E_{jm}.$$

We also let, for $X = (x, \xi) \in \mathbb{R}^n \times \mathbb{R}^n = \mathbb{R}^{2n}$,

$$\psi_j(X) := \frac{x_j + i\xi_j}{\sqrt{2}}, \quad 1 \le j \le n,$$

so that $\psi_j^{\mathrm{w}}(x, D)$ is the annihilation operator and $\psi_j^{\mathrm{w}}(x, D)^* = (\bar{\psi}_j)^{\mathrm{w}}(x, D)$ is the creation operator, with respect to the j-th variable. Hence, with $p_2(X) = |X|^2/2$ being the (standard) harmonic oscillator,

$$\sum_{j=1}^{n} \psi_j^{\mathrm{w}}(x, D)^* \psi_j^{\mathrm{w}}(x, D) = p_2^{\mathrm{w}}(x, D) - \frac{n}{2}.$$

We will also have to consider $2N \times 2N$ matrices of the form $\boldsymbol{\sigma}_j \otimes E_{jk}$, in which case the product is given by

$$(\boldsymbol{\sigma}_j \otimes E_{hk})(\boldsymbol{\sigma}_{j'} \otimes E_{h'k'}) = \boldsymbol{\sigma}_j \boldsymbol{\sigma}_{j'} \otimes E_{hk} E_{h'k'},$$

and the action on a vector $w \in \mathbb{C}^{2N}$, written as

$$w = \sum_{j=1}^{N} \begin{bmatrix} w_{2j-1} \\ w_{2j} \end{bmatrix} \otimes e_j,$$

given by

$$(\boldsymbol{\sigma}_m \otimes E_{hk}) w = \sum_{j=1}^{N} (\boldsymbol{\sigma}_m \begin{bmatrix} w_{2j-1} \\ w_{2j} \end{bmatrix}) \otimes (E_{hk} e_j).$$

We next list a few important models due to Jaynes and Cummings.

3.1 The JC-Model by Semiregular NCHOs

This is the model of a two-level atom in one cavity, given by the 2×2 system in one real variable $x \in \mathbb{R}$

$$A^{\mathrm{w}}(x, D) = \alpha p_2^{\mathrm{w}}(x, D) I_2 + \beta\Big(\boldsymbol{\sigma}_+ \psi^{\mathrm{w}}(x, D)^* + \boldsymbol{\sigma}_- \psi^{\mathrm{w}}(x, D)\Big) + \gamma \boldsymbol{\sigma}_3,$$

$$\alpha > 0, \beta, \gamma \in \mathbb{R},$$

where the atom levels are given by $\pm\gamma$.

3.2 The JC-Model for One Atom with N-Level and One Cavity-Mode in the Ξ-Configuration

In this case we consider, for $\alpha > 0$, $\beta_1, \ldots, \beta_{N-1} \in \mathbb{R} \setminus \{0\}$, $\gamma_1, \ldots \gamma_N \in \mathbb{R}$ with $\gamma_1 < \gamma_2 < \ldots < \gamma_N$, the $N \times N$ system in $\mathbb{R}$ given by

$$\begin{aligned} A^{\mathrm{w}}(x, D) =& \alpha p_2^{\mathrm{w}}(x, D) I_N + \frac{1}{2} \sum_{k=1}^{N-1} \beta_k \Big(\psi^{\mathrm{w}}(x, D)^* E_{k,k+1} + \psi^{\mathrm{w}}(x, D) E_{k+1,k} \Big) \\ &+ \sum_{k=1}^{N} \gamma_k E_{kk}. \end{aligned}$$

In this case the atom levels are given by the γ_k.

3.3 The JC-Model for an N-Level Atom and n = N − 1 Cavity-Modes in the Ξ-Configuration

In this case, for $\alpha > 0$, $\beta_1, \ldots \beta_{N-1} \in \mathbb{R} \setminus \{0\}$, $\gamma_1, \ldots \gamma_{N-1} \in \mathbb{R}$ with $\gamma_1 \leq \gamma_2 \leq \ldots \leq \gamma_{N-1}$, we consider the $N \times N$ system in $\mathbb{R}^n$, $n = N - 1$, given by

$$\begin{aligned} A^{\mathrm{w}}(x, D) &= \alpha p_2^{\mathrm{w}}(x, D) I_N \\ &+ \sum_{k=1}^{N-1} \beta_k \Big(\psi_k^{\mathrm{w}}(x, D)^* E_{k,k+1} + \psi_k^{\mathrm{w}}(x, D) E_{k+1,k} \Big) + \sum_{k=1}^{N-1} \gamma_k E_{k+1,k+1}. \end{aligned}$$

In this case, the levels of the atom are given by 0 and the γ_k.

3.4 The JC-Model for an N-Level Atom and n = N − 1 Cavity-Modes in the Λ-Configuration

In this case, for $\alpha > 0$, $\beta_1, \ldots \beta_{N-1} \in \mathbb{R} \setminus \{0\}$, $\gamma_1, \ldots \gamma_{N-1} \in \mathbb{R}$ with $\gamma_1 \leq \gamma_2 \leq \ldots \leq \gamma_{N-1}$, we consider the $N \times N$ system in $\mathbb{R}^n$, $n = N - 1$, given by

$$\begin{aligned} A^{\mathrm{w}}(x, D) &= \alpha p_2^{\mathrm{w}}(x, D) I_N \\ &+ \sum_{k=1}^{N-1} \beta_k \Big(\psi_k^{\mathrm{w}}(x, D)^* E_{k,N} + \psi_k^{\mathrm{w}}(x, D) E_{N,k} \Big) + \sum_{k=1}^{N-1} \gamma_k E_{k+1,k+1}. \end{aligned}$$

In this case, the levels of the atom are given by 0 and the γ_k.

3.5 The JC-Model for an N-Level Atom and $n = N - 1$ Cavity-Modes in the So-Called $\bigvee$-Configuration

In this case, for $\alpha > 0$, $\beta_1, \ldots \beta_{N-1} \in \mathbb{R} \setminus \{0\}$, $\gamma_1, \ldots \gamma_{N-1} \in \mathbb{R}$ with $\gamma_1 \leq \gamma_2 \leq \ldots \leq \gamma_{N-1}$, we consider the $N \times N$ system in $\mathbb{R}^n$, $n = N - 1$, given by

$$A^{\mathrm{w}}(x, D) = \alpha p_2^{\mathrm{w}}(x, D) I_N + \sum_{k=1}^{N-1} \beta_k \Big(\psi_k^{\mathrm{w}}(x, D)^* E_{1,k+1} + \psi_k^{\mathrm{w}}(x, D) E_{k+1,1}\Big) + \sum_{k=1}^{N-1} \gamma_k E_{k+1,k+1}.$$

In this case, the levels of the atom are given by 0 and the γ_k.

4 The Asymptotics of the Eigenvalues Counting Function

In this section we start defining the class of systems we will be concerned with here (see Sect. 4.2). Next, we state the decoupling theorem, which shows that for the class we consider here it is possible to obtain a pseudodifferential block-reduction of the system (see Sect. 4.2). This is fundamental in the study of a parametrix of the Schrödinger flow associated with our system, which in turns is the basic object to study for obtaining the Weyl asymptotics we are interested in. Finally, we state the Weyl asymptotic results: the first one extending to our class of systems the asymptotics due to Helffer and Robert [3], and the second presenting a better error term when the angular gradients of the X-ray transform of the semiprincipal symbol eigenvalues vanish to infinite order exactly on a subset of measure zero of $\mathbb{S}^{2n-1}$ (see Sect. 4.3).

4.1 The Semiregular Metric Globally Elliptic System Class (SMGES Class)

We will be using the following notation for the Hörmander metric and admissible weight (see Hörmander [4]): with $X = (x, \xi)$, $Y = (y, \eta)$ etc. belonging to the phase-space $\mathbb{R}^n \times \mathbb{R}^n$, and $m(X) := \langle X \rangle = (1 + |X|^2)^{1/2}$ the usual "Japanese bracket", we consider the Hörmander metric $g_X = |dX|^2/m(X)^2$. Then m is an admissible function (and so is m^μ for any given $\mu \in \mathbb{R}$), and we may exploit the full power of the Weyl-Hörmander pseudodifferential calculus.

We will write $\dot{\mathbb{R}}^{2n}$ for $\mathbb{R}^n \times \mathbb{R}^n \setminus \{(0,0)\}$.

Definitions 4.1 Let M_N denote the algebra of $N \times N$ complex matrices. A symbol $a \in S(m^\mu, g; \mathsf{M}_N)$ is said to be semiregular (see Remark 3.2.4 of [10]) if it possesses an asymptotic expansion $\sum_{j\geq 0} a_{\mu-j}$ in isotropic (i.e. positively homogenous and smooth outside the origin) terms $a_{\mu-j}$ positively homogeneous of degree $\mu - j$.
In other words, a matrix-symbol a of order μ is semiregular if $a = a^{(0)} + a^{(1)}$, where $a^{(0)} \in S_{\rm cl}(m^\mu, g; \mathsf{M}_N)$ and $a^{(1)} \in S_{\rm cl}(m^{\mu-1}, g; \mathsf{M}_N)$. We write $a \in S_{\rm sreg}(m^\mu, g; \mathsf{M}_N)$.

The terms a_μ, $a_{\mu-1}$ and $a_{\mu-2}$ are called the principal symbol, the semiprincipal symbol and the subprincipal symbol, respectively, of the operator $a^{\rm w}(x, D)$.

Hence, $a \in S_{\rm sreg}(m^\mu, g; \mathsf{M}_N)$ if and only if there exists a sequence $(a_{\mu-j})_{j\geq 0} \subset C^\infty(\dot{\mathbb{R}}^{2n}; \mathsf{M}_N)$ where $a_{\mu-j}$ is positively homogeneous of degree $\mu - j$ (in X) and, for an excision function χ,

$$a - \chi \sum_{j=0}^{N} a_{\mu-j} \in S(m^{\mu-(N+1)}, g; \mathsf{M}_N), \quad \forall N \in \mathbb{Z}_+.$$

As usual, we write

$$a \sim \sum_{j\geq 0} a_{\mu-j}.$$

We are now in a position to introduce the class of systems we are interested in.

Definitions 4.2 We say that an $N \times N$ symbol $a \in S_{\rm sreg}(m^\mu, g; \mathsf{M}_N)$ is a **semiregular metric globally elliptic system** (SMGES for short) of order μ, when

$$a(X) = a(X)^* = p_\mu(X) I_N + a_{\mu-1}(X) + a_{\mu-2}(X) + S_{\rm sreg}(m^{\mu-3}, g; \mathsf{M}_N), \quad X \neq 0,$$

where:

- $p_\mu \in C^\infty(\dot{\mathbb{R}}^{2n}; \mathbb{R})$ is positively homogeneous of degree μ and such that $|X|^\mu \approx p_\mu(X)$ for all $X \neq 0$;
- $a_{\mu-1} = a^*_{\mu-1}$ is such that there exists $r \geq 1$ and $e_0 \in C^\infty(\dot{\mathbb{R}}^{2n}; \mathsf{M}_N)$ **unitary** and positively homogeneous of degree 0 such that

  $$e_0(X)^* a_{\mu-1}(X) e_0(X) = \mathsf{diag}(\lambda_{\mu-1,j}(X) I_{N_j}; 1 \leq j \leq r), \quad X \neq 0$$

  where $N = N_1 + N_2 + \ldots + N_r$ and $\lambda_{\mu-1,j} \in C^\infty(\dot{\mathbb{R}}^{2n}; \mathbb{R})$ are positively homogeneous of degree $\mu - 1$ and such that

  $$j < k \Longrightarrow \lambda_{\mu-1,j}(X) < \lambda_{\mu-1,k}(X), \quad \forall X \neq 0.$$

4.2 The Decoupling Theorem

Let us state here a decoupling theorem for classes of semiregular global pseudodifferential systems from our class SMGES (that is, of the Jaynes-Cummings kind).

Theorem 4.3 *Let $\mu > 0$, and let $A = A^* \sim \sum_{j\geq 0} a_{\mu-j} \in S_{\rm sreg}(m^\mu, g; \mathsf{M}_N)$. Suppose $a_\mu = p_\mu I_N$ with $p_\mu \in S_{\rm cl}(m^\mu, g)$, and that $a_{\mu-1}$, for some $e_0 \in C^\infty(\mathbb{R}^{2n} \setminus \{0\}; \mathsf{M}_N)$ positively homogenous of degree 0 and such that $e_0 e_0^* = e_0^* e_0 = I_N$, $X \neq 0$, can be written as*

$$a_{\mu-1} = e_0 b_{\mu-1} e_0^*, \quad \textit{where } b_{\mu-1} = b_{\mu-1}^* = \left[\begin{array}{c|c} \lambda_{\mu-1,1} & 0 \\ \hline 0 & \lambda_{\mu-1,2} \end{array}\right], \quad X \neq 0,$$

where $\lambda_{\mu-1,j} \in C^\infty(\mathbb{R}^{2n} \setminus \{0\}; \mathsf{M}_{N_j})$, $j = 1, 2$ and $N = N_1 + N_2$, are positively homogeneous of degree $\mu - 1$, and are such that

$$\operatorname{Spec}(\lambda_{\mu-1,1}(X)) \cap \operatorname{Spec}(\lambda_{\mu-1,2}(X)) = \emptyset, \quad \forall X \in \mathbb{R}^{2n}, \ |X| = 1.$$

Then there exists $E \in S_{\rm sreg}(1, g; \mathsf{M}_N)$ with $E \sim \sum_{j\geq 0} e_{-j}$ and principal symbol e_0 (hence $e_{-k} \in C^\infty(\mathbb{R}^{2n} \setminus \{0\}; \mathsf{M}_N)$ is positively homogeneous of degree $-k$) such that

$$E^{\rm w}(x, D)^* E^{\rm w}(x, D) - I, \quad E^{\rm w}(x, D) E^{\rm w}(x, D)^* - I \in S(m^{-\infty}, g; \mathsf{M}_N), \tag{1}$$

and

$$E^{\rm w}(x, D)^* A^{\rm w}(x, D) E^{\rm w}(x, D) - B^{\rm w}(x, D) \in S(m^{-\infty}, g; \mathsf{M}_N), \tag{2}$$

where the symbol $B \sim \sum_{j\geq 0} b_{\mu-j} \in S_{\rm sreg}(m^\mu, g; \mathsf{M}_N)$ is blockwise diagonal, with

$$b_{\mu-j}(X) = \left[\begin{array}{c|c} b_{\mu-j,1}(X) & 0 \\ \hline 0 & b_{\mu-j,2}(X) \end{array}\right], \forall X \neq 0, \forall j \geq 0,$$

the blocks $b_{\mu-j,k}$ being of sizes $N_k \times N_k$, $k = 1, 2$, with

$$b_\mu = a_\mu = p_\mu I_N, \quad b_{\mu-1} = \left[\begin{array}{c|c} \lambda_{\mu-1,1} & 0 \\ \hline 0 & \lambda_{\mu-1,2} \end{array}\right], \quad X \neq 0.$$

4.3 The Weyl Law

Now we study the spectral Weyl asymptotics for a class of semiregular systems, extending to the vector-valued case results of Helffer and Robert [3], and more recently of Doll et al. [2]. Actually, the asymptotics by Doll, Gannot and Wunsch is more precise (that is why we call it *refined*) than the classical result by Helffer and Robert but deal with a less general class of systems since an hypothesis on the measure of the subset of $\mathbb{S}^{2n-1}$ on which the angular gradients of the X-ray transform of the semiprincipal symbol eigenvalues vanish to infinite order.

Theorem 4.4 (Weyl Law) *Let $A = A^*$, with $A \sim \sum_{j\geq 0} a_{2-j} \in S_{\mathrm{sreg}}(m^2, g; \mathsf{M}_N)$, be a second-order SMGES, with principal symbol $p_2 I_N$, p_2 being the harmonic oscillator. Adopting the notation used in Definition 4.2, we hence denote by $\lambda_{1,j}$, (with multiplicity N_j), $1 \leq j \leq r$, the eigenvalues of the semiprincipal part. Then, if $\rho \in \mathscr{S}(\mathbb{R})$ is chosen such that $\hat{\rho}$ has compact support in $(-\varepsilon, \varepsilon)$ for a sufficiently small $\varepsilon > 0$ and $\hat{\rho} = 1$ on a neighborhood of 0, as $\lambda \to +\infty$*

$$(\mathsf{N} * \rho)(\lambda) = \left(\sum_{j=1}^{r} \left(\frac{N_j}{(2\pi)^n} \int_{p_2+\lambda_{1,j}\leq\lambda} dX \right) - (2\pi)^{-n} \int_{p_2=\lambda} \mathsf{Tr}\,(a_0) \frac{ds}{|\nabla p_2|} \right) + O(\lambda^{n-3/2}), \tag{3}$$

(recall that Tr *is the matrix trace).*
Therefore, as $\lambda \to +\infty$

$$\mathsf{N}(\lambda) = \left(\frac{N}{(2\pi)^n} \int_{p_2\leq 1} dX \right) \lambda^n - \left((2\pi)^{-n} \int_{p_2=1} \mathsf{Tr}\,(a_1) \frac{ds}{|\nabla p_2|} \right) \lambda^{n-1/2} + O(\lambda^{n-1}). \tag{4}$$

We finally state the refined asymptotics of $\mathsf{N}(\lambda)$ for a positive ψdo system A^{w} satisfying the hypotheses of Theorem 4.4 and a condition on the measure of the subset of $\mathbb{S}^{2n-1}$ on which the angular gradients of the X-ray transform of the semiprincipal symbol eigenvalues of A^{w} vanish to infinite order.

Theorem 4.5 (Refined Weyl Law) *Let $A = A^* \in S_{\mathrm{sreg}}(m^2, g; \mathsf{M}_N)$ be a second-order SMGES satisfying the hypotheses of Theorem 4.4. If for all $1 \leq j \leq r$*

$$\Pi_{2\pi,j} := \left\{ \omega \in \mathbb{S}^{2n-1};\ \partial_\omega^\alpha \int_0^{2\pi} (\lambda_{1,j} \circ \exp tH_{p_2})(\omega) dt = 0,\ \forall \alpha \in \mathbb{N}^{2n-1} \setminus \{0\} \right\} \tag{5}$$

has measure zero.

then we have the following asymptotics, as $\lambda \to +\infty$

$$\mathsf{N}(\lambda) = (2\pi)^{-n} \left(\sum_{j=1}^{r} \left(N_j \int_{p_2+\lambda_{1,j} \leq \lambda} dX \right) - \int_{p_2=\lambda} \mathrm{Tr}\,(a_0) \frac{ds}{|\nabla p_2|} \right) + o(\lambda^{n-1}). \tag{6}$$

In particular,as $\lambda \to +\infty$

$$\begin{aligned} \mathsf{N}(\lambda) = (2\pi)^{-n} \Big(N\lambda^n \int_{p_2 \leq 1} dX - \lambda^{n-1/2} \int_{p_2=1} \mathrm{Tr}(a_1) \frac{ds}{|\nabla p_2|} \\ + \lambda^{n-1} \int_{p_2=1} \Big(\frac{n}{2} \mathrm{Tr}(a_1^2) - \mathrm{Tr}(a_0) \Big) \frac{ds}{|\nabla p_2|} \Big) + o(\lambda^{n-1}). \end{aligned} \tag{7}$$

Detailed proofs of all the above results can be found in [14].

References

1. M. Doll, S. Zelditch, Schrödinger trace invariants for homogeneous perturbations of the harmonic oscillator. J. Spectr. Theory **10**, 1303–1332 (2020). https://doi.org/10.4171/JST/328
2. M. Doll, O. Gannot, J. Wunsch, Refined Weyl law for homogeneous perturbations of the harmonic oscillator. Commun. Math. Phys. **362**(1), 269–294 (2018). https://doi.org/10.1007/s00220-018-3100-5
3. B. Helffer, D. Robert, Comportement asymptotique précisé du spectre d'opérateurs globalement elliptiques dans $\mathbb{R}^n$. Sém. Équ. aux dérivées Partielles (Polytechnique) **2**, 1–22 (1980–1981)
4. L. Hörmander, *The Analysis of Linear Partial Differential Operators III*. Classics in Mathematics (Springer, Berlin, 2007)
5. R. Howe, E.C. Tan, *Non-Abelian Harmonic Analysis. Applications of SL(2,* $\mathbb{R}$*)* (Springer, Berlin, 1992)
6. C. Parenti, A. Parmeggiani, Lower bounds for systems with double characteristics. J. Anal. Math. **86**, 49–91 (2002)
7. A. Parmeggiani, On the Fefferman-Phong inequality for systems of PDEs, in *Phase Space Analysis of Partial Differential Equations*. Progress in Nonlinear Differential Equations and their Applications, vol. 69 (Birkhäuser Verlag, Basel, 2006), pp. 247–266
8. A. Parmeggiani, On positivity of certain systems of partial differential equations. Proc. Natl. Acad. Sci. USA **104**, 723–726 (2007)
9. A. Parmeggiani, On the spectrum of certain non-commutative harmonic oscillators and semiclassical analysis. Commun. Math. Phys. **279**(2), 285–308 (2008). https://doi.org/10.1007/s00220-008-0436-2
10. A. Parmeggiani, *Spectral Theory of Non-commutative Harmonic Oscillators: An Introduction*. Lecture Notes in Mathematics, vol. 1992 (Springer, Berlin, 2010). xii+254pp. https://doi.org/10.1007/978-3-642-11922-4
11. A. Parmeggiani, M. Wakayama. Oscillator representations and systems of ordinary differential equations. Proc. Nat Acad. Sci. USA **98**, 2630 (2001). https://doi.org/10.1073/pnas.98.1.26
12. A. Parmeggiani, M. Wakayama, Non-commutative harmonic oscillators-I, -II. Forum Math. **14**, 539–604 ibid. 669–690 (2002). https://doi.org/10.1515/form.2002.025 and https://doi.org/10.1515/form.2002.029

13. B.W. Shore, P.L. Knight. The Jaynes-Cummings model. J. Modern Opt. **40**(7), 1195–1238 (1993). https://doi.org/10.1080/09500349314551321
14. M. Malagutti, A. Parmeggiani. *Spectral Asymptotic Properties of Semiregular Non-Commutative Harmonic Oscillators.* to appear in Commun. Math. Phys. (2024). https://doi.org/10.1007/s00220-024-04934-7

Global Compactness, Subcritical Approximation of the Sobolev Quotient, and a Related Concentration Result in the Heisenberg Group

Giampiero Palatucci, Mirco Piccinini, and Letizia Temperini

Abstract We investigate some effects of the lack of compactness in the critical Sobolev embedding in the Heisenberg group.

Keywords Sobolev embeddings · Heisenberg group · CR Yamabe · Global compactness · Profile decompositions · Green's function

2010 Mathematics Subject Classification 35R03, 46E35, 35J08, 35A15

1 Critical Sobolev Embeddings in the Heisenberg Group

Let $\mathbb{H}^n := (\mathbb{C}^n \times \mathbb{R}, \circ, \delta_\lambda)$ be the usual Heisenberg-Weyl group, endowed with the group multiplication law $\circ$,

$$\xi \circ \xi' := \Big(x + x',\ y + y',\ t + t' + 2\langle y, x'\rangle - 2\langle x, y'\rangle\Big)$$

for $\xi := (x + iy, t)$ and $\xi' := (x' + iy', t') \in \mathbb{R}^n \times \mathbb{R}^n \times \mathbb{R}$, whose group of non-isotropic *dilations* $\{\delta_\lambda\}_{\lambda>0}$ on $\mathbb{R}^{2n+1}$ is given by

$$\xi \mapsto \delta_\lambda(\xi) := (\lambda x,\ \lambda y,\ \lambda^2 t). \tag{1}$$

Consider the standard Folland-Stein-Sobolev space $S^1_0(\mathbb{H}^n)$ defined as the completion of $C^\infty_0(\mathbb{H}^n)$ with respect to the homogeneous subgradient norm $\|D_H \cdot\|_{L^2}$, where the horizontal (or intrinsic) gradient D_H is given by

$$D_H u(\xi) := \big(Z_1 u(\xi), \ldots, Z_{2n} u(\xi)\big),$$

G. Palatucci (✉) · M. Piccinini · L. Temperini
Dipartimento di Scienze Matematiche, Fisiche e Informatiche, Università di Parma, Parma, Italy
e-mail: giampiero.palatucci@unipr.it; mirco.piccinini@unipr.it; letizia.temperini@outlook.it

D. Cardona et al. (eds.), *Extended Abstracts 2021/2022*, Research Perspectives Ghent Analysis and PDE Center 3, https://doi.org/10.1007/978-3-031-48579-4_15

with $Z_j := \partial_{x_j} + 2y_j\partial_t$, $Z_{n+j} := \partial_{y_j} - 2x_j\partial_t$ for $1 \leq j \leq n$, and $T := \partial_t$ being the Jacobian base of the Heisenberg Lie algebra.

As well known, the following Sobolev-type inequality holds for some positive constant S^*,

$$\|u\|_{L^{2^*}}^{2^*} \leq S^* \|D_H u\|_{L^2}^{2^*}, \quad \forall u \in S_0^1(\mathbb{H}^n), \tag{2}$$

where $2^* = 2^*(Q) := 2Q/(Q-2)$ is the Folland-Stein-Sobolev critical exponent, depending on the *homogeneous dimension* $Q := 2n+2$ of the Heisenberg group $\mathbb{H}^n$.

The validity of (2) is equivalent to show that the constant S^* **defined in the** following maximization problem,

$$S^* := \sup\left\{\int_{\mathbb{H}^n} |u(\xi)|^{2^*}\,\mathrm{d}\xi \;:\; u \in S_0^1(\mathbb{H}^n), \int_{\mathbb{H}^n} |D_H u(\xi)|^2 \mathrm{d}\xi \leq 1\right\}, \tag{3}$$

is finite. The explicit form of the maximizers has been showed, amongst other results, in the breakthrough paper by Jerison and Lee [9], together with the computation of the optimal constant in (3).

For any bounded domain $\Omega \subset \mathbb{H}^n$, consider now

$$S_\Omega^* := \sup\left\{\int_\Omega |u(\xi)|^{2^*}\,\mathrm{d}\xi \;:\; u \in S_0^1(\Omega), \int_\Omega |D_H u(\xi)|^2 \mathrm{d}\xi \leq 1\right\}, \tag{4}$$

where the Folland-Stein-Sobolev space $S_0^1(\Omega)$ is given by the closure of $C_0^\infty(\Omega)$ with respect to the homogeneous subgradient norm in Ω. One can check that $S_\Omega^* \equiv S^*$ via a standard scaling argument, and thus—in view of the explicit form of the optimal functions in (3)—the variational problem (4) has no maximizers. The situation changes drastically for the subcritical embeddings: $S_0^1(\Omega) \hookrightarrow L^{2^*-\varepsilon}(\Omega)$ is compact (for each $0 < \varepsilon < 2^* - 2$), and this guarantees the existence of a maximizer $u_\varepsilon \in S_0^1(\Omega)$ for

$$S_\varepsilon^* := \sup\left\{\int_\Omega |u(\xi)|^{2^*-\varepsilon}\,\mathrm{d}\xi \;:\; u \in S_0^1(\Omega), \int_\Omega |D_H u(\xi)|^2 \mathrm{d}\xi \leq 1\right\}. \tag{5}$$

Such a dichotomy can be also found in the Euler-Lagrange equation for the energy functionals in (5); that is,

$$-\Delta_H u_\varepsilon = \lambda |u_\varepsilon|^{2^*-\varepsilon-2} u_\varepsilon \;\text{ in } (S_0^1(\Omega))', \tag{6}$$

where λ is a Lagrange multiplier, and $\Delta_H := \sum_{j=1}^{2n} Z_j^2$ is the standard Kohn Laplacian (or sub-Laplacian) operator. While when $\varepsilon > 0$ it has a solution u_ε, the problem above becomes very delicate when $\varepsilon = 0$: one falls in the CR Yamabe equation realm, and even the existence of the solutions is not granted. In view of such a qualitative change when $\varepsilon = 0$ (in both (5) and (6)), it sounds natural to analyze the asymptotic behavior as ε goes to 0 of both the subcritical Sobolev constant S_ε^* in the Heisenberg group given in (6) and of the corresponding optimal functions u_ε of the embedding $S_0^1(\Omega) \hookrightarrow L^{2^*-\varepsilon}(\Omega)$. This is the aim of papers [15] and [17], whose main results will be stated in the rest of the present note.

2 Subcritical Approximation of the Sobolev Quotient

Our first result is the subcritical approximation of the Sobolev embedding S^* in the Heisenberg group described below.

Theorem 1 (See Theorem 1.1 in [17]) *Let $\Omega \subseteq \mathbb{H}^n$ be a bounded domain, and denote by $\mathcal{M}(\overline{\Omega})$ the set of nonnegative Radon measures in Ω. Let $X = X(\Omega)$ be the space*

$$X := \Big\{(u, \mu) \in S_0^1(\Omega) \times \mathcal{M}(\overline{\Omega}) : \mu \geq |D_H u|^2 \mathrm{d}\xi,\ \mu(\overline{\Omega}) \leq 1\Big\},$$

endowed with the product topology $\mathcal{T}$ such that

$$(u_k, \mu_k) \xrightarrow{\mathcal{T}} (u, \mu) \overset{def}{\Leftrightarrow} \begin{cases} u_k \rightharpoonup u \text{ in } L^{2^*}(\Omega), \\ \mu_k \overset{*}{\rightharpoonup} \mu \text{ in } \mathcal{M}(\overline{\Omega}). \end{cases} \tag{7}$$

Let us consider the following family of functionals,

$$\mathcal{F}_\varepsilon(u, \mu) := \int_\Omega |u|^{2^*-\varepsilon} \mathrm{d}\xi \quad \forall (u, \mu) \in X.$$

Then, as $\varepsilon \to 0$, the Γ^+-limit of the family of functionals $\mathcal{F}_\varepsilon$ with respect to the topology $\mathcal{T}$ given by (7) is the functional $\mathcal{F}$ defined by

$$\mathcal{F}(u, \mu) = \int_\Omega |u|^{2^*} \mathrm{d}\xi + S^* \sum_{j=1}^{\infty} \mu_j^{\frac{2^*}{2}} \quad \forall (u, \mu) \in X.$$

Here S^* is the best Sobolev constant in $\mathbb{H}^n$, $2^* = 2Q/(Q-2)$ is the Folland-Stein-Sobolev critical exponent, and the numbers μ_j are the coefficients of the atomic part of the measure μ.

In order to prove such a result in the very general situation considered here, and thus requiring no additional regularity assumptions nor special geometric features on the domains, we attack the problem pursuing a new approach and for this we rely on De Giorgi's Γ-convergence techniques. This is in the same spirit of previous results regarding the classical Sobolev embedding in the Euclidean framework, as seen in [1, 12, 13], though the core of the proof in [17] goes in a very different line because the optimal recovery sequences have been concretely constructed whereas in all the aforementioned Euclidean papers such an existence result has been proven via compactness and locality properties of the Γ-limit energy functional. In this respect, the adopted strategy is surprisingly close to that in the *fractional Sobolev spaces* framework [18, 20], but various differences evidently arose because of the natural discrepancy between the involved frameworks.

It could be interesting to investigate whether or not the techniques introduced in [17] and [20] could be combined with the estimates involving the "nonlocal tail" in the Heisenberg framework firstly introduced in [14] in order to prove a similar result for fractional Folland-Stein-Sobolev spaces; see also [11, 16, 21].

As a corollary of Theorem 1, one can deduce that the sequences of maximizers $\{u_\varepsilon\}$ for the subcritical Sobolev quotient S^*_ε concentrates energy at one point $\xi_o \in \overline{\Omega}$, and this is in clear accordance with the analogous result in the Euclidean case.

Theorem 2 (See Theorem 1.2 in [17]) *Let $\Omega \subset \mathbb{H}^n$ be a bounded domain and let $u_\varepsilon \in S^1_0(\Omega)$ be a maximizer for S^*_ε. Then, as $\varepsilon = \varepsilon_k \to 0$, up to subsequences, we have that there exists $\xi_o \in \overline{\Omega}$ such that*

$$u_k = u_{\varepsilon_k} \rightharpoonup 0 \text{ in } L^{2^*}(\Omega),$$

and

$$|D_H u_k|^2 \mathrm{d}\xi \overset{*}{\rightharpoonup} \delta_{\xi_o} \text{ in } \mathcal{M}(\overline{\Omega}),$$

with δ_{ξ_o} being the Dirac mass at ξ_o.

3 Struwe's Global Compactness in the Heisenberg Group

Since the seminal paper [23] by Struwe, the celebrated Global Compactness in the Sobolev space H^1 have become a fundamental tool in Analysis which have been proven to be crucial in order to achieve various existence results, as e. g. for ground states solutions for nonlinear Schrödinger equations, for prescribing Q-curvature problems, for solutions of Yamabe-type equations in conformal geometry, for harmonic maps from Riemann surfaces into Riemannian manifolds, for Yang-Mills connections over four-manifolds, and many others. The involved literature is really too wide to attempt any reasonable account here. In Theorem 3 below, we will state the counterpart of Struwe's Global Compactness in the Heisenberg framework.

In order to precisely state such a result, consider for any fixed $\lambda \in \mathbb{R}$ the problem,

$$-\Delta_H u - \lambda u - |u|^{2^*-2}u = 0 \qquad \text{in } (S^1_0(\Omega))', \tag{P_λ}$$

together with its corresponding Euler–Lagrange energy functional $\mathcal{E}_\lambda : S_0^1(\Omega) \to \mathbb{R}$ given by

$$\mathcal{E}_\lambda(u) = \frac{1}{2}\int_\Omega |D_H u|^2 \,d\xi - \frac{\lambda}{2}\int_\Omega |u|^2 \,d\xi - \frac{1}{2^*}\int_\Omega |u|^{2^*} \,d\xi.$$

Consider also the following limiting problem,

$$-\Delta_H u - |u|^{2^*-2}u = 0 \qquad \text{in } (S_0^1(\Omega_o))', \tag{P_0}$$

where Ω_o is either a half-space or the whole $\mathbb{H}^n$; i.e., the Euler-Lagrange equation which corresponds to the energy functional $\mathcal{E}^* : S_0^1(\Omega_o) \to \mathbb{R}$,

$$\mathcal{E}^*(u) = \frac{1}{2}\int_{\Omega_o} |D_H u|^2 \,d\xi - \frac{1}{2^*}\int_{\Omega_o} |u|^{2^*} \,d\xi.$$

Theorem 3 (See Theorem 1.3 in [17]) *Let $\{u_k\} \subset S_0^1(\Omega)$ be a Palais-Smale sequence for $\mathcal{E}_\lambda$; i.e., such that*

$$\mathcal{E}_\lambda(u_k) \le c \quad \textit{for all } k,$$
$$d\mathcal{E}_\lambda(u_k) \to 0 \quad \textit{as } k \to \infty \quad \textit{in } (S_0^1(\Omega))'.$$

Then, there exists a (possibly trivial) solution $u^{(0)} \in S_0^1(\Omega)$ to (P_λ) such that, up to a subsequence, we have

$$u_k \rightharpoonup u^{(0)} \quad \textit{as } k \to \infty \quad \textit{in } S_0^1(\Omega).$$

Moreover, either the convergence is strong or there is a finite set of indexes $I = \{1, \ldots, J\}$ such that for all $j \in I$ there exist a nontrivial solution $u^{(j)} \in S_0^1(\Omega_o^{(j)})$ to (P_0) with $\Omega_o^{(j)}$ being either a half-space or the whole $\mathbb{H}^n$, a sequence of nonnegative numbers $\{\lambda_k^{(j)}\}$ converging to zero and a sequences of points $\{\xi_k^{(j)}\} \subset \Omega$ such that, for a renumbered subsequence, we have for any $j \in I$

$$u_k^{(j)}(\cdot) := \lambda_k^{(j)\frac{Q-2}{2}} u_k\big(\tau_{\xi_k^{(j)}}(\delta_{\lambda_k^{(j)}}(\cdot))\big) \rightharpoonup u^{(j)}(\cdot) \quad \textit{in } S_0^1(\mathbb{H}^n) \quad \textit{as} \to \infty.$$

In addition, as $k \to \infty$ we have

$$u_k(\cdot) = u^{(0)}(\cdot) + \sum_{j=1}^{J} \lambda_k^{(j)\frac{2-Q}{2}} u_k\big(\delta_{1/\lambda_k^{(j)}}(\tau^{-1}_{\xi_k^{(j)}}(\cdot))\big) + o(1) \ \textit{in } S_0^1(\mathbb{H}^n);$$

$$\left|\log \frac{\lambda_k^{(i)}}{\lambda_k^{(j)}}\right| + \left|\delta_{1/\lambda_k^{(j)}}\big(\xi_k^{(j)^{-1}} \circ \xi_k^{(i)}\big)\right|_{\mathbb{H}^n} \to \infty \quad \textit{for } i \neq j, \ i, j \in I;$$

$$\|u_k\|^2_{S^1_0} = \sum_{j=1}^{J} \|u^{(j)}\|^2_{S^1_0} + o(1);$$

$$\mathcal{E}_\lambda(u_k) = \mathcal{E}_\lambda(u^{(0)}) + \sum_{j=1}^{J} \mathcal{E}^*(u^{(j)}) + o(1)$$

In the display above, given $\xi' \in \mathbb{H}^n$, we denoted by $\tau_{\xi'}$ *the left translation* **defined** by $\tau_{\xi'}(\xi) := \xi' \circ \xi$ for all $\xi \in \mathbb{H}^n$.

The original proof by Struwe in [23] consists in a subtle analysis concerning how the Palais-Smale condition does fail for the functional $\mathcal{E}^*$, based on rescaling arguments, used in an iterated way to extract convergent subsequences with non-trivial limit, together with some slicing and extension procedures on the sequence of approximate solutions to (P_λ). **Such a proof revealed to be very difficult to** extend to different frameworks, and the aforementioned strategy seems even more cumbersome to be adapted to the Heisenberg framework considered here. For this, we completely changed the approach to the problem, and we proved how to deduce the results in Theorem 3 in quite a simple way by means of the so-called *Profile Decomposition*, **firstly proven by Gérard** [6] for bounded sequences in the fractional Euclidean space H^s, and extended to the Heisenberg framework by Benameur [2]. This is in clear accordance with the strategy in [19]; see the related result in the fractional Heisenberg framework in [7].

Remark 4 The limiting domain Ω_o in Theorem 3 can be either the whole $\mathbb{H}^n$ or a half-space. On the contrary, in the original proof in the Euclidean case by Struwe [23] one can exclude the existence of nontrivial solutions to the limiting problem in the half-space by Unique Continuation and Pohozaev's Identity. Such a possibility can not be a priori excluded in the sub-Riemannian setting, even in the very special case when a complete characterization of the limiting set is possible under further regularity assumptions on Ω. Indeed, in the Heisenberg framework, a very few nonexistence results are known, basically only in the case when the domain reduces to a half-plane parallel or perpendicular to the group center; see [4]. We also refer to the last paragraphs in [17, Section 5] for further details.

4 Asymptotics of the Optimal Functions

We present an asymptotic control of the maximizing sequence u_ε for S^*_ε in (5) via the Jerison and Lee extremals. This is shown in Theorem 5 below, which will be one of the key in the proof of the localization of the concentration result presented in Sect. 5 below and it could be also useful to investigate further properties related to subcritical Folland-Stein-Sobolev embeddings.

Theorem 5 (See Theorem 1.2 in [15]) *Let $\Omega \subset \mathbb{H}^n$ be a smooth bounded domain such that*

$$\liminf_{\rho \to 0} \frac{|(\mathbb{H}^n \setminus \Omega) \cap B_\rho(\xi)|}{|B_\rho(\xi)|} > 0 \ \forall \xi \in \partial\Omega.$$

Then, for each $0 < \varepsilon < 2^* - 2$ letting $u_\varepsilon \in S_0^1(\Omega)$ being a maximizer for S_ε^*, there exist $\{\eta_\varepsilon\} \subset \Omega$, $\{\lambda_\varepsilon\} \subset \mathbb{R}^+$ such that, up to choosing ε sufficiently small, we have that

$$u_\varepsilon \lesssim U_{\lambda_\varepsilon, \eta_\varepsilon} \qquad \text{on } \Omega,$$

where $U_{\lambda_\varepsilon,\eta_\varepsilon} = U\left(\delta_{1/\lambda_\varepsilon}\left(\tau_{\eta_\varepsilon}(\xi)\right)\right)$ are the Jerison and Lee *extremal functions, and the sequences $\{\eta_\varepsilon\}$ and $\{\lambda_\varepsilon\}$ satisfy*

$$\eta_\varepsilon \sim \xi_o \quad \text{and} \quad \lambda_\varepsilon^\varepsilon \sim 1 \quad \text{as } \varepsilon \searrow 0,$$

with ξ_o being the concentration point given in Theorem 2.

The result in Theorem 5 above reminds to the literature following the pioneering work in the Euclidean framework due to Aubin and Talenti, and in such a framework it is fundamental in the proof of a precise conjecture about the localization of the concentration point ξ_o given in Corollary 2 by Han in [8]. In the proof of Theorem 5 in [15] in the sub-Riemannian framework we are dealing with, one has also to deal with the fact that, in strong contrast with the Euclidean setting, the Jerison and Lee extremals cannot be reduced to functions depending only on the standard Korányi gauge. For this, such a proof will require a delicate strategy which makes use and refines the concentration result obtained via the Γ-convergence result in Theorem 1 in order to detect the right involved scalings η_ε and λ_ε. Also the Global Compactness-type result presented in Sect. 3 is needed.

5 Localization of the Energy Concentration

A natural question arises: can the blowing up be localized; i.e., is the concentration point ξ_o in Theorem 2 in Sect. 2 related in a specific way to the geometry of the domain Ω ?

In the Euclidean framework, under standard regularity assumptions, Han [8] and Rey [22] proved the connection with the Green function associated to the domain Ω by answering to a famous conjecture by Brezis and Peletier [3], who had previously investigated the spherical domains setting. The involved proofs strongly rely on the regularity of Euclidean domains, which is in clear contrast with the complexity of the underlying sub-Riemannian geometry here; as well-known, even if the domain Ω is smooth, the situation is drastically different because of the possible presence of characteristic points on the boundary $\partial\Omega$. From one side, near those characteristic points—as firstly discovered by Jerison—even harmonic functions on the Heisenberg group can encounter a sudden loss of regularity; from the other side, one did not want to work in the restricted class of domains not having characteristic points. In order to deal with those specific difficulties, it is thus quite natural to work under the assumption that the domain Ω is *geometrical regular near its characteristic set* as given by Definition 7 below. In forthcoming Theorem 8 we state the expected localization result for the concentration point ξ_o of the maximizing sequence u_ε in terms of the Green function associated with the domain Ω, in turn establishing the validity of the aforementioned Brezis-Peletier conjecture in the Heisenberg group.

As customary, denote by $\mathcal{D}$ the infinitesimal generator of the one-parameter group of non-isotropic dilations $\{\delta_\lambda\}_{\lambda>0}$ in (1); that is,

$$\mathcal{D} := \sum_{j=1}^{n} \left(x_j \partial_{x_j} + y_j \partial_{y_j}\right) + 2t\partial_t. \tag{8}$$

Definitions 6 (δ_λ-Starlike Sets) Let Ω be a C^1 connected open set of $\mathbb{H}^n$ containing the group identity $\mathfrak{e}$. We say that Ω is δ_λ-starlike (with respect to the identity $\mathfrak{e}$) along a subset $K \subseteq \partial\Omega$ if

$$\langle \mathcal{D}, \mathfrak{n}\rangle(\eta) \geq 0,$$

at every $\eta \in K$; in the display above $\mathfrak{n}$ indicates the exterior unit normal to $\partial\Omega$.

We say that Ω is uniformly δ_λ-starlike (with respect to the identity $\mathfrak{e}$) along K if there exists $\alpha_\Omega > 0$ such that, at every $\eta \in K$,

$$\langle \mathcal{D}, \mathfrak{n}\rangle(\eta) \geq \alpha_\Omega.$$

A domain as above Ω is δ_λ-starlike (uniformly δ_λ-starlike, respectively) with respect to one of its point $\zeta \in \Omega$ along K if $\tau_{\zeta^{-1}}(\Omega)$ is δ_λ-starlike (uniformly δ_λ-starlike, respectively) with respect to the origin along $\tau_{\zeta^{-1}}(K)$.

Given a domain $\Omega \subset \mathbb{H}^n$, we recall that its *characteristic set* Σ_{Ω,D_H}, the collection of all its characteristic point, is given by

$$\Sigma_{\Omega,D_H} := \Big\{\xi \in \partial\Omega \mid Z_j(\xi) \in T_\xi(\partial\Omega),\ \text{for } j = 1, \dots, 2n\Big\}.$$

We now recall the definition of regular domains in accordance with the by-now classical paper [5].

Definitions 7 (See Definition 2.2 in [15]) A smooth domain $\Omega \subset \mathbb{H}^n$ such that $\partial\Omega$ is an orientable hypersurface is "geometrical regular near its characteristic set" if the following conditions hold true,

(Ω1) There exist $\Phi \in C^\infty(\mathbb{H}^n)$, $c_\Omega > 0$ and $\rho_\Omega \in \mathbb{R}$ such that

$$\Omega := \{\Phi < \rho_\Omega\}, \quad \text{and} \quad |D\Phi| \geq c_\Omega.$$

(Ω2) For any $\xi \in \partial\Omega$ it holds

$$\liminf_{\rho \to 0^+} \frac{|(\mathbb{H}^n \setminus \Omega) \cap B_\rho(\xi)|}{|B_\rho(\xi)|} > 0.$$

(Ω3) There exist M_Ω such that

$$\Delta_H \Phi \geq \frac{4|z|}{M_\Omega} \langle D_H \Phi, D_H |z| \rangle \quad \text{in } \omega,$$

where ω is an interior neighborhood of Σ_{Ω, D_H}.

(Ω4) Ω is δ_λ-starlike with respect to one of its point $\zeta_o \in \Omega$ and uniformly δ_λ-starlike with respect to ζ_o along Σ_{Ω, D_H}.

We are finally in the position to state the localization result.

Theorem 8 (See Theorem 1.3 in [15]) *Consider a bounded domain $\Omega \subset \mathbb{H}^n$ geometrical regular near its characteristic set, and let $u_\varepsilon \in S_0^1(\Omega)$ be a maximizer for S^*_ε. Then, up to subsequences, u_ε concentrates at some point $\xi_o \in \Omega$ such that*

$$\int_{\partial\Omega} |D_H G_\Omega(\cdot, \xi_o)|^2 \langle \mathcal{D}, \mathfrak{n} \rangle \, d\mathcal{H}^{Q-2} = 0, \tag{9}$$

with $G_\Omega(\cdot; \xi_o)$ being the Green function associated to Ω with pole in ξ_o, and $\mathcal{D}$ being the infinitesimal generator of the one-parameter group of non-isotropic dilations in the Heisenberg group defined in (8).

The proof can be found in [15]; it involves all the results stated in the preceding sections together with other general tools in the sub-Riemaniann framework, as e.g., maximum principles, Caccioppoli-type estimates, H-Kelvin transform, boundary Schauder-type regularity estimates, as well as with a fine boundary analysis of the solutions to subcritical Yamabe equations. We refer also to the interesting related result in [10] in the case of domains with no characteristic points.

Acknowledgments The authors are member of Gruppo Nazionale per l'Analisi Matematica, la Probabilità e le loro Applicazioni (GNAMPA) of Istituto Nazionale di Alta Matematica "F. Severi" (INdAM), whose support is acknowledged. The authors are also supported by INdAM Project "Fenomeni non locali in problemi locali", CUP_E55F220 00270001. The second author is also supported by the Project "Local vs Nonlocal: mixed type operators and nonuniform ellipticity", CUP_D91B21005370003.

References

1. M. Amar, A. Garroni, Γ-convergence of concentration problems. Ann. Scuola Norm. Sup. Pisa Cl. Sci. **2**, 151–179 (2003)
2. J. Benameur, Description du défaut de compacité de l'injection de Sobolev sur le groupe de Heisenberg. Bull. Belg. Math. Soc. Simon Stevin **15**(4), 599–624 (2008)
3. H. Brezis, L.A. Peletier, Asymptotic for Elliptic equations involving critical growth, in *Partial Differential Equations and the Calculus of Variations. Essays in Honor of Ennio De Giorgi*. Progress in Differential Equations Applications, vol. 1 (Birkhäuser, Boston, 1989), pp. 149–192
4. G. Citti, F. Uguzzoni, Critical semilinear equations on the Heisenberg group: the effect of the topology of the domain. Nonlinear Anal. **46**, 399–417 (2001)
5. N. Garofalo, D. Vassilev, Regularity near the characteristic set in the non-linear Dirichlet problem and conformal geometry of sub-Laplacians on Carnot groups. Math. Ann. **318**, 453–516 (2000)
6. P. Gérard, Description du défaut de compacité de l'injection de Sobolev. ESAIM Control Optim. Calc. Var. **3**, 213–233 (1998)
7. C. Guidi, A. Maalaoui, V. Martino, Palais-Smale sequences for the fractional CR Yamabe functional and multiplicity results. Calc. Var. Partial Differ. Equ. **57**, Art. 152 (2018)
8. Z.-C. Han, Asymptotic approach to singular solutions for nonlinear elliptic equations involving critical Sobolev exponent. Ann. Inst. Henri Poincaré Anal. Non Linéaire **8**, 159–174 (1991)
9. D. Jerison, J.M. Lee, Extremals for the Sobolev inequality on the Heisenberg group and the CR Yamabe problem. J. Am. Math. Soc. **1**(1), 1–13 (1988)
10. A. Maalaoui, V. Martino, A. Pistoia, Concentrating solutions for a sub-critical sub-elliptic problem. Differ. Int. Equ. **26**(11–12), 1263–1274 (2013)
11. M. Manfredini, G. Palatucci, M. Piccinini, S. Polidoro, Hölder continuity and boundedness estimates for nonlinear fractional equations in the Heisenberg group. J. Geom. Anal. **33**(3), Art. 77 (2023)
12. G. Palatucci, Subcritical approximation of the Sobolev quotient and a related concentration result. Rend. Sem. Mat. Univ. Padova **125**, 1–14 (2011)
13. G. Palatucci, p-Laplacian problems with critical Sobolev exponent. Asymptot. Anal. **73**, 37–52 (2011)
14. G. Palatucci, M. Piccinini, Nonlocal Harnack inequalities in the Heisenberg group. Calc. Var. Partial Differ. Equ. **61**, Art. 185 (2022)
15. G. Palatucci, M. Piccinini, Asymptotic approach to singular solutions for the CR Yamabe equation, and a conjecture by H. Brezis and L. A. Peletier in the Heisenberg group. arXiv:2307.14933 (2024)
16. G. Palatucci, M. Piccinini, Nonlinear fractional equations in the Heisenberg group. Bruno Pini Math. Anal. Semin. **15**(1), (2024)
17. G. Palatucci, M. Piccinini, L. Temperini, Struwe's Global Compactness and energy approximation of the critical Sobolev embedding in the Heisenberg group. arXiv:2308.01153 (2024)
18. G. Palatucci, A. Pisante, Improved Sobolev embeddings, profile decomposition, and concentration-compactness for fractional Sobolev spaces. Calc. Var. Partial Differ. Equ. **50**(3–4), 799–829 (2014)
19. G. Palatucci, A. Pisante, A global compactness type result for palais-smale sequences in fractional Sobolev spaces. Nonlinear Anal. **117**, 1–7 (2015)
20. G. Palatucci, A. Pisante, Y. Sire, Subcritical approximation of a Yamabe-type nonlocal equation: a Gamma-convergence approach. Ann. Sc. Norm. Super. Pisa Cl. Sci. (5) **14**(3), 819–840 (2015)

21. M. Piccinini, The obstacle problem and the Perron Method for nonlinear fractional equations in the Heisenberg group. Nonlinear Anal. **222**, Art. 112966. (2022)
22. O. Rey, Proof of the conjecture of H. Brezis and L. A. Peletier. Manuscripta Math. **65**, 19–37 (1989)
23. M. Struwe, A global compactness result for elliptic boundary value problems involving limiting nonlinearities. Math. Z. **187**, 511–517 (1984)

A Note on Carleson-Hunt Type Theorems for Vilenkin-Fourier Series

L.-E. Persson, F. Schipp, G. Tephnadze, and F. Weisz

Abstract In this paper we discuss an analogy of the Carleson-Hunt theorem with respect to Vilenkin systems. In particular, we investigate the almost everywhere convergence of Vilenkin-Fourier series of $f \in L_p(G_m)$ for $p > 1$ in case the Vilenkin system is bounded. Moreover, we state an analogy of the Kolmogorov theorem for $p = 1$ and construct a function $f \in L_1(G_m)$ such that the partial sums with respect to Vilenkin systems diverge everywhere.

Keywords Fourier analysis · Vilenkin system · Vilenkin group · Vilenkin-Fourier series · Almost everywhere convergence · Carleson-Hunt theorem · Kolmogorov theorem

2000 Mathematics Subject Classification 42C10, 42B25

The research was supported by Shota Rustaveli National Science Foundation grant no. FR-21-2844.

L.-E. Persson
Department of Computer Science and Computational Engineering, UiT-The Arctic University of Norway, Narvik, Norway

Norway and Department of Mathematics and Computer Science, Karlstad University, Karlstad, Sweden
e-mail: lars.e.persson@uit.no; larserik.persson@kau.se

F. Schipp · F. Weisz
Department of Numarical Analysis, Eötvös University, Budapest, Hungary
e-mail: schipp@inf.elte.hu; weisz@inf.elte.hu

G. Tephnadze (✉)
School of science and technology, The University of Georgia, Tbilisi, Georgia
e-mail: g.tephnadze@ug.edu.ge

D. Cardona et al. (eds.), *Extended Abstracts 2021/2022*, Research Perspectives Ghent Analysis and PDE Center 3, https://doi.org/10.1007/978-3-031-48579-4_16

1 Introduction

In 1947 Vilenkin [53, 54] investigated a group G_m, which is a direct product of the additive groups $Z_{m_k} := \{0, 1, \dots, m_k - 1\}$ of integers modulo m_k, where $m := (m_0, m_1, \dots)$ are positive integers not less than 2, and introduced the Vilenkin systems $\{\psi_j\}_{j=0}^{\infty}$. These systems include as a special case the Walsh system, when $m \equiv 2$.

The classical theory of Hilbert spaces (see e.g the books [49] and [52]) certifies that if we consider the partial sums $S_n f := \sum_{k=0}^{n-1} \widehat{f}(k)\, \psi_k$, with respect to Vilenkin systems, then $\|S_n f\|_2 \leq \|f\|_2$. In the same year 1976 Schipp [37], Simon [43] and Young [58] (see also the book [41]) generalized this inequality for $1 < p < \infty$: there exists an absolute constant c_p, depending only on p, such that

$$\|S_n f\|_p \leq c_p \|f\|_p, \quad \text{when} \quad f \in L_p(G_m).$$

It follows that for every $f \in L_p(G_m)$ with $1 < p < \infty$, $\|S_n f - f\|_p \to 0$, as $n \to \infty$. The boundedness does not hold for $p = 1$, but Watari [55] (see also Gosselin[18], Young[58]) proved that there exists an absolute constant c such that, for $n = 1, 2, \dots$, the weak type estimate $y\mu\{|S_n f| > y\} \leq c \|f\|_1$, $f \in L_1(G_m)$, $y > 0$ holds.

The almost-everywhere convergence of Fourier series for L_2 functions was postulated by Luzin [30] in 1915 and the problem was known as Luzin's conjecture. Carleson's theorem is a fundamental result in mathematical analysis establishing the pointwise (Lebesgue) almost everywhere convergence of Fourier series of L_2 functions, proved by Carleson [8] in 1966. The name is also often used to refer to the extension of the result by Hunt [20] which was given in 1968 to L_p functions for $p \in (1, \infty)$ (also known as the Carleson-Hunt theorem).

Carleson's original proof is exceptionally hard to read, and although several authors have simplified the arguments there are still no easy proofs of his theorem. Expositions of the original Carleson's paper were published by Kahane [22], Mozzochi [31], Jorsboe and Mejlbro [21] and Arias de Reyna [35]. Moreover, Fefferman [14] published a new proof of Hunt's extension, which was done by bounding a maximal operator S^* of partial sums, defined by $S^* f := \sup_{n \in \mathbb{N}} |S_n f|$. This, in its turn, inspired a much simplified proof of the L_2 result by Lacey and Thiele [28], explained in more detail in Lacey [26]. In the books Fremlin [15] and Grafakos [17] it was also given proofs of the Carleson's theorem. An interesting extension of Carleson-Hunt result much more closer to L_1 space then L_p for any $p > 1$ was done by Carleson's student Sjölin [47] and later on, by Antonov [2]. Already in 1923, Kolmogorov [24] showed that the analogue of Carleson's result for L_1 is false by finding such a function whose Fourier series diverges almost everywhere (improved slightly in 1926 to diverging everywhere). This result indeed inspired many authors after Carleson proved positive results in 1966. In 2000, Kolmogorov's result was improved by Konyagin [25], by finding functions with everywhere-divergent Fourier series in a space smaller than L_1, but the candidate for such a space that is consistent with the results of Antonov and Konyagin is still an open problem.

The famous Carleson theorem was very important and surprising when it was proved in 1966. Since then this interest has remained and a lot of related research has been done. In fact, in recent years this interest has even been increased because of the close connections to e.g. scattering theory [32], ergodic theory [12, 13], the theory of directional singular integrals in the plane [3, 9, 11, 27] and the theory of operators with quadratic modulations [29]. We refer to [26] for a more detailed description of this fact. These connections have been discovered from various new arguments and results related to Carleson's theorem, which have been found and discussed in the literature. We mean that these arguments share some similarities, but each of them has also a distinct new ideas behind, which can be further developed and applied. **It is also interesting to note that, for almost every specific application of Carleson's theorem in the aforementioned fields, mainly only one of these new arguments was** used.

The analogue of Carleson's theorem for Walsh system was proved by Billard [4] for $p = 2$ and by Sjölin [46] and Demeter [10] for $1 < p < \infty$, while for bounded Vilenkin systems by Gosselin [18]. Schipp [38, 39] (see also [40, 56]) investigated the so called tree martingales and generalized the results about maximal function, quadratic variation and martingale transforms to these martingales and also gave a proof of Carleson's theorem for Walsh-Fourier series. A similar proof for bounded Vilenkin systems can be found in Schipp and Weisz [40, 56]. In each proof, it was proved that the maximal operator of the partial sums is bounded on $L_p(G_m)$, i.e.,

$$\left\|S^* f\right\|_p \leq c_p \left\|f\right\|_p, \quad \text{as } f \in L_p(G_m), \quad 1 < p < \infty.$$

A recent proof of almost everywhere convergence of Vilenkin-Fourier series was given by Persson, Schipp, Tephnadze and Weisz [33] (see also the book [34]) in 2022. Convergence of subsequences of Vilenkin-Fourier series were considered in [6, 7, 50, 51].

Stein [48] constructed an integrable function whose Walsh-Fourier series diverges almost everywhere. Later on Schipp [36, 41] proved that there exists an integrable function whose Walsh-Fourier series diverges everywhere. Kheladze [23] proved that for any set of measure zero there exists a function in $f \in L_p(G_m)$ $(1 < p < \infty)$ whose Vilenkin-Fourier series diverges on the set, while the result for continuous or bounded functions was proved by Harris [19] or Bitsadze [5]. Simon [44] constructed an integrable function such that its Vilenkin-Fourier series diverges everywhere. Generalization of results by Simon [44] and Kheladze [23] can be found in [33, 34].

2 Preliminaries

Denote by $\mathbb{N}_+$ the set of the positive integers, $\mathbb{N} := \mathbb{N}_+ \cup \{0\}$. Let $m := (m_0, m_1, \ldots)$ be a sequence of the positive integers not less than 2. Define the group G_m as the complete direct product of the the additive group $Z_{m_k} := \{0, 1, \ldots, m_k - 1\}$ of integers modulo with the product of the discrete topologies of Z_{m_j}'s. The direct product μ of the measures $\mu_k(\{j\}) := 1/m_k (j \in Z_{m_k})$ is the Haar measure on G_m with $\mu(G_m) = 1$. In this paper we discuss bounded Vilenkin groups, i.e. the case when $\sup_n m_n < \infty$. The elements of G_m are represented by sequences $x := (x_0, x_1, \ldots, x_j, \ldots) \left(x_j \in Z_{m_j}\right)$. It is easy to give a base for the neighborhood of G_m :

$$I_0(x) := G_m, \quad I_n(x) := \{y \in G_m \mid y_0 = x_0, \ldots, y_{n-1} = x_{n-1}\},$$

where $x \in G_m$, $n \in \mathbb{N}$. Denote $I_n := I_n(0)$ for $n \in \mathbb{N}_+$, and $\overline{I_n} := G_m \setminus I_n$.

If we define the so-called generalized number system based on m by

$$M_0 := 1, \quad M_{k+1} := m_k M_k \quad (k \in \mathbb{N}),$$

then every $n \in \mathbb{N}$ can be uniquely expressed as $n = \sum_{j=0}^{\infty} n_j M_j$, where $n_j \in Z_{m_j}$ $(j \in \mathbb{N}_+)$ and only a finite number of n_j's differ from zero.

We define the generalized Rademacher functions, by $r_k(x) : G_m \to \mathbb{C}$,

$$r_k(x) := \exp(2\pi \imath x_k/m_k), \quad \left(\imath^2 = -1, \quad x \in G_m, \quad k \in \mathbb{N}\right).$$

Now, define the Vilenkin system $\psi := (\psi_n : n \in \mathbb{N})$ on G_m as:

$$\psi_n(x) := \prod_{k=0}^{\infty} r_k^{n_k}(x), \quad (n \in \mathbb{N}).$$

The Vilenkin system is orthonormal and complete in $L_2(G_m)$ (see e.g. [1]).

If $f \in L_1(G_m)$, we can define the Fourier coefficients, the partial sums of the Fourier series, the Dirichlet kernels with respect to the Vilenkin system as:

$$\widehat{f}(n) := \int_{G_m} f \overline{\psi}_n d\mu, \quad (n \in \mathbb{N}), \quad S_n f := \sum_{k=0}^{n-1} \widehat{f}(k) \psi_k \quad \text{and}$$

$$D_n := \sum_{k=0}^{n-1} \psi_k, \quad (n \in \mathbb{N}_+)$$

respectively. Recall that (see e.g. Simon [42, 45] and Golubov et al. [16])

$$\sum_{s=0}^{m_k-1} r_k^s(x) = \begin{cases} m_k, & \text{if } x_k = 0, \\ 0, & \text{if } x_k \neq 0, \end{cases} \quad \text{and } D_{M_n}(x) = \begin{cases} M_n, & \text{if } x \in I_n, \\ 0, & \text{if } x \notin I_n. \end{cases} \tag{1}$$

A function P is called Vilenkin polynomial if $P = \sum_{k=0}^{n} c_k \psi_k$.

3 On Martingale Inequalities

The σ-algebra generated by the intervals $\{I_n(x) : x \in G_m\}$ will be denoted by $\mathcal{F}_n$ $(n \in \mathbb{N})$. If $\mathcal{F}$ denotes the set of Haar measurable subsets of G_m, then obviously $\mathcal{F}_n \subset \mathcal{F}$. By a Vilenkin interval we mean one of the form $I_n(x)$, $n \in \mathbb{N}$, $x \in G_m$. The conditional expectation operators relative to $\mathcal{F}_n$ are denoted by E_n. An integrable sequence $f = (f_n)_{n\in\mathbb{N}}$ is said to be a martingale if f_n is $\mathcal{F}_n$-measurable for all $n \in \mathbb{N}$ and $E_n f_m = f_n$ in the case $n \leq m$. We can see that if $f \in L_1(G_m)$, then $(E_n f)_{n\in\mathbb{N}}$ is a martingale. Martingales with respect to $(\mathcal{F}_n, n \in \mathbb{N})$ are called Vilenkin martingales. It is easy to prove (see e.g. Weisz [56, p.11]) that the sequence $(\mathcal{F}_n, n \in \mathbb{N})$ is regular, i.e., for all non-negative Vilenkin martingales (f_n),

$$f_n \leq R f_{n-1} \qquad \text{where} \qquad R := \max_{n\in\mathbb{N}} m_n, \qquad n \in \mathbb{N}. \tag{2}$$

Using (1), we can prove that $E_n f = S_{M_n} f$ for all $f \in L_p(G_m)$ with $1 \leq p \leq \infty$ (see e.g. [56]). By the well known martingale theorems, this implies that

$$\left\| S_{M_n} f - f \right\|_p \to 0, \text{ as } n \to \infty \text{ for all } f \in L_p(G_m) \text{ when } p \geq 1. \tag{3}$$

For a Vilenkin martingale $f = (f_n)_{n\in\mathbb{N}}$, the maximal function f^* is defined by $f^* := \sup_{n\in\mathbb{N}} |f_n|$. For a martingale $f = (f_n)_{n\geq 0}$ let $d_n f = f_n - f_{n-1}$ $(n \geq 0)$ denote the martingale differences, where $f_{-1} := 0$. The square function and the conditional square function of f are defined by

$$S(f) := \left(\sum_{n=0}^{\infty} |d_n f|^2 \right)^{1/2} \quad \text{and} \quad s(f) := \left(|d_0 f|^2 + \sum_{n=0}^{\infty} E_n |d_{n+1} f|^2 \right)^{1/2}.$$

We have shown the following theorem in [56]:

Theorem 9 *If* $0 < p < \infty$, *then* $\|f^*\|_p \sim \|S(f)\|_p \sim \|s(f)\|_p$. *If in addition* $1 < p \leq \infty$, *then* $\|f^*\|_p \sim \|f\|_p$.

4 a.e. Convergence of Vilenkin-Fourier Series

We introduce some notations. For $j, k \in \mathbb{N}$ we define the following subsets of $\mathbb{N}$:

$$I_{jM_k}^k := [jM_k, jM_k + M_k) \cap \mathbb{N} \quad \text{and} \quad \mathcal{I} := \{I_{jM_k}^k : j, k \in \mathbb{N}\}.$$

We introduce also the partial sums taken in these intervals:

$$s_{I_{jM_k}^k} f := \sum_{i \in I_{jM_k}^k} \widehat{f}(i)\psi_i.$$

For simplicity, we suppose that $\widehat{f}(0) = 0$. In [57] was proved that, for an arbitrary $n \in I_{jM_k}^k$, $s_{I_{jM_k}^k} f = \psi_n E_k(f\overline{\psi}_n)$. For $n = \sum_{j=0}^{\infty} n_j M_j \quad (0 \le n_j < m_j)$, we define

$$n(k) := \sum_{j=k}^{\infty} n_j M_j, \qquad I_{n(k)}^k = \big[n(k), n(k) + M_k\big) \qquad (n \in \mathbb{N}). \tag{4}$$

Let

$$T^I f := T^{I_{n(k)}^k} f := \sum_{\substack{[n(k+1), n(k)) \supset J \in \mathcal{I} \\ |J| = M_k}} s_J f, \quad \text{for} \quad I = I_{n(k)}^k.$$

Lemma 10 *For all $n \in \mathbb{N}$ and $I_{n(k)}^k$ defined in (4), we have that*

$$S_n f = \sum_{k=0}^{\infty} T^{I_{n(k)}^k} f = \psi_n \sum_{k=0}^{\infty} \sum_{l=0}^{n_k - 1} \overline{r}_k^{n_k - l} E_k \left(d_{k+1}(f\overline{\psi}_n) r_k^{n_k - l}\right),$$

Lemma 11 *For all $k, n \in \mathbb{N}$, the following inequality holds:*

$$|T^{I_{n(k)}^k} f| \le R E_k \left(|s_{I_{n(k+1)}^{k+1}} f - s_{I_{n(k)}^k} f|\right), \quad \textit{where} \quad R := \max(m_n, n \in \mathbb{N}).$$

Lemma 12 *For all $n \in \mathbb{N}$, $\left(\overline{\psi}_n T^{I_{n(k)}^k} f\right)_{k \in \mathbb{N}}$ is a martingale difference sequence with respect to $(\mathcal{F}_{k+1})_{k \in \mathbb{N}}$.*

Let I, J, K denote some elements of $\mathcal{I}$. Let $\mathcal{F}_K := \mathcal{F}_n$ and $E_K := E_n$ if $|K| = M_n$. Assume that $\epsilon = (\epsilon_K, K \in \mathcal{I})$ is a sequence of functions such that ϵ_K is $\mathcal{F}_K$ measurable. Set

$$T_{\epsilon; I, J} f := \sum_{I \subset K \subset J} \epsilon_K T^K f, \qquad T_{\epsilon; I}^* f := \sup_{I \subset J} |T_{\epsilon; I, J} f|, \qquad T_{\epsilon}^* f := \sup_{I \in \mathcal{I}} |T_{\epsilon; I}^* f|.$$

If $\epsilon_K(t) = 1$ for all $K \in \mathcal{I}$ and $t \in G_m$, then we omit the notation ϵ and write simply $T_{I,J}f$, T_I^*f and T^*f. For $I \in \mathcal{I}$ with $|I| = M_n$, let $I^+ \in \mathcal{I}$ such that $I \subset I^+$ and $|I^+| = M_{n+1}$. Moreover, let $I^- \in \mathcal{I}$ denote one of the sets $I^- \subset I$ with $|I^-| = M_{n-1}$. Note that $\mathcal{F}_{I^-} = \mathcal{F}_{n-1}$ and $E_{I^-} = E_{n-1}$ are well defined. We introduce the maximal functions s_I^* and s^* by $s_I^*f := \sup_{K \subset I} E_{K^-}|s_K f|$ and $s^*f := \sup_{I \in \mathcal{I}} s_I^* f$. Since $|s_{I^+}f|$ is $\mathcal{F}_{I^+}$ measurable, by the regularity condition (2), we conclude that $|s_{I^+}f| \le RE_I|s_{I^+}f| \le Rs_{I^+}^*f$.

Lemma 13 *For any real number $x > 0$ and $K \in \mathcal{I}$, let $\epsilon_K := \chi_{\{t \in G_m : x < s_{K^+}^* f(t) \le 2x\}}$ and $\alpha_K := \chi_{\{t \in G_m : s_K^* f(t) > x, s_I^* f(t) \le x, I \subset K\}}$. Then*

$$T_\epsilon^* f \le 2 \sup_{K \in \mathcal{I}} \alpha_K T_{\epsilon;K}^* f + 4R^2 x \chi_{\{t \in G_m : s^* f(t) > x\}}.$$

Now we introduce the quasi-norm $\|\cdot\|_{p,q}$ $(0 < p, q < \infty)$ by

$$\|f\|_{p,q} := \sup_{x>0} x \left(\int_{G_m} \left(\sum_{I \in \mathcal{I}} \alpha_I \right)^{p/q} d\mu \right)^{1/p},$$

where α_I is defined in Lemma 13. Observe that α_I can be rewritten as

$$\alpha_I := \chi_{\{t \in G_m : E_{I^-}|s_I f(t)| > x, E_{J^-}|s_J f(t)| \le x, J \subset I\}}. \tag{5}$$

Denote by $P^{p,q}$ the set of functions $f \in L_1$ which satisfy $\|f\|_{p,q} < \infty$. For $q = \infty$,

$$\|f\|_{p,\infty} := \sup_{x>0} x \left(\int_{G_m} \left(\sup_{I \in \mathcal{I}} \alpha_I \right)^p d\mu \right)^{1/p} \qquad (0 < p < \infty).$$

It is easy to see that

$$\|f\|_{p,\infty} \le \|f\|_{p,q} \quad (0 < q < \infty) \quad \text{and} \quad \|f\|_{p,\infty} = \sup_{x>0} x \mu(s^* f > x)^{1/p}.$$

Lemma 14 *Let $\max(1, p) < q < \infty$, $f \in P^{p,q}$ and $x, z > 0$. Then*

$$\mu\left(\sup_{I \in \mathcal{I}} \alpha_I T_{\epsilon;I}^* f > zx \right) \le C_{p,q} z^{-q} x^{-p} \|f\|_{p,q}^p, \text{ where } \alpha_I \text{ is defined in Lemma 13.}$$

Lemma 15 *Let $\max(1, p) < q < \infty$ and $f \in P^{p,q}$. Then*

$$\sup_{y>0} y^p \mu\Big(T^* f > (2 + 8R^2) y\Big) \le C_{p,q} \|f\|_{p,q}.$$

Let Δ denote the closure of the triangle in $\mathbb{R}^2$ with vertices $(0, 0)$, $(1/2, 1/2)$ and $(1, 0)$ except the points $(x, 1 - x)$, $1/2 < x \leq 1$.

Lemma 16 *Suppose that* $1 < p, q < \infty$ *satisfy* $(1/p, 1/q) \in \Delta$. *Then, for all* $f \in L_p$, *we have* $\|f\|_{p,q} \leq C_{p,q}\|f\|_p$.

Now we are ready to formulate our first main result.

Theorem 17 *Let* $f \in L_p(G_m)$, *where* $1 < p < \infty$. *Then*

$$\left\|S^* f\right\|_p \leq c_p \|f\|_p, \quad \text{where} \quad S^* f := \sup_{n\in\mathbb{N}} |S_n f| .$$

The next norm convergence result follow from Theorem 17.

Theorem 18 *Let* $f \in L_p(G_m)$, $1 < p < \infty$. *Then* $\|S_n f - f\|_p \to 0$, *as* $n \to \infty$.

Our announced Carleson-Hunt type theorem reads:

Theorem 19 *Let* $f \in L_p(G_m)$, *where* $p > 1$. *Then* $S_n f \to f$, *a.e., as* $n \to \infty$.

5 Almost Everywhere Divergence of Vilenkin-Fourier Series

A set $E \subset G_m$ is called a set of divergence for $L_p(G_m)$ if there exists a function $f \in L_p(G_m)$ whose Vilenkin-Fourier series diverges on E.

Lemma 20 *If* E *is a set of divergence for* $L_1(G_m)$, *then there is a function* $f \in L_1(G_m)$ *such that* $S^* f = \infty$ *on* E.

Lemma 21 *A set* $E \subseteq G_m$ *is a set of divergence for* $L_1(G_m)$ *if and only if there exist Vilenkin polynomials* $P_1, P_2, \ldots,$ *such that* $\sum_{j=1}^{\infty} \|P_j\|_1 < \infty$ *and*

$$\sup_{j\in\mathbb{N}_+} S^* P_j(x) = \infty \ (x \in E).$$

Corollary 22 *If* $E_1, E_2, \ldots$ *are sets of divergence for* $L_1(G_m)$, *then* $E := \cup_{n=1}^{\infty} E_n$ *is also a set of divergence for* $L_1(G_m)$.

Theorem 23 *If* $1 \leq p < \infty$ *and* $E \subseteq G_m$ *is a set of Haar measure zero, then* E *is a set of divergence for* $L_p(G_m)$.

Theorem 24 *There is a function* $f \in L_1(G_m)$ *whose Vilenkin-Fourier series diverges everywhere.*

Remark 25 For details of the above statements we refer to [33, 34].

Acknowledgments The research was supported by Shota Rustaveli National Science Foundation grant FR-21-2844.

References

1. G.N. Agaev, N.Y. Vilenkin, G.M. Dzhafarly, A.I. Rubinshtein, *Multiplicative Systems of Functions and Harmonic Analysis on Zero-dimensional Groups* (Elm, Baku 1981)
2. N.Y. Antonov, Convergence of Fourier series. East J. Approx. **2**(2), 187–196 (1996)
3. M. Bateman, C. Thiele, L_p estimates for the Hilbert transforms along a one-variable vector field. Anal. PDE **6**(7), 1577–1600 (2013)
4. P. Billard, Sur la convergence presque partout des séries de Fourier-Walsh des fonctions de l'espace $L^2[0, 1]$. Studia Math. **28**, 363–388 (1966/1967)
5. K. Bitsadze, On divergence of Fourier series with respect to multiplicative systems on the sets of measure zero. Georgian Math. J. **16**(3), 435–448 (2009)
6. I. Blahota, L.E. Persson, G. Tephnadze, Two-sided estimates of the Lebesgue constants with respect to Vilenkin systems and applications. Glasg. Math. J. **60**(1), 17–34 (2018)
7. I. Blahota, K. Nagy, L.E. Persson, G. Tephnadze, A sharp boundedness result concerning maximal operators of Vilenkin-Fourier series on martingale Hardy spaces. Georgian Math. J. **26**(3), 351–360 (2019)
8. L. Carleson, On convergence and growth of partial sums of Fourier series. Acta Math. **116**, 135–157 (1966)
9. C. Demeter, Singular integrals along N directions in R^2. Proc. Am. Math. Soc. **138**, 4433–4442 (2010)
10. C. Demeter, A guide to Carleson's theorem. Rocky Mountain J. Math. **45**(1), 169–212 (2015)
11. C. Demeter, F. Di Plinio, Logarithmic L^p bounds for maximal directional singular integrals in the plane. J. Geom. Anal. **24**(1), 375–416 (2014)
12. C. Demeter, C. Thiele, On the two-dimensional bilinear Hilbert transform. Am. J. Math. **132**(1), 201–256 (2010)
13. C. Demeter, M. Lacey, T. Tao, C. Thiele, Breaking the duality in the return times theorem. Duke Math. J. **143**(2), 281–355 (2008)
14. C. Fefferman, Pointwise convergence of Fourier series. Ann. Math. (2), **98**(3), 551–571 (1973)
15. D.H. Fremlin, *Measure Theory*. Broad Foundations, vol. 2. Corrected second printing of the 2001 original (Torres Fremlin, Colchester, 2003)
16. B.I. Golubov, A.V. Efimov, V.A. Skvortsov, *Walsh Series and Transforms, Theory and Application* (Kluwer Academic Publishers Group, Dordrecht, 1991)
17. L. Grafakos, Classical Fourier Analysis. Graduate Texts in Mathematics, vol. 249, 3rd edn. (Springer, New York, 2014)
18. J.A. Gosselin, Almost everywhere convergence of Vilenkin-Fourier series. Trans. Am. Math. Soc. **185**, 345–370 (1974)
19. D.C. Harris Compact sets of divergence for continuous functions on a Vilenkin group. Proc. Am. Math. Soc. **98**(3), 436–440 (1986)
20. R. Hunt, On the convergence of Fourier series, orthogonal expansions and their continuous analogues, in Proceedings of the Conference on Edwardsville, 1967) (Southern Illinois University Press, Carbondale, 1968), pp. 235–255
21. O.G. Jorsboe, L. Mejlbro, *The Carleson-Hunt Theorem on Fourier Series*. Lecture Notes in Mathematics, vol. 911 (Springer-Verlag, Berlin-New York, 1982)
22. J.P. Kahane, Sommes partielles des séries de Fourier (d'apres L. Carleson)", Sêminaire Bourbaki, 9 (Société Mathématique de France, Paris, 1995), pp. 491–507
23. S.V. Kheladze, On the everywhere divergence of Fourier-Walsh series, Sakharth. SSR Mecn. Acad. Moambe **77**, 305–307 (1975)

24. A.N. Kolmogorov, Une série de Fourier-Lebesgue divergente presque partout, Polska Akademii Nauk. Fund. Math. **4**, 324–328 (1923)
25. S.V. Konyagin, On the divergence everywhere of trigonometric Fourier series (Russian). Mat. Sb. **191**(1), 103–126 (2000)
26. M.T. Lacey, Carleson's theorem: proof, complements, variations. Publ. Mat. **48**(2), 251–307 (2004)
27. M. Lacey, X. Li, On a conjecture of E.M. Stein on the Hilbert transform on vector fields. Mem. Amer. Math. Soc. **205**(965), 1–80 (2010)
28. M. Lacey, C. Thiele, A proof of boundedness of the Carleson operator. Math. Res. Lett. **7**(4), 361–370 (2000)
29. V. Lie, The polynomial Carleson operator. Ann. Math. **192**(1), 47–163 (2020)
30. N.N. Luzin, Collected works, in *Metric Theory of Function and Functions of a Complex Variable*, vol. I (Russian) (Izdat. Acad. Nauk. SSSR, Moscow, 1953)
31. C.J. Mozzochi, *On the Pointwise Convergence of Fourier Series*. Lecture Notes in Mathematics, vol. 199 (Springer-Verlag, Berlin/New York, 1971)
32. C. Muscalu, T. Tao, C. Thiele, A Carleson type theorem for a Cantor group model of the scattering transform. Nonlinearity **16**(1), 219–246 (2003)
33. L.-E. Persson, F. Schipp, G. Tephnadze, F. Weisz, An analogy of the Carleson-Hunt theorem with respect to Vilenkin systems. J. Fourier Anal. Appl. **28**(48) 1–29 (2022)
34. L.-E. Persson, G. Tephnadze, F. Weisz, *Martingale Hardy Spaces and Summability of One-dimensional Vilenkin-Fourier Series* (Birkhäuser/Springer, Berlin, 2022). Book manuscript
35. J.A. Reyna, *Pointwise Convergence of Fourier Series.* Lecture Notes in Mathematics, vol. 1785 (Springer-Verlag, Berlin/New York, 2002)
36. F. Schipp, Über die divergenz der Walsh-Fourierreihen. Ann. Univ. Sci. Budapest Eötvös, Sect. Math. **12**, 49–62 (1969)
37. F. Schipp, On L^p-norm convergence of series with respect to product systems. Anal. Math. **2**(1), 49–64 (1976)
38. F. Schipp, Pointwise convergence of expansions with respect to certain product systems. Anal. Math. **2**(1), 65–76 (1976)
39. F. Schipp, Universal contractive projections and a.e. convergence, in *Probability Theory and Applications, Essays to the memory of József Mogyoródi*. Applied Mathematics, vol. 80, ed. by J. Galambos, I. Kátai (Kluwer Academic Publishers, Dordrecht/Boston/London, 1992), pp. 221–233
40. F. Schipp, F. Weisz, Tree martingales and almost everywhere convergence of Vilenkin-Fourier series. Math. Pannon. **8**(1), 17–35 (1997)
41. F. Schipp, W.R. Wade, P. Simon, J. Pál, Walsh series, in *An Introduction to Dyadic Harmonic Analysis* (Adam Hilger, Ltd., Bristol, 1990)
42. P. Simon, Verallgemeinerte Walsh-Fourierreihen. I. Ann. Univ. Sci. Budapest Eötvös, Sect. Math. **16**, 103–113 (1973)
43. P. Simon, Verallgemeinerte Walsh-Fourierreihen. II. Acta Math. Acad. Sci. Hung. **27**(3–4), 329–341 (1976)
44. P. Simon, On the divergence of Vilenkin-Fourier series. Acta. Math. Hungar. **41**(3–4), 359–370 (1983)
45. P. Simon, Investigations with respect to the Vilenkin system, Ann. Univ. Sci. Budapest Eötvös, Sect. Math. **27**, 87–101 (1984)
46. P. Sjölin, An inequality of Paley and convergence a.e. of Walsh-Fourier series. Ark. Mat. **7**, 551–570 (1969)
47. P. Sjölin, Convergence almost everywhere of certain singular integrals and multiple Fourier series. Ark. Mat. **9**, 65–90 (1971)
48. E.M. Stein, On limits of sequences of operators. Ann. Math. **74**(2), 140–170 (1961)
49. E.M. Stein, G. Weiss, *Introduction to Fourier Analysis on Euclidean Spaces* (Princeton University Press, Princeton, 1971)
50. G. Tephnadze, On the partial sums of Walsh-Fourier series. Colloq. Math. **141** (2), 227–242 (2015)

51. G. Tephnadze, On the convergence of partial sums with respect to Vilenkin system on the martingale Hardy spaces. J. Contemp. Math. Anal. **53** (5), 294–306 (2018)
52. A. Torchinsky, *Real-Variable Methods in Harmonic Analysis* (Dover Publications, Inc., Mineola, 2004)
53. N.Y. Vilenkin, On the theory of lacunary orthogonal systems (Russian) Izv. Akad. Nauk SSSR, Ser. Mat. **13**, 245–252 (1949)
54. N.Y. Vilenkin, On a class of complete orthonormal systems. Am. Math. Soc. Transi. **28**(2), 1–35 (1963)
55. C. Watari, On generalized Walsh-Fourier series. Tóhoku Math. J. **10**(2), 211–241 (1958)
56. F. Weisz, *Martingale Hardy Spaces and their Applications in Fourier Analysis*. Lecture Notes in Mathematics, vol. 1568 (Springer, Berlin/Heidelberg/New York, 1994)
57. F. Weisz, Convergence of Vilenkin-Fourier series in variable Hardy spaces. Math. Nachr., **295**(9), 1812–1839 (2022)
58. W.S. Young, Mean convergence of generalized Walsh-Fourier series. Trans. Am. Math. Soc. **218**, 311–320 (1976)

Self-Similar Gravitational Collapse for Polytropic Stars

Matthew Schrecker

Abstract We describe the construction of self-similar solutions to the gravitational Euler-Poisson equations for polytropic gases, providing exact self-similar profiles for the gravitational collapse of stars. These results are based on joint work with Guo et al. (Arch. Rat. Mech. Anal. 246:957–1066, 2022).

Keywords Singularity formation · Gravitational collapse · Euler-Poisson · Non-linear PDE

Mathematics Subject Classification 85A05, 35Q31, 35Q85

1 Introduction

The rigorous description of the collapse of a star under its own gravity is a fundamental mathematical and physical problem, described by the gravitational Euler-Poisson system. Stellar collapse is an important stage in understanding both the formation and the death of stars. The self-similarity hypothesis suggests that, in certain cases, on approach to collapse, the star should adopt an approximately self-similar form, with the intertwining of spatial and time scales dictated by the scaling symmetries of the underlying physical system (see, for example [9]). In recent work, jointly with Guo et al. [7], we have rigorously constructed exactly self-similar solutions to the Euler-Poisson system for the full range of supercriticial exponents.

M. Schrecker (✉)
Department of Mathematics, University of Bath, Bath, UK
e-mail: mris21@bath.ac.uk

D. Cardona et al. (eds.), *Extended Abstracts 2021/2022*, Research Perspectives Ghent Analysis and PDE Center 3, https://doi.org/10.1007/978-3-031-48579-4_17

In spherical symmetry, in three spatial dimensions, the (isentropic) gravitational Euler-Poisson equations take the form

$$\partial_t \rho + \partial_r(\rho u) + \frac{2}{r}\rho u = 0, \tag{1}$$

$$\rho\big(\partial_t u + u\partial_r u\big) + \partial_r p + \frac{1}{r^2}\rho m = 0, \tag{2}$$

where the density ρ and radial velocity u are functions only of time t and radial distance $r = |x|$. The mass m is determined through the relation

$$m(t,r) = 4\pi \int_0^r s^2 \rho(t,s)\,\mathrm{d}s, \tag{3}$$

corresponding to the radial component of the gravitational force field $\nabla\phi$, where ϕ is the gravitational potential determined through the Poisson equation

$$\Delta\phi = 4\pi\rho, \qquad \lim_{|x|\to\infty} \phi(t,x) = 0.$$

To close the system of equations, we require an equation of state for the pressure p, which we choose using the usual polytropic relation

$$p = p(\rho) = \kappa\rho^\gamma, \qquad \gamma \in \left(1, \frac{4}{3}\right), \quad \kappa > 0. \tag{4}$$

The main result of the paper [7], roughly stated, is then

Theorem 1 (Main Theorem, Rough Version) *For all $\gamma \in (1, \frac{4}{3})$, there exists a smooth initial data pair $(\rho_0(r), u_0(r))$, defined on $[0,\infty)$, with $\rho_0(r) \to 0$ as $r \to \infty$ such that the system (1)–(2) with initial data $(\rho, u)|_{t=-1} = (\rho_0, u_0)$ has a smooth solution $(\rho(t,r), u(t,r))$ for $t \in (-1, 0)$ such that, at the spatial origin $r = 0$, the density $\rho(t,0) \to \infty$ as $t \to 0^-$. For all $r > 0$, the limits of $\rho(t,r)$ and $u(t,r)$ exist as $t \to 0^-$ and define smooth functions $\rho(0,r), u(0,r)$ on $(0,\infty)$.*

As advertised above, we seek this claimed solution through self-similarity. To make this notion precise, we first observe that the system of Eqs. (1)–(2) is invariant under the scaling

$$\begin{aligned} \rho(t,r) &\mapsto \lambda^{-\frac{2}{2-\gamma}}\rho\Big(\frac{t}{\lambda^{\frac{1}{2-\gamma}}}, \frac{r}{\lambda}\Big), \\ u(t,r) &\mapsto \lambda^{-\frac{\gamma-1}{2-\gamma}}u\Big(\frac{t}{\lambda^{\frac{1}{2-\gamma}}}, \frac{r}{\lambda}\Big). \end{aligned} \tag{5}$$

This motivates the definition of a self-similar variable

$$y = \frac{r}{\sqrt{\kappa}(-t)^{2-\gamma}} \tag{6}$$

and the ansatz

$$\begin{aligned} \rho(t,r) &= (-t)^{-2}\tilde{\rho}(y), \\ u(t,r) &= \sqrt{\kappa}(-t)^{1-\gamma}\tilde{u}(y). \end{aligned} \tag{7}$$

Substituting this ansatz into the spherically symmetric Euler-Poisson system, defining a new unknown

$$\omega(y) = \frac{\tilde{u}(y)}{y} + 2 - \gamma, \tag{8}$$

and dropping the $\sim$ notation yields, after rearrangement, the self-similar system

$$\begin{aligned} \rho' &= \frac{y\rho\left(2\omega^2 + (\gamma-1)\omega - \frac{4\pi\rho\omega}{4-3\gamma} + (\gamma-1)(2-\gamma)\right)}{\gamma\rho^{\gamma-1} - y^2\omega^2}, \\ \omega' &= \frac{4-3\gamma-3\omega}{y} - \frac{y\omega\left(2\omega^2 + (\gamma-1)\omega - \frac{4\pi\rho\omega}{4-3\gamma} + (\gamma-1)(2-\gamma)\right)}{\gamma\rho^{\gamma-1} - y^2\omega^2}. \end{aligned} \tag{9}$$

It is then clear that any smooth solution of (9) with $\rho(0) > 0$ and $\rho(y) \to \infty$ as $y \to \infty$ gives a collapsing solution of the original system (1)–(2) with density blowup at the origin at time $t = 0$.

Seminal work of Larson and Penston [10, 14] offered a numerical solution to this system in the case $\gamma = 1$ (so-called isothermal stars), describing self-similar stellar collapse. However, two fundamental quantities formally conserved along solutions of the system are the total mass and energy, defined by

$$\begin{aligned} M[\rho] &= 4\pi \int_0^\infty \rho r^2 \, \mathrm{d}r, \\ E[\rho,u] &= 4\pi \int_0^\infty \left(\frac{1}{2}\rho u^2 + \frac{1}{\gamma-1}\rho^\gamma - \frac{1}{8\pi}|\partial_r\phi|^2\right) r^2 \, \mathrm{d}r. \end{aligned}$$

In the isothermal setting, the self-similar ansatz of Larson and Penston leads to solutions of infinite mass and energy, as can be seen either directly from the asymptotics of the Larson–Penston solution, or predicted from the scaling relation (5). Indeed, one checks easily that

$$M[\rho_\lambda] = \lambda^{\frac{4-3\gamma}{2-\gamma}} M[\rho], \quad E[\rho_\lambda, u_\lambda] = \lambda^{\frac{6-5\gamma}{2-\gamma}} E[\rho, u].$$

Thus $\gamma = \frac{4}{3}$ is the *mass-critical* exponent and $\gamma = \frac{6}{5}$ is the *energy-critical* exponent. This observation led Yahil [15] to construct finite energy numerical solutions in the range $\gamma \in (\frac{6}{5}, \frac{4}{3})$ and, for this range of γ, these are the solutions rigorously constructed in Theorem 1.

A key difficulty in solving this ODE system rigorously is the presence of singularities in the system. As well as the regular singular point at the origin (due to the radial symmetry assumption), there is an *a priori* unknown further singularity whenever $\gamma\rho^{\gamma-1} - y^2\omega^2 = 0$. At such a point, the relative speed $y\omega$ is exactly the speed of sound in the gas, $\sqrt{p'(\rho)}$, motivating the following definition.

Definitions 2 Let (ρ, ω) be a C^1 solution of system (9) on an open interval $I \subset (0, \infty)$. A point $y_* \in I$ is called a *sonic point* if

$$\gamma\rho(y_*)^{\gamma-1} - y_*^2\omega(y_*)^2 = 0.$$

A sonic point y_* for the self-similar system corresponds to a backwards acoustic cone emanating from the spatio-temporal origin $(t, r) = (0, 0)$ in physical variables. Although the location of a sonic point is *a priori* unknown, the necessity of the existence of at least one such point is given by the physical asymptotic and boundary conditions at infinity and the origin. As we are looking for smooth solutions with positive density at the origin and density tending to zero at infinity, a simple Taylor expansion shows that we require

$$\begin{aligned} &\rho(0) > 0, \quad \omega(0) = \frac{4-3\gamma}{3}, \\ &\rho(y) \sim y^{-\frac{2}{2-\gamma}} \text{ as } y \to \infty, \quad \lim_{y\to\infty} \omega(y) = 2-\gamma. \end{aligned} \tag{10}$$

The intermediate value theorem then immediately gives the existence of at least one sonic point for any smooth solution.

We can now state the main result of [7] rigorously.

Theorem 3 ([7, Theorem 1.3]) *Let $\gamma \in \left(1, \frac{4}{3}\right)$. Then there exists a global, real-analytic solution (ρ, ω) of (9), (10) with a single sonic point $y_* \in (0, \infty)$ and satisfying the natural, physical conditions*

$$\rho(y) > 0 \textit{ for all } y \in [0, \infty), \quad -\frac{2}{3}y < u(y) < 0 \textit{ for all } y \in (0, \infty). \tag{11}$$

It should be noted that such collapse solutions are not expected for $\gamma > \frac{4}{3}$, the mass-subcritical regime. In this range, it has been shown that no collapsing solutions of finite mass and energy can exist, see [3]. In the mass-critical case, $\gamma = \frac{4}{3}$, there is a famous family of solutions due to Goldreich and Weber, [4], which can either collapse or expand. These solutions are found using an effective separation of variables, allowing for the solution to be found as a time-modulated spatial profile satisfying a Lane-Emden type equation. In contrast, the solutions found in this work involve the careful balancing of all three main forces in the system: inertia, pressure and gravity.

In the isothermal case, $\gamma = 1$, the problem of existence of the Larson–Penston collapsing solutions was solved by Guo–Hadžić–Jang [6], who developed a delicate shooting argument based on the existence of local, smooth solutions around a candidate sonic point. In this case, the system (9) simplifies, and a less refined analysis is required to demonstrate the existence of the solutions.

More recently, the same authors have constructed exact self-similar blowup solutions to the Einstein-Euler equations in general relativity, [8]. This work constructs a smooth, self-similar spacetime with singularity (in both curvature and fluid variables) at a centre of symmetry. From this singularity, a null geodesic emanates and escapes to null infinity. To ensure the spacetime is physically meaningful, the spacetime is flattened far from the centre of symmetry to ensure that it is asymptotically flat. This therefore gives an example of a naked singularity for the Einstein-Euler equations.

We mention also the existence of collapsing (or imploding) self-similar solutions of the Euler equations without gravity, which were found recently by Merle et al. [12]. These solutions were constructed using a careful phase portrait analysis, based on an autonomous self-similar ODE system. The same authors proved the finite co-dimension stability of these solutions within the class of radially symmetric solutions, [13]. Later numerical work, [1] suggests that the finite co-dimension is positive, i.e. these solutions are unstable to generic radial perturbations.

In contrast, it is widely expected that the Larson-Penston and Yahil solutions are in fact stable in the class of radial solutions, based on numerical investigations, see [11]. The smoothness of the underlying self-similar profile appears to be essential for the stability properties, both for the full stability of the gravitational collapse and the finite co-dimensional stability of the gas flows.

2 Strategy of Proof

Before offering an outline of the proof, we first observe that there are two explicit solutions to the system (9), the Friedmann solution

$$\omega_F = \frac{4-3\gamma}{3}, \quad \rho_F = \frac{1}{6\pi},$$

and the far-field solution

$$\omega_f = 2-\gamma, \quad \rho_f = k y^{-\frac{2}{2-\gamma}}, \text{ where } k = \left(\frac{\gamma(4-3\gamma)}{2\pi(2-\gamma)^2}\right)^{\frac{1}{2-\gamma}}.$$

The Friedmann solution satisfies the boundary condition at the origin, but not at infinity, and the far-field solution satisfies the asymptotic condition as $y \to \infty$, but fails to be regular at the origin, compare (10). We find that these solutions each have a unique sonic point, $y_F(\gamma)$, $y_f(\gamma)$, respectively, with $0 < y_f(\gamma) < y_F(\gamma) < \infty$, defined by

$$y_F(\gamma) = \frac{3}{4-3\gamma}\sqrt{\frac{\gamma}{(6\pi)^{(\gamma-1)}}}, \quad y_f(\gamma) = \frac{\sqrt{\gamma}}{2-\gamma}\Big(\frac{4-3\gamma}{2\pi}\Big)^{\frac{\gamma-1}{2}}. \tag{12}$$

Thus, for each $\gamma \in (1, \frac{4}{3})$, the interval $[y_f(\gamma), y_F(\gamma)]$ is compact and we search for our solution with a sonic point in the open range (y_f, y_F) (we henceforth drop explicit dependence on γ).

The proof broadly proceeds in four steps. The first step is to construct local solutions around any candidate sonic point $y_* \in [y_f, y_F]$. This is based on a Taylor expansion argument to solve for

$$\rho(y; y_*) = \sum_{n=0}^{\infty} \rho_n(y_*)(y - y_*)^n, \quad \omega(y; y_*) = \sum_{n=0}^{\infty} \omega_n(y_*)(y - y_*)^n. \tag{13}$$

The order zero coefficients are determined from (9) by solving the pair of nonlinear equations

$$\gamma\rho_0^{\gamma-1} - y_*^2\omega_0^2 = 0, \quad 2\omega_0^2 + (\gamma-1)\omega_0 - \frac{4\pi\rho_0\omega_0}{4-3\gamma} + (\gamma-1)(2-\gamma) = 0.$$

This gives, for each $y_* \in [y_f, y_F]$, a unique choice $(\rho_0(y_*), \omega_0(y_*))$, with $\omega_0(y_f) = 2-\gamma$ and $\omega_0(y_F) = \frac{4-3\gamma}{3}$. A selection principle is necessary to determine the first order coefficients (for which there are two possible choices for every $y_* \in [y_f, y_F]$) and then a recurrence relation is used to determine the higher order coefficients. Through combinatorial bounds, these formal series are shown to converge in some neighbourhood of y_* and give a local, real-analytic solution of (9).

The second step is to extend the local solution to the right on the interval (y_*, ∞). This can be done for all $y_* \in [y_f, y_F]$, and the argument is based on the construction of dynamical invariances for the flow, using the precise structure of the non-linearities. To close some of the estimates, it is necessary to verify the sign of certain explicit (but high order) polynomial functions of γ and ω. This is achieved via the use of *interval arithmetic*, a rigorous computer-assisted form of proof that has attracted increasing attention in the PDE community in recent years (see, for example, [2, 5]).

The third step is the most difficult, and contains the key new ideas of the paper. This is to extend the local solution to the left onto the interval $(0, y_*)$. For a general y_*, the solution will *not* extend smoothly all the way to the origin, and so we here develop a shooting argument in order to find a critical $\bar{y}_*$ for which the solution does connect. The central difficulty is that the invariant region arguments used in extending to the right all fail in this direction, which is an unstable direction for the flow. Instead, we rely on a new *monotonicity lemma*, which allows us to obtain the necessary control on the solutions.

To be more precise, we observe that the solution needs to connect to the value $\frac{4-3\gamma}{3}$ at the origin in order to extend smoothly, and, for y_* close to y_F, the solution obtained by Taylor expansion around y_* will quickly decrease below this value. We therefore define the shooting set Y to be the set of candidate sonic points y_* for which the solution to the left intersects this value:

$$Y = \Big\{ y_* \in (y_f, y_F) \mid \text{for all } \tilde{y}_* \in [y_*, y_F), \\ \text{there exists } y \text{ such that } \omega(y; \tilde{y}_*) = \frac{4-3\gamma}{3} \Big\}$$

and search for the critical $\bar{y}_*$ as the infimum of Y.

Due to the instability of the flow to the left and the possibility of hitting a second sonic point, it is hard to achieve uniform convergence estimates. Instead, we prove the key monotonicity lemma. First, for each $y_* \in Y$, we define the critical time $y_c(y_*)$ as the first touching time

$$y_c(y_*) = \inf \Big\{ y \mid \omega(\tilde{y}; y_*) > \frac{4-3\gamma}{3} \text{ for all } \tilde{y} \in (y, y_*) \Big\}.$$

Lemma 4 *Let $\gamma \in (1, \frac{4}{3})$. For all $y_* \in Y$, the solution $(\rho(\cdot; y_*), \omega(\cdot; y_*))$ defined by the formal Taylor expansion* (13) *and extended to the left on the interval $[y_c(y_*), y_*]$ satisfies the strict monotonicity condition*

$$\omega'(y; y_*) > 0 \textit{ for all } y \in [y_c(y_*), y_*].$$

With this monotonicity, we achieve sufficient control of the flow in order to establish that the solution associated to $\bar{y}_*$ exists on the interval $(0, \bar{y}_*)$, in particular ruling out the existence of another sonic point on this interval.

The final step is to show that this solution connects *smoothly* to the origin. The monotonicity simplifies many arguments, avoiding the need for the complicated topological upper- and lower-solution arguments required in [6]. This allows us to show that $\omega(y; \bar{y}_*) \to \frac{4-3\gamma}{3}$ as $y \to 0$, while the density remains bounded. By a further Taylor expansion at the origin and a local uniqueness result, we show that the solution is in fact locally analytic, completing the proof of Theorem 3.

Acknowledgments M. Schrecker's research is supported by the EPSRC Post-doctoral Research Fellowship EP/W001888/1.

References

1. A. Biasi, Self-similar solutions to the compressible Euler equations and their instabilities, Commun. Nonlinear Sci. Numer. Simul. **103**, Paper No. 106014 (2021)
2. J. Chen, T. Hou, Stable nearly self-similar blowup of the 2D Boussinesq and 3D Euler equations with smooth data I: Analysis. Preprint (2022) ArXiv , arXiv:2210.07191
3. Y. Deng, T.P. Liu, T. Yang, Z. Yao, Solutions of Euler-Poisson equations for gaseous stars. Arch. Rat. Mech. Anal. **164**, 261–285 (2002)
4. P. Goldreich, S. Weber, Homologously collapsing stellar cores. Astrophys. J. **238**, 991–997 (1980)
5. J. Gómez-Serrano, Computer-assisted proofs in PDE: a survey. Preprint (2018). arxiv:1810.00745
6. Y. Guo, M. Hadžić, J. Jang, Larson-Penston self-similar gravitational collapse. Commun. Math. Phys. **386**, 1551–1601 (2021)
7. Y. Guo, M. Hadžić, J. Jang, M. Schrecker, Gravitational collapse for polytropic gaseous stars: self-similar solutions. Arch. Rat. Mech. Anal. **246**, 957–1066 (2022)
8. Y. Guo, M. Hadžić, J. Jang, Naked singularities in the Einstein-Euler system. Ann. PDE **9** Paper No. 4, 182 pp. (2023)
9. T. Harada, self-similar solutions, critical behavior and convergence to attractor in gravitational collapse, in *12th Workshop on General Relativity and Gravitation, Nov 25–28, 2002, Tokyo, Japan* (2003)
10. R.B. Larson, Numerical calculations of the dynamics of a collapsing protostar. Mon. Not. R. Astr. Soc. **145**, 271–295 (1969)
11. H. Maeda, T. Harada, Critical phenomena in Newtonian gravity. Phys. Rev. D **64**, 124024 (2001)
12. F. Merle, P. Raphaël, I. Rodnianski, J. Szeftel, On the implosion of a compressible fluid I: Smooth self-similar inviscid profiles. Ann. Math. **196**, 567–778, (2022)
13. F. Merle, P. Raphaël, I. Rodnianski, J. Szeftel, On the implosion of a compressible fluid II: Singularity formation. Ann. Math. **196**, 779–889, (2022)
14. M.V. Penston, Dynamics of self-gravitating spheres III. Mon. Not. R. Astr. Soc. **144**, 425–448 (1969)
15. A. Yahil, Self-similar stellar collapse. Astrophys. J. **265**, 1047–1055 (1983)

Control of Parabolic Equations with Inverse Square Infinite Potential Wells

Arick Shao

Abstract This survey summarises a presentation recently given by the author at the Ghent Methusalem Junior Analysis Seminar. The talk discussed the recent result of Enciso et al. (Controllability of parabolic equations with inverse square infinite potential wells via global Carleman estimates. Preprint, 2021), joint with Alberto Enciso (ICMAT) and Bruno Vergara (Brown), as well as the main ideas of its proof.

In Enciso et al. (Controllability of parabolic equations with inverse square infinite potential wells via global Carleman estimates. Preprint, 2021), we consider heat operators on a bounded convex domain, with a critically singular potential diverging as the inverse square of the distance to the boundary of the domain. We address the problem of boundary null controllability—whether one can drive the solution from any initial data to zero via suitable boundary data. We establish a null control result for such operators in all spatial dimensions, in particular providing the first result in more than one spatial dimension. The key step in the proof is a novel global Carleman estimate that captures both the relevant boundary asymptotics and the appropriate energy for this problem.

Keywords Parabolic · Heat · Singular · Potential · Null · Control · Carleman

Mathematics Subject Classification 35K05, 93B05, 93B07

A. Shao (✉)
School of Mathematical Sciences, Queen Mary University, London, UK
e-mail: a.shao@qmul.ac.uk

D. Cardona et al. (eds.), *Extended Abstracts 2021/2022*, Research Perspectives Ghent Analysis and PDE Center 3, https://doi.org/10.1007/978-3-031-48579-4_18

1 Critically Singular Heat Equations

Let $\Omega \subseteq \mathbb{R}^n$ be a bounded domain, with a sufficiently regular boundary $\Gamma := \partial\Omega$, and fix $T > 0$. In this article, we will primarily consider the following PDE and data:

$$\begin{aligned} -\partial_t v + (\Delta + \sigma d^{-2})v &= Y \cdot \nabla v + W v && \text{on } (0,T) \times \Omega, \\ v|_{t=0} &= v_0 && \text{on } \Omega, \\ \mathcal{D}_\sigma v &= f && \text{on } (0,T) \times \Gamma. \end{aligned} \tag{1}$$

Here, $d := d(\cdot, \Gamma)$ denotes the distance to the boundary Γ, so the first line of (1) gives a heat equation with a singular potential diverging as an inverse square at the hypersurface Γ. Moreover, $\sigma \in \mathbb{R}$ is a parameter measuring the strength of the singular potential, while $Y : \Omega \to \mathbb{R}^n$ and $W : \Omega \to \mathbb{R}$ represent (time-independent) lower-order coefficients that are less singular at Γ. The second part of (1) represents the initial data for the singular heat equation, while $\mathcal{D}_\sigma v$ is an appropriately defined Dirichlet data, which will be described in more detail later.

Our main objective is to investigate the following control-theoretic problem for (1):

Problem 1 (Boundary Null Controllability) *Given any $T > 0$ and initial data v_0, does there exist boundary data f such that the solution v of (1) satisfies $v|_{t=T} = 0$? In other words, can solutions of the singular heat equation of (1) be driven, through an appropriate Dirichlet boundary control, from any initial state to the equilibrium state in any finite time?*

Remark 2 A closely related question is *interior controllability*, which is similar to Problem 1, except that the control is imposed through an additional forcing term supported on subdomain $(0,T) \times \omega$, for some $\omega \subseteq \Omega$, rather than via boundary data.

When $\sigma = 0$, our setting (1) reduces to classical heat equations, for which there is extensive literature on Problem 1; see, for instance, [15, 33] for wide-ranging surveys. Roughly speaking, the methods used split into two categories. First, there are *spectral* or *Fourier methods*, which tend to yield the strongest results, but only for very specific lower-order coefficients Y, W (for which one has detailed information on the spectrum of the full linear operator). Next, there are also results obtained via *Carleman estimates*, which, while slightly less optimal, are robust in that they apply to a wide class of Y, W. As a result, the latter also has the benefit of being applicable to nonlinear equations through an iteration procedure.

On the other hand, when $\sigma \neq 0$, the factor σd^{-2} in (1) can be seen as an infinite potential well. Such a quantity also show up naturally in geometric settings—for example, the Laplace-Beltrami operator on a hyperbolic or asymptotically hyperbolic manifold gains such a singular potential after a conformal compactification that makes its boundary finite [24, 32].

Such a potential σd^{-2} introduces numerous novel difficulties, since it has the same scaling as Δ, and hence it has to be treated as a "principal" term (unlike Y, W); thus, we refer to σd^{-2} as *critically singular*. One consequence of this is that solutions v to (1) have radically different boundary asymptotics. More specifically, v behaves near Γ like specific powers of d:

$$v \sim_{\Gamma} d^{\kappa} v_D + d^{1-\kappa} v_N, \qquad \kappa := \tfrac{1-\sqrt{1-4\sigma}}{2}, \quad \sigma \le \tfrac{1}{4}. \tag{2}$$

Consequently, our notions of Dirichlet and Neumann data for (1) must be amended to capture the expected branches v_D and v_N. For this, the appropriate quantities are given by

$$\mathcal{D}_\sigma \phi := d^{-\kappa} \phi|_{(0,T)\times\Gamma}, \qquad \mathcal{N}_\sigma \phi := d^{2\kappa} \nabla d \cdot \nabla (d^{-\kappa} \phi)|_{(0,T)\times\Gamma}, \tag{3}$$

representing the Dirichlet and Neuman traces, respectively.

Remark 3 A more subtle issue is that for $\sigma \neq 0$, the natural energy spaces now have fractional regularity depending on σ. However, we will avoid this issue in this article.

Remark 4 Note it is especially natural to consider nonzero Y and W in our context, since d generally fails to be regular away from Γ, hence nontrivial (and irregular) lower-order coefficients are needed merely to study smooth operators of the form (1).

Remark 5 Of particular importance are the thresholds $\sigma = \frac{1}{4}$ and $\sigma = -\frac{3}{4}$, as well-posedness for (1) breaks down when $\sigma > \frac{1}{4}$ [3, 5], while the Dirichlet branch $d^{\kappa} v_D \notin L^2(\Omega)$ when $\sigma \le -\frac{3}{4}$. It is also known—see [2]—that null boundary controllability fails for $\sigma = \frac{1}{4}$.

2 Existing Results

The literature for (1) is much sparser than for the classical heat equation. Moreover, a vast majority of results for singular heat equations have focused instead on potentials $\sigma |x - x_0|^{-2}$ that diverge only at a single point $x_0 \in \bar{\Omega}$; see, for instance, [1, 8, 11, 14, 31] and references within. In order to keep this survey concise, here we will restrict our discussions only to the relatively few null controllability results that exist for (1) itself and its special cases.

In one spatial dimension, where we can set $\Omega := (0, 1)$ without loss of generality, there are numerous results treating the singular heat operator

$$-\partial_t + \partial_x^2 + \sigma x^{-2}. \tag{4}$$

(Note this is not quite a special case of (1), since the potential diverges only at $x = 0$ and not at $x = 1$.) For instance, interior null controllability results for (4) were established in [6, 7, 10, 23], while boundary null controllability results for (4) were obtained in [2, 9, 10, 17].

Of particular relevance is the result of Biccari [2], which established boundary null controllability at $x = 0$ for (4) with $-\frac{3}{4} < \sigma < \frac{1}{4}$. As the proof applied the moment method, a Fourier-based technique that relied strongly on the precise spectral decomposition of the operator, the results of [2] do not readily extend to more general settings, for instance:

- Singular heat Eq. (1) in higher spatial dimensions.
- Equation (4) with lower-order coefficients $Y, W \not\equiv 0$.
- Equation (4), but with a potential that is singular at both $x = 0$ and $x = 1$.

These were listed as open problems in [2, Section 8] and serve as key motivations for our result. Finding Carleman estimates for (4) was also highlighted as an especially challenging problem.

Now, for higher dimensions $n > 1$, the only null controllability results for (1) have been with interior controls. In particular, Biccari and Zuazua [3] proved interior null controllability for (1) using Carleman estimates. However, the results of [3] do not yield boundary controllability, as their Carleman estimates do not capture the natural Neumann trace (3) at the boundary.

3 The Main Result

Before stating the main result of [13] and this article, let us first roughly describe the assumptions we impose on the lower-order coefficients Y, W in (1):

- Y, W should be regular near Γ (in particular, $Y \in C^3$ and $dW \in C^2$).
- Y, W can be rougher away from Γ (in particular, $Y \in C^1$ and $W \in L^\infty$).

In particular, regularity near Γ is needed for our setting to be well-posed in the context of the Hilbert uniqueness method. On the other hand, since σd^{-2} can be rough away from Γ, then we must require this also from Y and W in order to treat smooth operators.

The main boundary null control result, joint with Enciso and Vergara, is then the following:

Theorem 6 ([13], Theorem 4.6) *Assume the setting of* (1), *with Y and W satisfying the above. In addition, suppose $-\frac{3}{4} < \sigma < 0$, and suppose Γ is a C^4 and convex (having strictly positive second fundamental form) hypersurface. Then, given any $T > 0$ and $v_0 \in H^{-1}(\Omega)$, there exists $f \in L^2((0, T) \times \Gamma)$ such that the solution v of* (1) *satisfies $v|_{t=T} = 0$.*

To the author's best knowledge, Theorem 6 is the first boundary null controllability result for (1) when $n > 1$, as well as the first result in any dimension for nontrivial Y, W. When $n = 1$, Theorem 6 also positively addresses the variant of (4) in which the potential diverges at both $x = 0$ and $x = 1$. Furthermore, [13] provides the first proof of boundary null control for (1) using Carleman estimates, which hence allows us to treat general Y, W.

Remark 7 On the other hand, the proof of Theorem 6 breaks down if $\sigma \leq -\frac{3}{4}$ or $\sigma \geq 0$. While $\sigma = 0$ is just the classical heat equation, Theorem 6 does fail to treat the range $0 < \sigma < \frac{1}{4}$, within which one still has boundary null control for (4) from [2].

4 The Hilbert Uniqueness Method

The overall framework for proving Theorem 6 is based on duality and is now widely used in the control theory of PDEs. Early versions of this approach were due to Russell [26, 27]. More modern treatments, usually known as the *Hilbert uniqueness method* (*HUM*) of Lions [22], also allow for one to characterise the desired control variationally; see [33].

Though there is not enough space to cover this approach in detail, here we simply stress that the key observation behind this framework is that the null controllability of (1) is equivalent to a quantitative uniqueness, or *observability*, statement for a dual problem. More specifically, this dual problem consists of the adjoint equation—a backwards singular heat equation:

$$\begin{aligned} \partial_t u + (\Delta + \sigma d^{-2}) u &= X \cdot \nabla u + V u && \text{on } (0, T) \times \Omega, \\ u|_{t=T} &= u_T && \text{on } \Omega, \\ \mathcal{D}_\sigma u &= 0 && \text{on } (0, T) \times \Gamma. \end{aligned} \tag{5}$$

Here, (5) is solved backwards in time, with homogeneous Dirichlet data. Furthermore, we can impose the same assumptions for X and V as for Y and V in Theorem 6.

The first task at hand is to obtain a viable well-posedness theory for both (1) and (5) in dual spaces. Such theories are well-known for $\sigma = 0$, see [25, 33], but for Theorem 6, we must further adapt the theory to treat the cases $\sigma \neq 0$. Much of the treatment of (5) is based on the approach taken in [3], however in [13] we further extend the analysis to (1) in dual spaces. Without going into details, the upshot of this well-posedness analysis is as follows:

- The dual theories for (5) and (1) hold for $-\frac{3}{4} < \sigma < \frac{1}{4}$.
- For the observability side (5), one has well-posedness with $u_T \in H_0^1(\Omega)$.
- For the control side (1), one has well-posedness with $v_0 \in H^{-1}(\Omega)$ and $f \in L^2((0, T) \times \Gamma)$. (In particular, here $H^{-1}(\Omega)$ is the Hilbert space dual of $H_0^1(\Omega)$.)

In particular, these well-posedness theories lead to the specific spaces $H^{-1}(\Omega)$ and $L^2((0,T) \times \Gamma)$ in the statement of Theorem 6. For details, see the discussions in [13, Sections 3–4].

With the well-posedness theories in place, the standard HUM machinery then yields that null controllability of (1) would follow—and hence the proof of Theorem 6 would be complete—if one can establish the following estimates on the observability side:

Proposition 8 *For any solution u of* (5)*, the Neumann trace $\mathcal{N}_\sigma u$ is well-defined and finite as an element of $L^2((0,T) \times \Gamma)$. Furthermore, the following estimates hold for u:*

$$\|u|_{t=0}\|_{H^1(\Omega)} \lesssim \|\mathcal{N}_\sigma u\|_{L^2((0,T)\times\Gamma)} \lesssim \|u_T\|_{H^1(\Omega)}. \tag{6}$$

(Here, the constants of the inequalities are independent of u_T and u.)

See, for instance, [33] for details of this HUM machinery. For our purposes, to complete the proof of Theorem 6, it suffices to establish the two estimates in (6).

The second estimate in (6), the upper bound for $\mathcal{N}_\sigma u$, is a consequence of the regularity theory of (5). When $\sigma = 0$, such an estimate is standard and follows by applying a trace estimate,

$$\|\mathcal{N}_0 u\|_{L^2((0,T)\times\Gamma)} \lesssim \|u\|_{L^2(0,T;H^2(\Omega))},$$

and then applying the usual smoothing and energy estimates to control the above by u_T. With a bit of work, the above can be adapted to (5) in the range $-\frac{3}{4} < \sigma < \frac{1}{4}$ by using the appropriate weighted spaces; for details of this process, see [13, Section 3.3],

The more interesting part is the first estimate in (6)—the so-called *observability estimate*. (The meaning is that one can determine everything about the solution u by observing $\mathcal{N}_\sigma u$.) When $\sigma = 0$, a well-known approach is to derive a global Carleman estimate for (5) to bound

$$\|\mathcal{N}_0 u\|_{L^2((0,T)\times\Gamma)} \gtrsim \|u\|_{L^2(0,T;H^1(\Omega))},$$

from which the observability estimate follows via a standard energy inequality.

The key step of the proof of Theorem 6 is to obtain, for the first time, a global Carleman estimate for (5) that yields the above observability estimate in the range $-\frac{3}{4} < \sigma < 0$. This serves as the main novelty of [13] and opens the doors for treating (1) using Carleman estimate methods. In the following section, we narrow our focus to this Carleman estimate.

5 The Global Carleman Estimate

Carleman estimates have been an indispensible tool in the study of unique continuation for a wide variety of PDEs; see, for instance, [18, 21]. They have also been applied to a variety of topics within PDEs, such as control theory [15, 16, 30] and inverse problems [4, 19].

In the context of Theorem 6, in particular of observability estimates for (5), we wish to prove a global Carleman estimate that has, very roughly, the following form:

$$\begin{aligned} C'\lambda \int_{(0,T)\times\Gamma} (\mathcal{N}_\sigma u)^2 + \int_{(0,T)\times\Omega} e^{-2\lambda F} (\partial_t u + \Delta u + \sigma d^{-2} u)^2 \\ \geq C\lambda \int_{(0,T)\times\Omega} e^{-2\lambda F} (|\nabla u|^2 + d^{-2} u^2). \end{aligned} \tag{7}$$

Note (7) is a spacetime integral estimate with some additional features. Here, $F : (0, T) \times \Omega \to \mathbb{R}$ is a weight function—depending on the PDE, the geometry of the domain, and the problem at hand—that is specially chosen so that (7) holds. Moreover, $\lambda \gg 1$ is an additional parameter in (7) that must be sufficiently large but can otherwise be freely chosen. Finally, C and C' are positive constants that are, crucially, independent of u and λ.

To see how (7) leads to observability, we apply (7) to a solution u of (5), which yields

$$\begin{aligned} C'\lambda \int_{(0,T)\times\Gamma} (\mathcal{N}_\sigma u)^2 + \int_{(0,T)\times\Omega} e^{-2\lambda F} (X \cdot \nabla u + V u)^2 \\ \geq C\lambda \int_{(0,T)\times\Omega} e^{-2\lambda F} (|\nabla u|^2 + d^{-2} u^2). \end{aligned}$$

By taking λ sufficiently large, the second term on the left-hand side of the above can be absorbed into the right, and we obtain an estimate of the form

$$\|\mathcal{N}_\sigma u\|_{L^2((0,T)\times\Gamma)} \gtrsim \|u\|_{L^2(0,T;H^1(\Omega))},$$

from which observability follows via an additional energy estimate.

As for how such an estimate (7) is proved, one begins this process by conjugating the PDE with the exponential weight $e^{-\lambda F}$, starting with an expression of the form

$$\int_{(0,T)\times\Omega} [e^{-\lambda F} (\partial_t + \Delta + \sigma d^{-2})(e^{\lambda F} w)] \cdot S w, \qquad w := e^{-\lambda F}, \tag{8}$$

where S is an appropriately chosen multiplier approximately of the form

$$S w := \partial_t w + \lambda \nabla F \cdot \nabla w + \ldots.$$

One then expands (8) and integrates by parts repeatedly. Roughly speaking, if an appropriate F is chosen, then after an extensive amount of computations, the boundary terms obtained will capture the Neumann trace $(\mathcal{N}_\sigma u)^2$, and the bulk terms will be dominated by a positive H^1-norm of u, as in the right-hand side of (7). By taking λ large enough, then this leading positive term will be large enough to absorb other lower-order terms that have no sign.

Remark 9 See [13, Section 2] for detailed computations behind the Carleman estimate for Theorem 6, as well as for the precise form of the Carlmeman estimate.

Below, we outline the main ideas and novelties behind the proof of our new global Carleman estimate. We begin with the Carleman estimate of Biccari and Zuazua [3], which was used to prove interior controllability for (1). There, the authors used a weight F roughly of the form

$$F(t,x) \approx \tfrac{1}{t^3(T-t)^3}\left(C - [d(x)]^2 - \left[d(x)e^{d(x)}\right]^s\right), \tag{9}$$

where C is a constant and $s \gg 1$ is a large exponent. (The factors of t^{-1} and $(T-t)^{-1}$ in (9) are standard; these force the weight $e^{-\lambda F}$ to vanish at $t = 0$ and $t = T$, thereby avoiding any boundary terms there.) However, this choice (9) does not suit our purposes, as performing the above-mentioned computations with this F fails to capture the Neumann trace at the boundary. Moreover, using this F only leads to control for the spacetime L^2-norm of u, but not the full H^1-norm, as terms containing $|\nabla u|^2$ are accompanied by a weight vanishing at Γ.

Thus, the first idea is that F *needs to contain a special power of d in order to capture the Neumann data*. For this, we instead consider F of the form

$$F(t,x) := \tfrac{1}{t(T-t)}\left(\tfrac{1}{1+2\kappa}[d(x)]^{1+2\kappa} + \beta\right), \qquad \beta > 0, \tag{10}$$

where κ is as in (2). Then, integrating by parts using F as in (10), we can see (after many computations) that one indeed recovers the L^2-norm of $\mathcal{N}_\sigma u$ at the boundary.

Remark 10 Technically, we must also ensure that only the Neumann data is captured at the boundary. For this, one must also show in the setting of (5) that (see [13, Section 3.3])

$$d^{-1+\kappa} u|_\Gamma = \mathcal{N}_\sigma u, \qquad \int_{(0,T)\times\Gamma} e^{-2\lambda F}\partial_t(\mathcal{D}_\sigma u)\mathcal{N}_\sigma u = 0.$$

Taking F as in (10), however, exposes an even more basic issue—that d, and hence F, fails to be regular away from Γ. This would prevent us from performing the needed integrations by parts away from Γ. Thus, we must also *alter d so that it is better behaved away from Γ*. Such a function is constructed [13, Section 2.1], and we summarise its main features below:

Proposition 11 *There exists a* boundary-defining function $0 < y \in C^4(\Omega)$ *such that:*

- *y coincides with d near Γ.*
- *$-y$ is almost everywhere convex on Ω (i.e. $-\nabla^2 y \geq -\varepsilon$).*
- *y has a unique critical point $x_* \in \Omega$.*

The idea is then to replace d by y in our definition of F:

$$F(t,x) := \frac{1}{t(T-t)}\left(\frac{1}{1+2\kappa}[y(x)]^{1+2\kappa} + \beta\right), \qquad \beta > 0, \tag{11}$$

For similar reasons, one also replaces the singular heat operator by a smoother variant:

$$\partial_t + \Delta + \sigma d^{-2} \quad \rightarrow \quad \partial_t + \Delta + \sigma y^{-2}.$$

Using F as in (11), one still recovers the Neumann trace at the boundary, since $y = d$ near Γ. Also, since y is sufficiently regular, the computations can now be performed on all of Ω.

Remark 12 The C^4-regularity of y is due to the assumption that Γ is C^4. The only significance of C^4 is that one takes 4 derivatives of y in the ensuing computations, although this is unlikely to be optimal. Similarly, the near-convexity of $-y$ is a consequence of Γ being convex.

Furthermore, using the near-convexity of $-y$, we see (after many computations) that the full $\dot{H}^1$-norm of u can be controlled when $-\frac{3}{4} < \sigma < 0$, and we obtain an inequality of the form

$$\begin{aligned} C'\lambda \int_{(0,T)\times\Gamma} (\mathcal{N}_\sigma u)^2 + \int_{(0,T)\times\Omega} e^{-2\lambda F}(\partial_t u + \Delta u + \sigma y^{-2} u)^2 \\ \geq C\lambda \int_{(0,T)\times\Omega} e^{-2\lambda F} \dots |\nabla u|^2 + C\lambda^3 \int_{(0,T)\times[\Omega\setminus B_\delta(x_*)]} e^{-2\lambda F} \dots u^2 \\ - C_*\lambda^2 \int_{(0,T)\times B_\delta(x_*)} e^{-2\lambda F} \dots u^2. \end{aligned} \tag{12}$$

(Here, ... denotes some additional weights that are omitted for brevity, and $B_\delta(x_*)$ is a sufficiently small ball around x_*.) Note, most alarmingly, that (12) fails to control the L^2-norm, since one has L^2-positivity only away from the critical point of y. The reason is that the weights "..." in the positive L^2-terms contain a factor of $|\nabla y|^2$, which vanishes at x_*.

The above yields what turns out to be the most signficant obstacle. The idea for overcoming this is to *construct instead two boundary-defining functions y_1 and y_2, with distinct critical points $x_{*,1} \neq x_{*,2}$*. One can then *obtain partial Carleman* *estimates from both* y_1 *and* y_2 *via the weight* (11), and then *sum both estimates together*. Observe then that:

- Near $x_{*,1}$, the L^2-contribution from the y_1-estimate is negative, but this is overtaken by a positive L^2-contribution from the y_2-estimate.
- Near $x_{*,2}$, the L^2-contribution from the y_2-estimate is negative, but this is overtaken by a positive L^2-contribution from the y_1-estimate.

In particular, in both points above, the positive term dominates, since it comes with a larger power of λ. Yet another concern lies with the exponentials $e^{-2\lambda F_1}$ and $e^{-2\lambda F_2}$ in the partial estimates (12). However, by a careful choice of constants β_1, β_2 from (11), one can ensure that near each critical point $x_{*,j}$, the exponential weight from the positive term is the largest.

Remark 13 Similar methods involving adding two Carleman estimates with different weights were used in [20, 28], in the context of controllability for wave equations.

From the above analysis, we now obtain our desired Carleman estimate:

Theorem 14 (Global Carleman Estimate) *If* $-\frac{3}{4} < \sigma < 0$, *then the following holds:*

$$C'\lambda \int_{(0,T)\times\Gamma} (\mathcal{N}_\sigma u)^2 + \sum_{j=1}^{2} \int_{(0,T)\times\Omega} e^{-2\lambda F_j} (\partial_t u + \Delta u + \sigma y_j^{-2} u)^2 \tag{13}$$

$$\geq C\lambda \sum_{j=1}^{2} \int_{(0,T)\times\Omega} e^{-2\lambda F_j} (|\nabla u|^2 + y_j^{-2} u^2).$$

See [13, Theorem 2.9] for the precise statement and proof. In particular, (13) is sufficient to obtain the desired observability estimate for (5) and complete the proof of Theorem 6.

6 Additional Remarks

We conclude the article with a few final remarks related to Problem 1 and Theorem 6:

Remark 15 A natural question is whether some variant of Theorem 6 could still hold when Γ is not convex. Although convexity was crucial in the proof of Theorem 14 in order to control the $\dot{H}^1$-norm, it turns out that *one can still obtain a similar Carleman estimate localised near* Γ *even when* Γ *fails to be convex*. One consequence of this is that for general Γ, one can still obtain an *approximate boundary controllability* result (i.e. that one can steer the solution arbitrarily close to any final state); this has recently been established, jointly with Vergara, in [29].

Remark 16 We conjecture that boundary null controllability for (1) still holds when $0 < \sigma < \frac{1}{4}$. The Carleman estimate of (13) seemingly fails in this range, while the natural energy space for (5) is a fractional H^s, for some $s < 1$. Thus, one likely needs to obtain an analogous Carleman estimate in weaker, fractional Sobolev norms, which significantly complicates the analysis.

Remark 17 Finally, [12] obtained a similar observability estimate for singular wave equations,

$$-\partial_t^2 u + (\Delta + \sigma d^{-2})u = X \cdot \nabla u + V\, u,$$

but only when the domain Ω is a disk. It would be interesting to see whether the techniques here can be used to extend the result of [12] to more general convex domains.

References

1. P. Baras, J.A. Goldstein, The heat equation with a singular potential, Trans. Am. Math. Soc. **284**, 121–139 (1984)
2. U. Biccari, Boundary controllability for a one-dimensional heat equation with a singular inverse-square potential. Math. Control Relat. Fields **9**, 191–219 (2019)
3. U. Biccari, E. Zuazua, Null controllability for a heat equation with a singular inverse-square potential involving the distance to the boundary function. J. Differ. Equ. **261**, 2809–2853 (2016)
4. A.L. Bukhgeĭm, M.V. Klibanov, Uniqueness in the large of a class of multidimensional inverse problems. Dokl. Akad. Nauk SSSR **260**, 269–272 (1981)
5. X. Cabré, Y. Martel, Existence versus explosion instantée pour des équations de la chaleur linéaires avec potentiel singulier. C. R. Math. Sci. Acad. Sci. Paris **329**, 973–978 (1999)
6. P. Cannarsa, P. Martinez, J. Vancostenoble, Null controllability of degenerate heat equations. Adv. Differ. Equ. **10**, 153–190 (2005)
7. P. Cannarsa, P. Martinez, J. Vancostenoble, Carleman estimates for a class of degenerate parabolic operators. SIAM J. Control Optim **47**, 1–19 (2008)
8. P. Cannarsa, P. Martinez, J. Vancostenoble, Carleman estimates and null controllability for boundary-degenerate parabolic operators. C. R. Acad. Sci. Paris, Serie I **347**, 147–152 (2009)
9. P. Cannarsa, P. Martinez, J. Vancostenoble, The cost of controlling weakly degenerate parabolic equations by boundary controls. Mat. Control Relat. Fields **7**, 171–211 (2017)
10. P. Cannarsa, P. Martinez, J. Vancostenoble, The cost of controlling strongly degenerate parabolic equations by boundary controls. ESAIM: Control Optim. Calc. Var. **26**, 50pp. (2020)
11. C. Cazacu, Controllability of the heat equation with an inverse-square potential localized on the boundary. SIAM J. Control Optim. **52**, 2055–2089 (2014)
12. A. Enciso, A. Shao, B. Vergara, Carleman estimates with sharp weights and boundary observability for wave operators with critically singular potentials. J. Eur. Math. Soc. **23**, 3459–3495 (2021)
13. A. Enciso, A. Shao, B. Vergara, Controllability of parabolic equations with inverse square infinite potential wells via global Carleman estimates. Preprint (2023). arXiv:2112.04457
14. S. Ervedoza, Control and stabilization properties for a singular heat equation with an inverse-square potential. Commun. Partial Differ. Equ. **33**, 1996–2019 (2008)

15. A.V. Fursikov, O.Y. Imanuvilov, *Controllability of Evolution Equations*. Lecture Notes Series, vol. 34 (Seoul National University, Research Institute of Mathematics, Global Analysis Research Center, Seoul, 1996)
16. X. Fu, Q. Lü, X. Zhang, *Carleman Estimates for Second Order Partial Differential Operators and Applications* (Springer, Berlin, 2019)
17. M. Gueye, Exact boundary controllability of 1-D parabolic and hyperbolic degenerate equations. SIAM J. Control Optim. **52**, 2037–2054 (2014)
18. L. Hörmander, *The Analysis of Linear Partial Differential Operators IV* (Springer-Verlag, Berlin, 1985)
19. M.V. Klibanov, Inverse problems and Carleman estimates. Inverse Problems **8**, 575 (1992)
20. V.K. Jena, Carleman estimate for ultrahyperbolic operators and improved interior control for wave equations. J. Differ. Equ. **302**, 273–333 (2021)
21. N. Lerner, *Carleman Inequalities: An Introduction and More* (Springer, Berlin, 2019)
22. J.L. Lions, *Controlabilité exacte perturbations et stabilisation de systèmesdistribués* (Masson, Paris, 1988)
23. P. Martinez, J. Vancostenoble, Carleman estimates for one-dimensional degenerate heat equations. J. Evol. Equ. **6**, 325–362 (2006)
24. R. Mazzeo, R. Melrose, Meromorphic extension of the resolvent on complete spaces with asymptotically constant negative curvature. J. Funct. Anal. **75**, 260–310 (1987)
25. J.-P. Raymond, Optimal control of partial differential equations. Lecture Notes (2004). https://www.math.univ-toulouse.fr/~raymond/book-ficus.pdf
26. D.L. Russell, Boundary value control of the higher-dimensional wave equation. SIAM J. Control **7**, 29–42 (1971)
27. D.L. Russell, Boundary value control theory of the higher-dimensional wave equation. II. SIAM J. Control **7**, 401–419 (1971)
28. A. Shao, On Carleman and observability estimates for wave equations on time-dependent domains. Proc. Lond. Math. Soc. **119**, 998–1064 (2019)
29. A. Shao, B. Vergara, Approximate boundary controllability for parabolic equations with inverse square infinite potential wells. Preprint (2023). arXiv:2311.01628
30. D. Tataru, A-priori estimates of Carleman's type in domains with boundaries. J. Math. Pures Appl. **73**, 355–387 (1994)
31. J. Vancostenoble, E. Zuazua, Null controllability for the heat equation with singular inverse-square potential. J. Funct. Anal. **254**, 1864–1902 (2008)
32. C.M. Warnick, The massive wave equation in asymptotically AdS spacetimes. Commun. Math. Phys. **321**, 85–111 (2013)
33. E. Zuazua, *Controllability of Partial Differential Equations*, 3rd cycle (Castro Urdiales, Espagne, 2006)

On Geometric Estimates for Some Problems Arising from Modeling Pull-in Voltage in MEMS

Durvudkhan Suragan and Dongming Wei

Abstract In this paper, we prove that the pull-in voltage of the multidimensional MEMS (micro-electro mechanical systems) problem on the whole space $\mathbb{R}^d$, $d \geq 3$, is minimized by symmetrizing the permittivity profile. The proof relies on Talenti's comparison principle.

Keywords MEMS problem · Pull-in voltage · Newton potential · Talenti's comparison principle · Geometric estimate

2000 Mathematics Subject Classification 47J10, 35J60

1 Introduction

Let us recall the second order differential equation with the singular nonlinearity modeling stationary MEMS (micro-electro mechanical systems):

$$\begin{cases} -\Delta u(x) = \lambda \frac{f(x)}{(1-u(x))^2}, \; 0 \leq u(x) < 1, \; x \in \Omega \subset \mathbb{R}^d, d \geq 1, \\ u(x) = 0, \;\; x \in \partial\Omega. \end{cases} \tag{1}$$

Here f describes the varying permittivity profile of the elastic membrane with $f \in C^\alpha(\bar{\Omega})$ for some $\alpha \in (0, 1]$, $0 \leq f \leq 1$, and $f \not\equiv 0$. The Dirichlet pull-in voltage is defined as

$$\lambda^*(\Omega, f) = \sup\{\lambda > 0 \mid (1.1) \text{ possesses at least one classical solution}\}.$$

For the Dirichlet pull-in voltage the following inequality holds:

D. Suragan (✉) · D. Wei
Department of Mathematics, Nazarbayev University, Astana, Kazakhstan
e-mail: durvudkhan.suragan@nu.edu.kz; dongming.wei@nu.edu.kz

D. Cardona et al. (eds.), *Extended Abstracts 2021/2022*, Research Perspectives Ghent Analysis and PDE Center 3, https://doi.org/10.1007/978-3-031-48579-4_19

Theorem 1 (Proposition 2.2.1 [1]) *A ball B (with the symmetrized permittivity profile) is a minimizer of the Dirichlet pull-in voltage among all domains of given volume, i.e.,*

$$\lambda^*(B, f^*) \leq \lambda^*(\Omega, f)$$

for an arbitrary domain $\Omega \subset \mathbb{R}^d$ with $|\Omega| = |B|$, where $|\cdot|$ is the Lebesgue measure in $\mathbb{R}^d$, $d \geq 1$. Here f^* is the symmetric decreasing rearrangement of f.

In the present paper, we consider a similar stationary MEMS problem, but in the infinity domain, that is, on the whole space $\mathbb{R}^d$:

$$\begin{cases} -\Delta u(x) = \lambda \frac{f}{(1-u(x))^2}, & 0 \leq u(x) < 1, \quad x \in \mathbb{R}^d,\ d \geq 3, \\ u(x) \longrightarrow 0, & |x| \longrightarrow \infty, \end{cases} \tag{2}$$

where $\lambda > 0$ and $f = 1$ with $\operatorname{supp} f \subset \Omega \subset \mathbb{R}^d$.

To analyse the main difference between the problems (1) and (2), let us briefly discuss linear analogues of these problems. A linear analogue of Theorem 1 is so called the Rayleigh-Faber-Krahn inequality. To recall it let us consider the minimization problem of the first eigenvalue of the Laplacian with the Dirichlet boundary condition (among domains of a given volume):

$$\begin{cases} -\Delta u(x) = \lambda^D u(x), \ x \in \Omega \subset \mathbb{R}^d, \\ u(x) = 0, \ x \in \partial\Omega. \end{cases} \tag{3}$$

The famous Rayleigh-Faber-Krahn inequality asserts that

$$\lambda_1^D(B) \leq \lambda_1^D(\Omega),$$

for any Ω with $|\Omega| = |B|$, where $B \subset \mathbb{R}^d$ is a ball and $|\cdot|$ is the Lebesgue measure in $\mathbb{R}^d$. Note that an analogue of the Rayleigh-Faber-Krahn inequality for general convolution type integral operators were given in [5] (see also [6]).

Similarly, we can consider a linear version of the problem (2):

$$-\Delta u(x) = \mu u(x), \quad x \in \Omega \subset \mathbb{R}^d, \tag{4}$$

with the nonlocal integral boundary condition

$$-\frac{1}{2}u(x) + \int_{\partial\Omega} \frac{\partial \varepsilon_d(x-y)}{\partial n_y} u(y) dS_y - \int_{\partial\Omega} \varepsilon_d(x-y) \frac{\partial u(y)}{\partial n_y} dS_y = 0, \ x \in \partial\Omega, \tag{5}$$

where ε_d is the fundamental solution of the Laplacian and $\frac{\partial}{\partial n_y}$ denotes the outer normal derivative at a point y on the boundary $\partial\Omega$. The spectral problem (4)–(5) is equivalent (see [3]) to

$$u(x) = \mu \int_{\Omega} \varepsilon_d(x - y)u(y)dy,\ x \in \Omega \subset \mathbb{R}^d. \tag{6}$$

This also means

Lemma 2 *[3] The problem (2) is equivalent to the nonlinear integral problem*

$$u(x) = \lambda \int_{\Omega} \varepsilon_d(x - y)\frac{1}{(1 - u(y))^2}dy,\ 0 \le u(x) < 1. \tag{7}$$

In Sect. 2 we briefly discuss some preliminary results, in particular, we recall the celebrated Talenti comparison principle [7], which states that the symmetric decreasing rearrangement (Schwarz rearrangement) of the Newtonian potential of a charge distribution is pointwise smaller than the potential resulting from symmetrizing the charge distribution itself. Main results of this paper and their proofs will be given in Sect. 3. Talenti's comparison principle plays a key role in the proofs.

2 Preliminaries

Let Ω be a measurable bounded domain of $\mathbb{R}^d$. An open ball (with origin 0) Ω^* is called a symmetric rearrangement of Ω if $|B| = |\Omega|$ and

$$\Omega^* = B = \left\{x \in \mathbb{R}^d \mid \sigma_d |x|^d < |\Omega|\right\},$$

where $\sigma_d = \frac{2\pi^{\frac{d}{2}}}{\Gamma(\frac{d}{2})}$ is the surface area of the unit ball in $\mathbb{R}^d$. Let u be a nonnegative measurable function vanishing at infinity in the sense that all of its positive level sets have a finite measure, i.e.,

$$\mathrm{Vol}\,(\{x|u(x) > t\}) < \infty, \quad \forall t > 0.$$

To define a symmetric decreasing rearrangement of u one uses (see, for example [4]) the layer-cake decomposition, which expresses a nonnegative function u in terms of its level sets in the following way

$$u(x) = \int_0^{\infty} \chi_{\{u(x)>t\}} dt,$$

where χ is the characteristic function. *Let u be a nonnegative measurable function vanishing at infinity. Then*

$$u^*(x) = \int_0^\infty \chi_{\{u(x)>t\}^*} dt \tag{8}$$

is called a symmetric decreasing rearrangement of the function u. Note that the symmetric decreasing rearrangement is also sometimes called the Schwarz rearrangement. The simple definition (8) can be useful in many proofs, for example, if $0 \le v(x) - u(x), \ \forall x \in \mathbb{R}^d$, then we directly get

$$u^*(x) = \int_0^\infty \chi_{\{u(x)>t\}^*} dt \le \int_0^\infty \chi_{\{v(x)>t\}^*} dt = v^*(x), \quad \forall x \in \mathbb{R}^d.$$

That is, if

$$0 \le u(x) \le v(x), \quad \forall x \in \mathbb{R}^d,$$

then

$$0 \le v^*(x) - u^*(x), \quad \forall x \in \mathbb{R}^d.$$

Theorem 3 (Talenti's Comparison Principle for the Laplacian) *Consider a (smooth) nonnegative function f with supp$f \subset \Omega \subset \mathbb{R}^d$, $d \ge 3$, for a bounded set Ω, and its symmetric decreasing rearrangement f^*. If solutions u and v of*

$$-\Delta u = f, \quad -\Delta v = f^*,$$

vanish at infinity, then

$$u^*(x) \le v(x), \quad \forall x \in \mathbb{R}^d.$$

Note that u and v exist, and are uniquely determined by the equation, i.e.

$$u(x) = \int_\Omega \varepsilon_d(x-y) f(y) dy$$

and

$$v(x) = \int_B \varepsilon_d(x-y) f^*(y) dy,$$

where $\varepsilon_d(\cdot)$ is the fundamental solution of the Laplacian, that is,

$$\varepsilon_d(x-y) = \frac{1}{(d-2)\sigma_d |x-y|^{d-2}}, \quad d \ge 3,$$

and σ_d is the surface area of d-dimensional unit ball. They are nonnegative since the fundamental solution is nonnegative. The inequality also holds for nonnegative measurable functions f vanishing when $|x| \to \infty$. It is also known that the fundamental solution σ_d, $d \geq 3$, does not change its formula under the symmetric decreasing rearrangement, see e.g. Lieb and Loss [4].

3 The Pull-in Voltage for the Newtonian Potential

We consider the pull-in voltage for the stationary deflection of an infinity elastic membrane satisfying

$$\begin{cases} -\Delta u(x) = \lambda \frac{f(x)}{(1-u(x))^2}, & 0 \leq u(x) < 1, \quad x \in \mathbb{R}^d,\ d \geq 3, \\ u(x) \longrightarrow 0, & |x| \longrightarrow \infty, \end{cases} \tag{9}$$

where $\lambda > 0$ is the applied voltage and the permittivity profile f is a constant with finite support, that is, $f = 1$ in Ω with $\operatorname{supp} f \subset \Omega \subset \mathbb{R}^d$.

As usual, the pull-in voltage is defined as

$$\lambda^*(\Omega) = \sup\{\lambda > 0 \mid (3.1)\,\text{possesses at least one classical solution}\}.$$

Theorem 4 *There exists a positive pull-in voltage $\lambda^* < \infty$ such that*

(a) For any $\lambda < \lambda^*$, there exists at least one solution of (9).
(b) For any $\lambda > \lambda^*$, there is no solution of (9).

Proof of Theorem 4 By using Lemma 2, problem (9) is equivalent to the nonlinear integral problem (7). Thus, since (7) has the trivial solution $u = 0$ with $\lambda = 0$, by the implicit function theorem (7) has a solution. In addition, since the fundamental solution ε_d is positive, the integral on the right hand sight of (7) is positive. This means that λ must be positive, that is, $0 < \lambda < \lambda^*$. Now we need to show that $\lambda^* < \infty$. Let $0 \leq u(x) < 1$ be a solution of (7). We also use the following known fact: The first eigenvalue μ_1 of the spectral problem

$$\phi_1(x) = \mu_1 \int_\Omega \varepsilon_d(x-y)\phi_1(y)dy \tag{10}$$

is simple and positive as well as the corresponding eigenfunction ϕ_1 can be chosen positive. Thus, let us multiply (7) by ϕ_1 and integrate over Ω, then we have

$$\int_\Omega u(x)\phi_1(x)dx = \lambda \int_\Omega \int_\Omega \varepsilon_d(x-y)\frac{1}{(1-u(y))^2}dy\phi_1(x)dx, \ 0 \leq u(x) < 1.$$

By (10) we obtain

$$\int_{\Omega} u(x)\phi_1(x)dx = \frac{\lambda}{\mu_1}\int_{\Omega} \frac{\phi_1(y)}{(1-u(y))^2}dy,\ 0 \le u(x) < 1,$$

that is,

$$\lambda = \frac{\mu_1 \int_{\Omega} u(x)\phi_1(x)dx}{\int_{\Omega} \frac{\phi_1(y)}{(1-u(y))^2}dy} \le \frac{\mu_1 \int_{\Omega} \phi_1(x)dx}{\int_{\Omega} \phi_1(y)dy}.$$

This means

$$\lambda^* \le \frac{\mu_1 \int_{\Omega} \phi_1(x)dx}{\int_{\Omega} \phi_1(y)dy} < \infty, \tag{11}$$

and there is no solution of (7) for any $\lambda > \lambda^*$. By the definition of λ^* for any $\lambda \in (0, \lambda^*)$ there exists $\tilde{\lambda} \in (\lambda, \lambda^*)$ for which (7) has a solution $u_{\tilde{\lambda}}$, that is,

$$u_{\tilde{\lambda}}(x) = \tilde{\lambda}\int_{\Omega} \varepsilon_d(x-y)\frac{1}{(1-u_{\tilde{\lambda}}(y))^2}dy \ge \lambda \int_{\Omega} \varepsilon_d(x-y)\frac{1}{(1-u_{\tilde{\lambda}}(y))^2}dy, \tag{12}$$

This also means that $u_{\tilde{\lambda}}$ is a supsolution of (12) for the parameter λ. On the other hand, since

$$0 \le \lambda \int_{\Omega} \varepsilon_d(x-y)dy, \tag{13}$$

$u \equiv 0$ is a subsolution of

$$u_{\tilde{\lambda}}(x) \le \lambda \int_{\Omega} \varepsilon_d(x-y)\frac{1}{(1-u_{\tilde{\lambda}}(y))^2}dy. \tag{14}$$

Therefore, by the method of sub- and supsolutions (see the proof of [1, Theorem 2.1.1]) we prove existence of a solution u_λ of (7) for any $\lambda \in (0, \lambda^*)$. □

Now we are ready to prove the following result.

Theorem 5 *We have*

$$\lambda^*(f^*) \le \lambda^*(f)$$

for the constant permittivity profile $f = 1$ in a smooth bounded domain Ω satisfying the assumption supp $f \subset \Omega \subset \mathbb{R}^d$, $d \ge 3$.

Proof of Theorem 5 Let u be any positive solution of (7). Define the sequence (the Picard iteration scheme)

$$u_m(x) = \lambda \int_{\mathbb{R}^d} \varepsilon_d(x-y)\frac{f(y)}{(1-u_{m-1}(y))^2}dy,\ u_0(x) \equiv 0,\ m = 1, 2, \ldots, \tag{15}$$

with $f = 1$ and $\operatorname{supp} f \subset \Omega$. We have $u > u_0 \equiv 0$ and whenever $u \geq u_{m-1}$, then

$$u(x) - u_m(x) = \lambda \int_{\mathbb{R}^d} \varepsilon_d(x-y) f(y) \left(\frac{1}{(1-u(y))^2} - \frac{1}{(1-u_{m-1}(y))^2}\right) dy \geq 0,$$

for all $x \in \mathbb{R}^d$, that is, $1 > u \geq u_m$ in $\mathbb{R}^d$ for each $m \geq 0$. Moreover, from (15) it is straightforward to see that the sequence $\{u_m\}$ is monotone increasing. Thus, it converges uniformly to a positive solution u_λ satisfying $u \geq u_\lambda$ in $\mathbb{R}^d$. Consider the following two sequences

$$u_n(x) = \lambda \int_{\mathbb{R}^d} \varepsilon_d(x-y)\frac{f(y)}{(1-u_{n-1}(y))^2}dy,\ u_0(x) \equiv 0,\ n = 1, 2, \ldots, \tag{16}$$

with $\operatorname{supp} f \subset \Omega$, and

$$v_n(x) = \lambda \int_{\mathbb{R}^d} \varepsilon_d(x-y)\frac{f^*(|y|)}{(1-v_{n-1}(y))^2}dy,\ v_0(x) \equiv 0,\ n = 1, 2, \ldots. \tag{17}$$

We have

$$-\Delta u_1 = \lambda f, \quad -\Delta v_1 = \lambda f^*,$$

therefore, by Talenti's comparison principle for the Laplacian (see Theorem 3) we obtain

$$u_1^*(r) \leq v_1(r), \quad \forall r \in [0, \infty). \tag{18}$$

Note that here we have radial v_1 since f^* is radial. We also have

$$-\Delta u_2 = \lambda \frac{f}{(1-u_1)^2} \tag{19}$$

and

$$-\Delta v_2 = \lambda \frac{f^*}{(1-v_1)^2}. \tag{20}$$

Furthermore, by induction as in the proof of [1, Proposition 2.2.1], for all n we get

$$u_n^*(r) \leq v_n(r), \quad \forall r \in [0, \infty). \tag{21}$$

Since $\max_B u_n^* = \max_\Omega u_n$, it means that for a given λ if $\{v_n\}$ converges, then the sequence $\{u_n\}$ is also convergent. Theorem 5 is proved. □

We have the following upper bound for the pull-in voltage (for the non-constant permittivity profile):

Theorem 6 *Let f be an integrable function with supp $f \subset \Omega \subset \mathbb{R}^d$, $d \geq 3$. Let $\mu_1(\Omega)$ be the first eigenvalue of the Newtonian potential* (10) *in Ω. Then*

$$\lambda^*(f) \leq \frac{4\mu_1(\Omega)}{27}(\inf_\Omega f)^{-1}. \tag{22}$$

Proof of Proposition 22 As in the proof of Theorem 4, for any $\lambda \in (0, \lambda^*)$ we have

$$\int_\Omega u(x)\phi_1(x)dx = \frac{\lambda}{\mu_1}\int_\Omega \frac{\phi_1(y)f(y)}{(1-u(y))^2}dy, \ 0 \leq u < 1.$$

Since $u(1-u)^2 \leq \frac{4}{27}$ we obtain

$$\int_\Omega u(x)\phi_1(x)dx = \frac{\lambda}{\mu_1}\int_\Omega \frac{u(y)\phi_1(y)f(y)}{u(y)(1-u(y))^2}dy \geq \frac{27\lambda \inf_\Omega f}{4\mu_1}\int_\Omega u(y)\phi_1(y)dy$$

proving the inequality (22). □

Note that, moreover, one can prove other upper estimates of the pull-in voltage that depends on the global properties of the (non-constant) permittivity profile. For instance, for the Dirichlet case (see, e.g. [2]) we have the estimate

$$(\lambda^D)^*(f) \leq \frac{4\mu_1^D(\Omega)}{3}\frac{\int_\Omega \phi_1^D(x)dx}{\int_\Omega \phi_1^D(y)f(y)dy}, \tag{23}$$

where μ_1^D and ϕ_1^D are the first eigenvalue and the first eigenfunction of the Dirichlet Laplacian, respectively.

Acknowledgments This research was funded by the Committee of Science of the Ministry of Science and Higher Education of Kazakhstan (Grant number BR21882172). The first author was also funded by NU CRP 20122022CRP1601. The second author was funded by NU CRP 20122022FD4126.

References

1. P. Esposito, N. Ghoussoub, Y. Guo, *Mathematical Analysis of Partial Differential Equations Modeling Electrostatic MEMS*. Courant Lecture Notes in Mathematics, vol. 20 (American Mathematical Society, Providence, 2009)
2. Y. Guo, Z. Pan, M. J. Ward, Touchdown and pull-in voltage behavior of a MEMS device with varying dielectric properties. SIAM J. Appl. Math. **66**, 309–338 (2005)
3. T.S. Kal'menov, D. Suragan, On spectral problems for the volume potential. Dokl. Math. **428**(1), 16–19 (2009)
4. E.H. Lieb, M. Loss, *Analysis*. Graduate Studies in Mathematics, vol. 14, 2nd edn. (American Mathematical Society, Providence, 2001)
5. M. Ruzhansky, D. Suragan, Communications of the Moscow mathematical society: on Schatten norm for convolution-type integral operator. Russ. Math. Surv. **71**, 157–158 (2016)
6. M. Ruzhansky, D. Suragan, On first and second eigenvalues of Riesz transforms in spherical and hyperbolic geometries. Bull. Math. Sci. **6**(2), 325–334 (2016)
7. G. Talenti, Elliptic equations and rearrangements. Ann. Sc. Norm. Super. Pisa Ñ Cl. Sci. **3**, 697–718 (1976)

A Note on Fractional Powers of the Hermite Operator

Sundaram Thangavelu

Abstract We give a very short proof of a result proved by Cappiello-Rodino-Toft on the Weyl symbol of the inverse of the Harmonic oscillator. We also extend their results to fractional powers.

Keywords Hermite operator · Fractional powers · Weyl transform · Pseudo-differential operators

2010 Mathematics Subject Classification Primary: 35Q40, 35S05, 46F05; Secondary: 33C10, 30G

1 Introduction

As is well known the spectrum of the operator H consists of $(2k+n), k \in \mathbb{N}$ and hence it is invertible. The formal inverse can be written in terms of the spectral theorem by

$$H^{-1} = \sum_{k=0}^{\infty} (2k+n)^{-1} P_k$$

This work is supported by the J. C. Bose Fellowship of the author from the Department of Science and Technology, Government of India.

S. Thangavelu (✉)
Department of Mathematics, Indian Institute of Science, Bangalore, India
e-mail: veluma@iisc.ac.in

D. Cardona et al. (eds.), *Extended Abstracts 2021/2022*, Research Perspectives Ghent Analysis and PDE Center 3, https://doi.org/10.1007/978-3-031-48579-4_20

where P_k are the orthogonal projections associated to the eigenspaces corresponding to the eigenvalues $(2k+n)$. However, it is known that H^{-1} is a pseudo-differential operator with a symbol $b(x,\xi)$ in the Weyl calculus. Thus

$$H^{-1}\varphi(\xi) = (2\pi)^{-n}\int_{\mathbb{R}^n}\int_{\mathbb{R}^n} e^{i(\xi-\eta)\cdot y} b(\frac{\xi+\eta}{2}, y)\varphi(\eta)dyd\eta$$

for $\varphi \in L^2(\mathbb{R}^n)$. In [1] the authors have obtained the following explicit expression for the symbol $b(x,\xi)$ when the dimension n is even.

Theorem 1 (Cappiello-Rodino-Toft) *Let $b_{2n}(x,\xi)$ stand for the Weyl symbol of H^{-1} on $\mathbb{R}^{2n}$. Then one has the explicit formula*

$$b_{2n}(x,\xi) = \sum_{j=0}^{n-1} \frac{(n+j-1)!}{(n-1)!j!}(-1)^j(2j)!\frac{1-p_{2j}(|x|^2+|\xi|^2)e^{-(|x|^2+|\xi|^2)}}{(|x|^2+|\xi|^2)^{2j+1}}$$

where $p_j(t)$ are the Taylor polynomials of the function e^{-t} about $t=0$.

The proof given in [1] is quite long and based on the fact that the symbol b satisfies a partial differential equation. In this note, the above theorem becomes an easy consequence of an integral representation for the symbol b which is based on the formula

$$H^{-1} = \int_0^\infty e^{-tH}dt$$

and the fact that e^{-tH} is a pseudodifferential operator with an explicit symbol. In the same paper [1] the authors have proved the following result giving estimates on the derivatives of the symbol b of H^{-1}.

Theorem 2 *The following estimates on the Weyl symbol $b(x,\xi)$ of the operator H^{-1} are valid: there exists a constant $C>0$ such that for any $\alpha \in \mathbb{N}^{2n}$ and $r \in [0,1]$*

$$|\partial_{x,\xi}^\alpha b(x,\xi)| \le C^{|\alpha|+1}(\alpha!)^{(r+1)/2}(|x|^2+|\xi|^2)^{-1-(r/2)|\alpha|}.$$

In this note we give a short proof of the above theorem. Actually we can consider H^{-s} for any $s>0$ and prove similar estimates for the Weyl symbol b_s of the operator H^{-s}. We will also say something about conformally invariant fractional powers H_{-s} studied in the literature.

2 Fractional Powers H^{-s} of the Hermite Operator

In this section we consider fractional powers of the Hermite operator $H = -\Delta+|x|^2$ on $\mathbb{R}^n$. We first consider the negative powers H^{-s} where $s \geq 0$ which are given in terms of the Hermite semigroup e^{-tH} via the Gamma integral:

$$H^{-s} f(x) = \frac{1}{\Gamma(s)} \int_0^{\infty} e^{-tH} f(x) t^{s-1} dt.$$

The kernel of the semigroup e^{-tH} is explicitly known and is given in terms of the Mehler's formula for the Hermite functions, see [6]. However, we can also write e^{-tH} as the Weyl transform of a function on $\mathbb{C}^n$ which allows us to realise e^{-tH} and hence H^{-s} as a pseudo-differential operator. Recall that the Weyl transform $W(F)$ of a function F on $\mathbb{C}^n$ is defined by

$$W(F)\varphi = \int_{\mathbb{C}^n} F(z)\pi(z)\varphi dz$$

for $\varphi \in L^2(\mathbb{R}^n)$. Here, $\pi(z)$ is the projective representation of $\mathbb{C}^n$ which is closely related to the Schrödinger representations of the Heisenberg group. It is given explicitly by

$$\pi(x+iy)\varphi(\xi) = e^{i(x\cdot\xi+\frac{1}{2}x\cdot y)}\varphi(\xi+y).$$

It turns out that $W(F)$ is an integral operator with kernel

$$K_F(\xi,\eta) = \int_{\mathbb{R}^n} e^{\frac{i}{2}x\cdot(\xi+\eta)} F(x,\eta-\xi) dx$$

where by abuse of notation we have written $F(x, y)$ in place of $F(x+iy)$. If $\tilde{F}(\xi, y)$ stands for the inverse Fourier transform of $F(x, y)$ in the first set of variables, then we have $K_F(\xi,\eta) = \tilde{F}(\frac{\xi+\eta}{2}, \eta-\xi)$. By letting $b(\xi,\eta)$ stand for the full inverse Fourier transform of F in both variables we can write $W(F)$ as

$$W(F)\varphi(\xi) = (2\pi)^{-n} \int_{\mathbb{R}^n}\int_{\mathbb{R}^n} e^{i(\xi-\eta)\cdot y} b(\frac{\xi+\eta}{2}, y)\varphi(\eta) dy d\eta.$$

Thus we see that the Weyl transform $W(F)$ is a pseudo-differential operator in the Weyl calculus with symbol $b(x,\xi)$.

We now make use of the well known fact that $e^{-tH} = W(p_t)$ where

$$p_t(z) = c_n(\sinh t)^{-n} e^{-\frac{1}{4}(\coth t)|z|^2}$$

is the heat kernel associated to the so called special Hermite opereator, see e.g. [6]. In view of the relation between a function F and the Weyl symbol of $W(F)$, we observe that the Weyl symbol of the Hermite semigroup e^{-tH} is given by the function $a_t(x,\xi) = c_n(\cosh t)^{-n} e^{-(\tanh t)(|x|^2+|\xi|^2)}$. As $\Gamma(s)H^{-s} = \int_0^\infty t^{s-1} e^{-tH} dt$ the Weyl symbol of H^{-s} is given by

$$b_s(x,\xi) = \frac{c_n}{\Gamma(s)} \int_0^\infty t^{s-1} (\cosh t)^{-n} e^{-(\tanh t)(|x|^2+|\xi|^2)} dt.$$

By taking $s = 1$ and making a change of variables we see that the Weyl symbol $b(x,\xi)$ of H^{-1} is given by

$$b(x,\xi) = c_n \int_0^1 (1-t^2)^{n/2-1} e^{-t(|x|^2+|\xi|^2)} dt.$$

It is an easy matter to prove Theorem 1.

Proof of Theorem 1 Let b_{2n} stands for the Weyl symbol of H^{-1} on $\mathbb{R}^{2n}$ given by the above expression. Then expanding $(1-t^2)^{n-1}$ and making a change of variables we get

$$b_{2n}(x,\xi) = \sum_{j=0}^{n-1} \frac{(n+j-1)!}{j!(n-1)!} (-1)^j \Big(\int_0^{(|x|^2|+|\xi|^2)} t^{2j} e^{-t} dt \Big) (|x|^2 + |\xi|^2)^{-2j-1}.$$

The proof is completed by showing that $\frac{1}{j!} \int_0^a t^j e^{-t} dt = 1 - e^{-a} p_j(a)$ where p_j are the Taylor polynomials of e^{-t}. But this follows immediately by induction.

In [1] the authors have studied H^{-1} as a pseudo-differential operator. For the Weyl symbol $b(x,\xi)$ of H^{-1} the authors have proved the estimate

$$|\partial_{x,\xi}^\alpha b(x,\xi)| \leq C^{|\alpha|+1} (\alpha!)^{(r+1)/2} (|x|^2 + |\xi|^2)^{-1-(r/2)|\alpha|}$$

for some constant C which is independent of $\alpha \in \mathbb{N}^{2n}$ and $r \in [0,1]$. The proof given in [1] is quite long and uses several results from microlocal analysis. Here we give a very short proof of the same. □

Theorem 3 *For* $0 < s \leq 1$ *we have the following estimates on the Weyl symbol* $b_s(x,\xi)$ *of the operator* H^{-s}*: there exists a constant constant* $C > 0$ *such that for all* $\alpha \in \mathbb{N}^{2n}$ *and* $r \in [0,1]$

$$|\partial_{x,\xi}^\alpha b_s(x,\xi)| \leq C^{|\alpha|+1} (\alpha!)^{(r+1)/2} (|x|^2 + |\xi|^2)^{-s-(r/2)|\alpha|}.$$

Proof We make use of some properties of the Hermite functions on $\mathbb{R}^n$. Recall that Hermite polynomials $H_k(t)$ on the real line are defined by the equation

$$H_k(t) = (-1)^k e^{t^2} \frac{d^k}{dt^k} e^{-t^2}$$

and the normalised Hermite functions are given by $h_k(t) = (2^k k! \sqrt{\pi})^{-1/2} H_k(t) e^{-\frac{1}{2}t^2}$. It is then well known that $h_k(t)$ are bounded functions uniformly in k. The multi-dimensional Hermite functions $H_\alpha(x), x \in \mathbb{R}^n, \alpha \in \mathbb{N}^n$ are defined by taking tensor products. Thus the $2n$-dimensional Hermite polynomials $H_\alpha, \alpha \in \mathbb{N}^{2n}$ are defined by the equation

$$H_\alpha(x, \xi) e^{-(|x|^2+|\xi|^2)} = (-1)^{|\alpha|} \partial_{x,\xi}^\alpha e^{-(|x|^2+|\xi|^2)}.$$

Therefore, from the integral representation for b_s we obtain the relation

$$\partial_{x,\xi}^\alpha b_s(x, \xi) = (-1)^{|\alpha|} \frac{c_n}{\Gamma(s)} \int_0^\infty t^{s-1} \times (\cosh t)^{-n} (\tanh t)^{\frac{1}{2}|\alpha|} H_\alpha((\tanh t)^{1/2}(x, \xi)) e^{-(\tanh t)(|x|^2+|\xi|^2)} dt.$$

We now make use of the fact that the normalised Hermite functions $\Phi_\alpha(x, \xi)$ defined by

$$\Phi_\alpha(x, \xi) = (2^{|\alpha|}(\alpha!)\pi^n)^{-1/2} H_\alpha(x, \xi) e^{-\frac{1}{2}(|x|^2+|\xi|^2)}$$

are uniformly bounded (which follows from the fact that $h_k(t)$ are uniformly bounded). This leads to the estimate

$$|\partial_{x,\xi}^\alpha b_s(x, \xi)| \leq C_n 2^{\frac{1}{2}|\alpha|} (\alpha!)^{1/2} \int_0^\infty t^{s-1} (\cosh t)^{-n} (\tanh t)^{\frac{1}{2}|\alpha|} e^{-\frac{1}{2}(\tanh t)(|x|^2+|\xi|^2)} dt.$$

In order to estimate the integral appearing above, we write it as

$$I = \int_0^\infty \Pi_{j=1}^n t^{(s-1)/n} (\cosh t)^{-1} (\tanh t)^{\frac{1}{2}\alpha_j} e^{-\frac{1}{2n}(\tanh t)(|x|^2+|\xi|^2)} dt.$$

Applying generalised Holder's inequality, we are led to estimating $I \leq \Pi_{j=1}^n I_j^{1/n}$ where

$$I_j = \int_0^\infty t^{s-1} (\cosh t)^{-n} (\tanh t)^{\frac{n}{2}\alpha_j} e^{-\frac{1}{2}(\tanh t)(|x|^2+|\xi|^2)} dt.$$

Assuming $s = 1$ and making a change of variables, we have to estimate the integral

$$J = \int_0^1 (1 - t^2)^{n/2-1} t^{nk/2} e^{-\frac{n}{2}ta^2} dt.$$

Further assuming that $n \geq 2$ we get two kinds of estimates for J. Namely, $J \leq Ca^{-2}$ and $J \leq C\Gamma(1 + (nk)/2)a^{-2-nk}$. These estimates immediately lead to the estimates $I \leq C(|x|^2 + |\xi|^2)^{-1}$ and

$$I \leq C^{|\alpha|}(\alpha!)^{1/2}(|x|^2 + |\xi|^2)^{-1-(1/2)|\alpha|}$$

where we have used Stirling's formula to estimate the Gamma function. Thus we have proved

$$|\partial^\alpha_{x,\xi} b_s(x, \xi)| \leq C^{|\alpha|}(\alpha!)^{1/2}(|x|^2 + |\xi|^2)^{-1}$$

as well as

$$|\partial^\alpha_{x,\xi} b_s(x, \xi)| \leq C^{|\alpha|}(\alpha!)(|x|^2 + |\xi|^2)^{-1-(1/2)|\alpha|}.$$

Interpolation now gives the required estimate when $s = 1$. When $0 < s < 1$, we are led to estimate the integrals

$$\int_0^\infty t^{s-1}(\cosh t)^{-n}(\tanh t)^{\frac{1}{2}|\alpha|} e^{-\frac{1}{2}(\tanh t)(|x|^2+|\xi|^2)} dt.$$

As $\tanh t$ behaves like t for t small and is dominated by t for $t \geq 1$ and since $s - 1 < 0$ we can bound the above integral by

$$\int_0^\infty (\tanh t)^{s-1}(\cosh t)^{-n}(\tanh t)^{\frac{1}{2}|\alpha|} e^{-\frac{1}{2}(\tanh t)(|x|^2+|\xi|^2)} dt.$$

This can be estimated as before yielding the required estimate.

□

3 More on Fractional Powers of the Hermite Operator

As noted elsewhere, it is sometimes more convenient to use a variant of the fractional power. At least in the case of the sublaplacian $\mathcal{L}$ on the Heisenberg group $\mathbb{H}^n$, it has turned out to be more natural and fruitful to use the conformally invariant fractional power $\mathcal{L}_s$ instead of the pure fractional power $\mathcal{L}^s$, see [5] for the definition. For the case of the Hermite operator it amounts to replace H^s by the operator defined by

$$H_s\varphi = \sum_{k=0}^{\infty} \frac{\Gamma(\frac{2k+n+1+s}{2})}{\Gamma(\frac{2k+n+1-s}{2})} P_k\varphi$$

where P_k are the spectral projections associated to H. In view of Stirling's formula for the Gamma function, it follows that H_s differs from the pure power H^s by a bounded operator U_s. Indeed, if we let

$$U_s\varphi = \sum_{k=0}^{\infty} \frac{\Gamma(\frac{2k+n+1+s}{2})}{\Gamma(\frac{2k+n+1-s}{2})} (2k+n)^{-s} P_k\varphi$$

then clearly, U_s is bounded on $L^2(\mathbb{R}^n)$ and $H_s = U_s H^s$. We also note that $H_s^{-1} = H_{-s}$. Using the connection between $\mathcal{L}_s$ and H_s we can obtain an explicit formula for the Weyl symbol of H_s^{-1}.

We make use of several known facts: first of all we recall (see [6]) that $P_k = (2\pi)^{-n} W(\varphi_k)$ where $\varphi_k(z) = L_k^{n-1}(\frac{1}{2}|z|^2)e^{-\frac{1}{4}|z|^2}$ are the Laguerre functions of type $(n-1)$ on $\mathbb{C}^n$. Here $L_k^\alpha(r)$ are Laguerre polynomials of type α. Thus if we let

$$F_s(z) = (2\pi)^{-n} \sum_{k=0}^{\infty} \frac{\Gamma(\frac{2k+n+1-s}{2})}{\Gamma(\frac{2k+n+1+s}{2})} \varphi_k(z)$$

then it follows that $H_s^{-1} = W(F_s)$. The function F_s is known explicitly. To see this, let $\varphi_k^\alpha(r) = L_k^\alpha(r^2)e^{-\frac{1}{2}r^2}$ be Laguerre functions of type α. Let $K_\nu(r)$ stands for the Macdonald function of type ν defined by the Sommerfeld integral (see [4] p.226)

$$K_\nu(r) = \frac{1}{2}(\frac{r}{2})^\nu \int_0^\infty e^{-(t+\frac{r^2}{4t})} t^{-\nu-1} dt.$$

Then the function $G_{\alpha,\sigma}(r)$ defined by

$$G_{\alpha,\sigma}(r) = \frac{2^{\alpha+\sigma}\Gamma(\frac{\alpha-\sigma}{2})}{\sqrt{\pi}\Gamma(\sigma)} r^{-\alpha-1+\sigma} K_{(\alpha+1-\sigma)/2}(\frac{1}{2}r^2)$$

can be expanded in terms of the functions φ_k^α. In [2] the authors have shown that

$$G_{\alpha,\sigma}(r) = \frac{2}{\Gamma(\alpha+1)} \sum_{k=0}^{\infty} \frac{\Gamma(\frac{2k+\alpha+1+1-\sigma}{2})}{\Gamma(\frac{2k+\alpha+1+1+\sigma}{2})} \varphi_k^\alpha(r).$$

Thus we see that, by choosing $\alpha = n - 1$ and $\sigma = s$, the function F_s is explicitly given by

$$F_s(z) = c_{n,s}|z|^{-n+s} K_{(n-s)/2}(\frac{1}{4}|z|^2)$$

where $c_{n,s}$ is an explicit constant. Finally the Weyl symbol of H_s^{-1} is given by

$$b_s(x, \xi) = (2\pi)^{-n} \int_{\mathbb{R}^{2n}} F_s(u + iv) e^{-i(x \cdot u + \xi \cdot v)} du dv.$$

Theorem 4 *For $0 < s \leq 1$ the Weyl symbol of H_s^{-1} is given explicitly by*

$$b_s(x, \xi) = c_{n,s} \int_0^1 e^{-a(|x|^2 + |\xi|^2)} a^{s-1} (1 - a^2)^{\frac{(n-s-1)}{2}} da.$$

Moreover, the following estimates are valid:

$$|\partial_{x,\xi}^{\alpha} b_s(x, \xi)| \leq C^{|\alpha|+1} (\alpha!)^{(r+1)/2} (|x|^2 + |\xi|^2)^{-s-(r/2)|\alpha|}$$

for some constant C which is independent of $\alpha \in \mathbb{N}^{2n}$ and $r \in [0, 1]$.

Proof In order to get the integral representation for $b_s(x, \xi)$ we make use of the Poisson integral representation of K_ν: (see [4], p.223)

$$K_\nu(r) = \frac{\sqrt{\pi}}{\sqrt{2}\Gamma(\nu + 1/2)} r^{-1/2} e^{-r} \int_0^\infty e^{-t} t^{\nu-1/2} (1 + t/(2r))^{\nu-1/2} dt.$$

Recalling the formula for $F_s(z)$ in terms of $K_{(n-s)/2}(\frac{1}{4}|z|^2)$ and using the fact that the Fourier transform of $e^{-t|z|^2}$ is a constant multiple of $t^{-n} e^{-\frac{1}{4t}|z|^2}$ we see that, after a change of variables,

$$b_s(x, \xi) = c_{n,s} \int_0^1 e^{-a(|x|^2 + |\xi|^2)} a^{s-1} (1 - a^2)^{\frac{(n-s-1)}{2}} da.$$

We observe that the above expression coincides with the formula we got for b_1 earlier. Estimating derivatives of b_s is done as in the case of $s = 1$. We leave the details to the reader.

□

Remark 5 The integral representation for b_s can also be obtained easily by making use of the numerical identity (see [3, p. 382, 3.541.1])

$$\int_0^\infty e^{-\mu t} \sinh^\nu \beta t \, dt = \frac{1}{2^{\nu+1}} \frac{\Gamma(\frac{\mu}{2\beta} - \frac{\nu}{2}) \Gamma(\nu + 1)}{\Gamma(\frac{\mu}{2\beta} + \frac{\nu}{2} + 1)},$$

which is valid for $\operatorname{Re}\beta > 0$, $\operatorname{Re}\nu > -1$, $\operatorname{Re}\mu > \operatorname{Re}\beta\nu$. The proof given above has the added advantage that the Fourier transform of b_s is given explicitly. Indeed, we have

$$\int_{\mathbb{R}^{2n}} b_s(x,\xi)e^{-i(y\cdot x+\xi\cdot\eta)}dxd\xi = C_{n,s}(|y|^2+|\eta|^2)^{-(n-s)/2}K_{(n-s)/2}(\frac{1}{4}(|y|^2+|\eta|^2)).$$

Since K_ν is a linear combination of the modified Bessel functions I_ν and $I_{-\nu}$, (see [4], p.224), the above is an explicit formula for the Fourier transform of b_s.

References

1. M. Cappiello, L. Rodino, J. Toft, On the inverse to the harmonic oscillator. Commun. Partial Differ. Equ. **40**, 1096–1118 (2015)
2. Ó. Ciaurri, L. Roncal, S. Thangavelu, Hardy-type inequalities for fractional powers of the Dunkl-Hermite operator. Proc. Edinb. Math. Soc. **61**, 513–544 (2018)
3. I.S. Gradshteyn, I.M. Ryzhik, *Table of Integrals, Series and Products*, 7th edn. (Elsevier/Academic Press, Amsterdam, 2007)
4. A.F. Nikiforov, V.B. Uvarov, *Special Functions of Mathematical Physics* (Birkhauser, Basel, 1988)
5. L. Roncal, S. Thangavelu, Hardy's inequality for fractional powers of the sublaplacian on the Heisenberg group. Adv. Math. **302**, 106–158 (2016)
6. S. Thangavelu, *Lectures on Hermite and Laguerre expansions*. Mathematical Notes, vol. 42 (Princeton University Press, Princeton, 1993)

Non-Standard Version of Egorov Algebra of Generalized Functions

Todor D. Todorov

Abstract We consider a non-standard version of Egorov's algebra of generalized functions, with improved properties of the generalized scalars and the embedding of Schwartz distributions compared with the original standard Egorov's version. The embedding of distributions is similar to, but different from author's works in the past and independently done by Hans Vernaeve.

Keywords Schwartz distributions · Generalized functions · Colombeau algebra · Egorov algebra · Multiplication of distributions · Sheaf of differential functional spaces · Partial differential equations · Non-standard analysis · Infinitesimals · Non-standard real numbers · Non-standard complex numbers · Transfer principle · Saturation principle · Underflow principle

2020 Mathematics Subject Classification Primary: 46F30; Secondary: 46F05, 46F10, 46S10, 46S20, 03H05, 03C50 35D05, 35D10, 35R05, 54B40

1 Introduction

Egorov's article (Egorov [2]) on algebra $\mathcal{G}(\Omega)$ on generalized functions was published relatively soon after the arrival of Colombeau theory of generalized functions (Colombeau [1]) and from the very beginning it was treated from the mathematical community in close comparison with Colombeau theory. One striking difference in this comparison is the *simplicity of Egorov construction*: Unlike in Colombeau construction, all representatives of the generalized functions are *moderate* and the *ideal is relatively simple* (even *trivial* in the case of generalized scalars). This simplicity of Egorov's construction is particularly advantageous when one is trying to define *composition between generalized functions* or *generalized*

T. D. Todorov (✉)
Professor Emeritus of California Polytechnic State University, San Luis Obispo, CA, USA
e-mail: ttodorov@calpoly.edu; https://math.calpoly.edu/todor-todorov

D. Cardona et al. (eds.), *Extended Abstracts 2021/2022*, Research Perspectives Ghent Analysis and PDE Center 3, https://doi.org/10.1007/978-3-031-48579-4_21

functions on a manifold. In addition to the general theory, numerous interesting applications of the theory to partial differential equations appear in (Egorov [2]). In spite of all of these, Egorov's theory was mostly ignored from mathematical community dealing with non-linear theories of generalized functions (standard and non-standard alike) - for one and one reason only: the embedding of Schwartz distribution into Egorov algebra $\mathcal{G}(\Omega)$ is not of *Colombeau type* in the sense that the product on $\mathcal{G}(\Omega)$, if restricted to smooth functions in $\mathcal{E}(\Omega)$, reduces to classical product only for constant functions. We summarize this in $\mathbb{C} \subset \mathcal{D}'(\Omega) \subset \mathcal{G}(\Omega)$ (compared with $\mathcal{E}(\Omega) \subset \mathcal{D}'(\Omega) \subset \mathcal{G}(\Omega)$ in Colombeau theory). The purpose of this article is to improve the properties of the generalized scalars and the embedding of the distributions as much as possible, while preserving the rest of the attractiveness features of Egorov approach including its simplicity.

The non-standard version of Egorov algebra $\widehat{{}^*\mathcal{E}}(\Omega)$ of generalized functions had been studied in the past under the notation $\mathcal{A}(\Omega)$ in (Todorov [11], p. 680–684) and under the notation ${}^*\mathcal{C}^\infty(\Omega)|\text{ns}({}^*\Omega)$ in (Vernaeve [13]). For convenience of the reader we give an independent presentation in Sect. 3. One reason to involve non-standard analysis into non-linear theory of generalized functions (Egorov or Colombeau theories alike) is to improve the properties of generalized scalars (defined as generalized functions with zero gradient). The sets of generalized scalars in both Egorov and Colombeau algebras are rings with zero divisors. In contrast, the sets of scalars, ${}^*\mathbb{R}$ and ${}^*\mathbb{C}$, of the algebra $\widehat{{}^*\mathcal{E}}(\Omega)$ are fields, real closed and algebraically closed, respectively.

In Sect. 4 we discuss the existence of a particular non-standard delta-function in the space ${}^*\mathcal{D}(\mathbb{R}^d)$, slightly modifying some results in Todorov [9, 10]. Here ${}^*\mathcal{D}(\mathbb{R}^d)$ stands for the non-standard extension of the space of test functions $\mathcal{D}(\mathbb{R}^d)$.

In Sect. 5 we define a particular embedding $\iota_\Omega : \mathcal{D}'(\Omega) \to \widehat{{}^*\mathcal{E}}(\Omega)$ of the space of Schwartz distributions. The embedding ι_Ω is similar to, but different the embedding of distributions in (Vernaeve [13]); we make a short comparison below. The properties of ι_Ω are described in Theorem 6, but the differences with the previous works are best visible in Corollary 8. In short, the product in $\widehat{{}^*\mathcal{E}}(\Omega)$ reduces to the classical (pointwise) product on the ring $\mathbb{C}[\Omega]$ of polynomials (not on the whole $\mathcal{E}(\Omega)$) and in a weaker sense on the ring $\mathcal{C}(\Omega)$ of continuous functions. This is an improvement relative to Egorov theory (Egorov [2]), where the product in Egorov's algebra $\mathcal{G}(\Omega)$ reduces to the classical (pointwise) product only on $\mathbb{C}$ (if complex numbers are treated as constant functions in $\mathcal{E}(\Omega)$). We should mention that in both $\mathcal{G}(\Omega)$ and $\widehat{{}^*\mathcal{E}}(\Omega)$ there is one more embedding (in addition to what we discussed above) of $\mathcal{E}(\Omega)$ as a differential subalgebra; in our text it appears under notation σ (see the end of Definition 1 and Corollary 8).

In Sect. 6 we introduce a differential subalgebra $\widehat{\mathcal{R}}_\rho(\Omega)$ of $\widehat{{}^*\mathcal{E}}(\Omega)$, which is similar to, but different from the algebra $\mathcal{G}^\infty(\Omega)$ of regular generalized functions introduced and study in (Oberguggenberger [6, 7]) within (standard) Colombeau theory (Colombeau [1]). The algebra $\widehat{\mathcal{R}}_\rho$ does not contain the counterexample constructed in Vernaeve [15]) and in that way we remove the obstacles to developing a regularity methods in non-standard setting.

We shortly compare our approach based on the algebra $^*\widehat{\mathcal{E}}(\Omega)$ with Hans Vernaeve's work on his algebra $^*\mathcal{E}(\Omega)|\text{ns}(^*\Omega)$ in (Vernaeve [13]). We should mention that Vernaeve translated his theory also in standard setting (Vernaeve [14]).

- The algebra $^*\widehat{\mathcal{E}}(\Omega)$ defined here and Vernaeve's algebra $^*\mathcal{E}(\Omega)|\text{ns}(^*\Omega)$ (Vernaeve [13]) are the same: $^*\widehat{\mathcal{E}}(\Omega) = {}^*\mathcal{E}(\Omega)|\text{ns}(^*\Omega)$ and $\mu(\Omega) = \text{ns}(^*\Omega)$. So, the scalars, $^*\mathbb{C}$ and $^*\mathbb{R}$, are also the same. *The difference is only in the embeddings* of the space of distributions.
- Vernaeve's embedding is of *Colombeau type* in the sense that, the product on $^*\mathcal{E}(\Omega)|\text{ns}(^*\Omega)$ *reduces on* $\mathcal{E}(\Omega)$ *to the usual classical product between smooth functions* - very much like in Colombeau theory (Colombeau [1]) as well as in its non-standard versions (Oberguggenberger and Todorov [8] and Todorov and Vernaeve [12]). Unfortunately, Vernaeve's embedding is defined *for convex open sets* Ω *only*. Hence, the family of spaces of distributions $\{\mathcal{D}'(\Omega)\}_{\Omega\in\mathcal{T}^d}$ fails to be a subsheaf of $\{^*\mathcal{E}(\Omega)|\text{ns}(^*\Omega)\}_{\Omega\in\mathcal{T}^d}$, where $\mathcal{T}^d$ stands for the usual topology on $\mathbb{R}^d$. Consequently, the embedding in Vernaeve's approach does not, in general, preserve the supports of distributions.
- In contrast to the above, our embedding ι_Ω is *not of Colombeau type* - the product in $^*\widehat{\mathcal{E}}(\Omega)$ *generalizes the classical product on the ring of polynomials* $\mathbb{C}[\Omega]$ *only*, not on the whole space $\mathcal{E}(\Omega)$ (and also on the ring of continuous functions $\mathcal{C}(\Omega)$ in a weaker sense - after the restriction of the functions to Ω). However, our embedding is well-defined for *any open set* Ω *of* $\mathbb{R}^d$, the family $\{\mathcal{D}'(\Omega)\}_{\Omega\in\mathcal{T}^d}$ is a subsheaf of $\{^*\widehat{\mathcal{E}}(\Omega)\}_{\Omega\in\mathcal{T}^d}$. Consequently, ι_Ω preserves the support of distributions (very much like in Egorov and Colombeau algebras). Whether the "trade-off" is worth doing, remains to be seen.
- *The question* for defining a *Colombeau type of embedding* of Schwartz's distributions into $^*\widehat{\mathcal{E}}(\Omega)$ or $^*\mathcal{E}(\Omega)|\text{ns}(^*\Omega)$ *for every open set* Ω of $\mathbb{R}^d$, which preserves the support of distributions, *remains open* (with or without the requirement on the generalized scalars to be fields).

2 Notations and Set-Theoretical Framework

- If Ω is an open subset of $\mathbb{R}^d$, we denote by $\mathcal{C}(\Omega)$ the space continuous functions from Ω to $\mathbb{C}$. Similarly, we write $\mathcal{E}(\Omega) = \mathcal{C}^\infty(\Omega)$, $\mathcal{D}(\Omega) = \mathcal{C}_0^\infty(\Omega)$, $\mathcal{L}_{loc}(\Omega)$, $\mathcal{D}'(\Omega)$ and $\mathcal{E}'(\Omega)$ for the popular classes of functions and distributions. The *Schwartz embedding* $\mathcal{S}_\Omega : \mathcal{L}_{loc}(\Omega) \mapsto \mathcal{D}'(\Omega)$ is defined by $\langle \mathcal{S}_\Omega(f), \varphi\rangle = \int_\Omega f(x)\varphi(x)\,dx$ for all $\varphi \in \mathcal{D}(\Omega)$ (Vladimirov [16]). In addition, we let $\mathbb{C}[\Omega] = \mathbb{C}[x_1, \dots, x_d] \restriction \Omega$, where $\mathbb{C}[x_1, \dots, x_d]$ stands for the ring of polynomials in d-many variables with coefficients in $\mathbb{C}$ and $\restriction$ stands for the point-wise *restriction*.
- Our framework is a $\mathfrak{c}_+$-saturated ultrapower non-standard model with the set of individuals $\mathbb{R}$, where $\mathfrak{c} = \text{card}\,\mathbb{R}$. For a presentation of the topic we refer to (Lindstrøm [4], Loeb and Wolff [5]) and/or (the Appendix in Todorov [11]). If S is a set (in the superstructure of $\mathbb{R}$), we write $^*S$ for the non-standard extension of

S. In particular, ${}^*\mathbb{N}$, ${}^*\mathbb{R}$, ${}^*\mathbb{C}$, ${}^*\mathbb{R}^d$, ${}^*\Omega$, ${}^*\mathcal{L}_{loc}(\Omega)$, ${}^*\mathcal{C}(\Omega)$, ${}^*\mathcal{D}(\Omega)$, ${}^*\mathcal{E}(\Omega)$, ${}^*\mathcal{D}'(\Omega)$. etc., are the non-standard extensions of $\mathbb{N}$, $\mathbb{R}$, $\mathbb{C}$, $\mathbb{R}^d$, Ω, $\mathcal{L}_{loc}(\Omega)$, $\mathcal{C}(\Omega)$, $\mathcal{D}(\Omega)$, $\mathcal{E}(\Omega)$, $\mathcal{D}'(\Omega)$. etc., respectively. Recall that ${}^*\mathbb{R}$ is $\mathfrak{c}_+$–saturated *real closed (non-Archimedean) field* of cardinality $\mathfrak{c}_+$, which contains $\mathbb{R}$ as a subfield. Also, ${}^*\mathbb{C}$ is an *algebraically closed field* containing $\mathbb{C}$ as a subfield and we have the usual connection ${}^*\mathbb{C} = {}^*\mathbb{R}(i)$. Notice that ${}^*\mathcal{E}(\Omega)$ and ${}^*\mathcal{D}(\Omega)$ are *differentia algebras* over the field ${}^*\mathbb{C}$. Also, ${}^*\mathbb{C}$ is a differential subring of ${}^*\mathcal{E}(\Omega)$ (if the elements of ${}^*\mathbb{C}$ are treated as *constant functions*). We should mention that the functions in ${}^*\mathcal{E}(\Omega)$ are mapping from ${}^*\Omega$ to ${}^*\mathbb{C}$ (not from Ω to ${}^*\mathbb{C}$) and similarly for the rest of the spaces.

- If $X \subseteq \mathbb{R}^d$, we let $\mu(X) = \{x + dx : x \in X,\ dx \in {}^*\mathbb{R}^d, dx \approx 0\}$ for the set of *near-standard points of* ${}^*X$. Here $dx \approx 0$ means that $||dx||$ is an infinitesimal in ${}^*\mathbb{R}$. If $f \in {}^*\mathcal{E}(\Omega)$, then the following are equivalent: (a) $f \upharpoonright \mu(\Omega) = 0$; (b) $f \upharpoonright {}^*K = 0$ for all $K \Subset \Omega$.

3 Non-Standard Version of Egorov Algebra $\widehat{{}^*\mathcal{E}}(\Omega)$

Definition 1 (Non-Standard Version of Egorov Algebra) Let Ω be an open set of $\mathbb{R}^d$.

1. We let $\widehat{{}^*\mathcal{E}}(\Omega) = {}^*\mathcal{E}(\Omega)/\mathcal{N}(\Omega)$, where $\mathcal{N}(\Omega) = \{f \in {}^*\mathcal{E}(\Omega) : f \upharpoonright \mu(\Omega) = 0\}$.
2. We supply $\widehat{{}^*\mathcal{E}}(\Omega)$ with the operations of a differential algebra over the field ${}^*\mathbb{C}$ with the operations inherited from ${}^*\mathcal{E}(\Omega)$.
3. For every $f \in {}^*\mathcal{E}(\Omega)$ we let $\widehat{f} = f + \mathcal{N}(\Omega)$ or $\widehat{f} = f \upharpoonright \mu(\Omega)$ and refer to $\widehat{f}$ as a *generalized function on* Ω. By exception, we shall write simply c instead of $\widehat{c}$ in the particular case $c \in {}^*\mathbb{C}$ (if c is treated as a constant function in ${}^*\mathcal{E}(\Omega)$). If $\mathcal{S} \subseteq {}^*\mathcal{E}(\Omega)$, we let $\widehat{\mathcal{S}} = \{\widehat{f} : f \in S\}$. In particular, $\widehat{{}^*\mathcal{E}(\Omega)} = \widehat{{}^*\mathcal{E}}(\Omega)$, $\widehat{\mathcal{N}(\Omega)} = \{0\}$, $\widehat{{}^*\mathbb{C}} = {}^*\mathbb{C}$ and $\widehat{{}^*\mathbb{R}} = {}^*\mathbb{R}$.
4. For every $\widehat{f} \in \widehat{\mathcal{E}}(\Omega)$ we define $\widehat{f} : \mu(\Omega) \mapsto {}^*\mathbb{C}$ by $\widehat{f}(\xi) = {}^*f(\xi)$.
5. Let $\mathcal{O}$ be an open subset of Ω and $\widehat{f} \in \widehat{{}^*\mathcal{E}}(\Omega)$. We define the *restriction* $\widehat{f} \upharpoonright \mathcal{O} \in \widehat{{}^*\mathcal{E}}(\mathcal{O})$ by $\widehat{f} \upharpoonright \mathcal{O} = \widehat{f \upharpoonright {}^*\mathcal{O}}$. We say that $\widehat{f}$ *vanishes on* $\mathcal{O}$ if $\widehat{f} \upharpoonright \mathcal{O} = 0$ in $\widehat{{}^*\mathcal{E}}(\mathcal{O})$. The *support* $\mathrm{supp}(\widehat{f})$ of $\widehat{f}$ is the the complement to Ω of the largest open subset of Ω, on which $\widehat{f}$ vanishes.
6. Let X be a Lebesgue measurable subset of $\mathbb{R}^d$ whose closure is a compact subset of Ω. We define a (Lebesgue) *integral of* $\widehat{f} \in \widehat{{}^*\mathcal{E}}(\Omega)$ *over* X with values in ${}^*\mathbb{C}$ by the formula $\int_X \widehat{f}(x)\,dx = \int_{{}^*X} f(\xi)\,d\xi$.
7. We define the *pairing between* $\widehat{{}^*\mathcal{E}}(\Omega)$ and $\mathcal{D}(\Omega)$ by $\langle \widehat{f}, \varphi \rangle = \int_{{}^*\Omega} f(\xi)\, {}^*\varphi(\xi)\,d\xi$ for all $\widehat{f} \in \widehat{{}^*\mathcal{E}}(\Omega)$ and all $\varphi \in \mathcal{D}(\Omega)$, where ${}^*\varphi$ stands for the non-standard extension of φ.
8. We say that two generalized functions $\widehat{f},\ \widehat{g} \in \widehat{{}^*\mathcal{E}}(\Omega)$ are *weakly equal* or *associated*, and write $\widehat{f} \cong \widehat{g}$, if $\langle \widehat{f}, \varphi \rangle = \langle \widehat{g}, \varphi \rangle$ for all $\varphi \in \mathcal{D}(\Omega)$.
9. We define *standard embedding* $\sigma : \mathcal{E}(\Omega) \to \widehat{{}^*\mathcal{E}}(\Omega)$ by $\sigma(f) = \widehat{{}^*f} = {}^*f \upharpoonright \mu(\Omega)$.

Theorem 2 (Basic Properties of $^*\widehat{\mathcal{E}}(\Omega)$)

(i) *$^*\widehat{\mathcal{E}}(\Omega)$ is a differential algebra over the field $^*\mathbb{C}$. Also, the mapping $f + \mathcal{N}(\Omega) \mapsto f \upharpoonright \mu(\Omega)$ from $^*\mathcal{E}(\Omega)/\mathcal{N}(\Omega)$ onto $\{f \upharpoonright \mu(\Omega) : f \in {}^*\mathcal{E}(\Omega)\}$ is a differential algebra isomorphism (justifying the notation $\widehat{f} = f \upharpoonright \mu(\Omega)$ used in advance).*

(ii) $^*\mathbb{C} = \{\widehat{f} \in {}^*\widehat{\mathcal{E}}(\Omega) : \nabla \widehat{f} = 0\}$ *for every open connected subset Ω of $\mathbb{R}^d$.*

(iii) *The family $\{^*\widehat{\mathcal{E}}(\Omega)\}_{\Omega \in \mathcal{T}^d}$, is a sheaf of differential algebras on $\mathbb{R}^d$, where $\mathcal{T}^d$ stands for the usual topology on $\mathbb{R}^d$.*

(iv) *$\sigma[\mathcal{E}(\Omega)]$ is a differential $\mathbb{C}$-subalgebra of $^*\widehat{\mathcal{E}}(\Omega)$, isomorphic of $\mathcal{E}(\Omega)$. Also, $\int_X \sigma(f)(v)\, dx = \int_X f(x)\, dx$ for all $f \in \mathcal{E}(\Omega)$ and all Lebesgue measurable subset X of $\mathbb{R}^d$ with compact closure in Ω. Moreover, $\{\sigma[\mathcal{E}(\Omega)]\}_{\Omega \in \mathcal{T}^d}$ is a subsheaf of $\{^*\widehat{\mathcal{E}}(\Omega)\}_{\Omega \in \mathcal{T}^d}$.*

Proof For the proof we refer to (Todorov [11], §5) or/and (Vernaeve [13]). □

4 Non-Standard Delta-Function

We discuss the existence of a particular non-standard delta-function Δ in the space $^*\mathcal{D}(\mathbb{R}^d)$, the non-standard extension $\mathcal{D}(\mathbb{R}^d)$. In this section we slightly modify similar results in Todorov [9, 10].

Lemma 3 (Non-Standard Delta-Function) *For every $d \in \mathbb{N}$ there exists (not necessarily unique) $\Delta \in {}^*\mathcal{D}(\mathbb{R}^d)$ such that $\Delta(\xi) = 0$ for all infinitely large and for all finite, but non-infinitesimal $\xi \in {}^*\mathbb{R}^d$ and such that $\int_{^*\mathbb{R}^d} \Delta(\xi)\, {}^*\varphi(\xi)\, d\xi = \int_{^*\mathbb{R}^d} \Delta(-\xi)\, {}^*\varphi(\xi)\, d\xi = \varphi(0)$ for all continuous functions $\varphi \in C(\mathbb{R}^d)$. We let $\rho = {}^*\sup\{||\xi|| : \xi \in {}^*\mathbb{R}^d,\ \Delta(\xi) \neq 0\}$ for the radius of support of Δ. Moreover, for each open $\Omega \subseteq \mathbb{R}^d$ and each $\varepsilon \in {}^*\mathbb{R}_+$ we let*

$$\Omega_\varepsilon = \{\xi \in {}^*\Omega : \text{dist}(\xi, \partial\Omega) \geq \varepsilon \ \&\ \text{dist}(\xi, 0) \leq 1/\varepsilon\},$$

and define $\Pi_\Omega : {}^*\mathbb{R}^d \mapsto {}^*\mathbb{C}$ by the formula $\Pi_\Omega(\xi) = \int_{\Omega_{3\rho}} \Delta(\xi - \eta)\, d\eta$.

Theorem 4 (Regularization in $^*\mathcal{E}(\Omega)$) *Let Ω and Δ be as in Lemma 3. Then:*

(i) *$^*f \star \Delta \in {}^*\mathcal{E}(\mathbb{R}^d)$ for every $f \in C(\mathbb{R}^d)$. Here $^*f \star \Delta : {}^*\mathbb{R}^d \to {}^*\mathbb{C}$ is defined by $(^*f \star \Delta)(\xi) = \int_{^*\mathbb{R}^d} {}^*f(\eta)\, \Delta(\xi - \eta)\, d\eta$. Moreover, $^*f \star \Delta$ is an extension of f from $\mathbb{R}^d$ to $^*\mathbb{R}^d$, in symbol, $(^*f \star \Delta) \upharpoonright \mathbb{R}^d = f$.*

(ii) *$^*P \star \Delta = {}^*P$ for all polynomials $P \in \mathbb{C}[\Omega]$.*

(iii) *ρ is a positive infinitesimal in $^*\mathbb{R}$ and $\mu(\Omega) \subseteq \Omega_{2\rho} \subset \Omega_\rho \subseteq {}^*\Omega$.*

(iv) *$\Pi_\Omega \in {}^*\mathcal{D}(\Omega)$, $\Pi_\Omega \upharpoonright \mu(\Omega) = 1$. Moreover, $\text{supp}(\Pi_\Omega) = \Omega_{2\rho}$.*

(v) *$\Pi_\Omega(^*T \star \Delta) \in {}^*\mathcal{D}(\Omega)$ for all $T \in \mathcal{D}'(\Omega)$.*

Proof For (i) and (ii) we refer to (Todorov [10], where the results are based on infinite-dimensional linear algebra and saturation principle in non-standard analysis

(Lindstrøm [4] and/or Loeb and Wolff [5]). The fact that ρ **is an infinitesimal follows** by **underflow principle (Lindstrøm** [4]). For the standard counterpart of the rest we refer to (Vladimirov [16], §4.6). □

5 Embedding of Distributions Into $\widehat{{}^*\mathcal{E}(\Omega)}$

Although the mapping $T \to \Pi_\Omega({}^*T \star \Delta)$ from $\mathcal{D}'(\Omega)$ to ${}^*\mathcal{E}(\Omega)$ (Theorem 4) is injective, it does not commute with the partial derivatives ∂^α in ${}^*\mathcal{E}(\Omega)$. Moreover, the family $\{{}^*\mathcal{E}(\Omega)\}_{\Omega\in\mathcal{T}^d}$ is not a sheaf on $\mathbb{R}^d$. Thus ${}^*\mathcal{E}(\Omega)$ cannot be treated as an algebra of generalized functions on $\mathbb{R}^d$ we are looking for; we return to the algebra $\widehat{{}^*\mathcal{E}(\Omega)}$ **defined in Sect.** 3.

Definition 5 (Embedding of Distributions in $\widehat{{}^*\mathcal{E}(\Omega)}$) Let Ω and Δ be chosen (and fixed) as in Lemma 3. **We define** $\iota_\Omega : \mathcal{D}'(\Omega) \mapsto \widehat{{}^*\mathcal{E}(\Omega)}$ by $\iota_\Omega(T) = \widehat{\Pi_\Omega({}^*T \star \Delta)}$ or equivalently, by $\iota_\Omega(T) = \Pi_\Omega({}^*T \star \Delta) \upharpoonright \mu(\Omega)$. Here ${}^*T \star \Delta : {}^*\Omega \to {}^*\mathbb{C}$ **is defined by** $({}^*T \star \Delta)(\xi) = \langle {}^*T(\eta), \Delta(\xi - \eta)\rangle$ on the ground of transfer principle (Lindstrøm [4] and/or Loeb and Wolff [5]).

Theorem 6 (Properties of the Embedding)

(i) *ι_Ω commutes with the partial derivatives on $\mathcal{D}'(\Omega)$. Moreover, $\langle \iota_\Omega(T), \varphi\rangle = \langle T, \varphi\rangle$ for all $\varphi \in \mathcal{D}(\Omega)$ (Definition 1). Consequently, ι_Ω is injective and $\iota_\Omega[\mathcal{D}'(\Omega)]$ is a differential $\mathbb{C}$-vector subspace of $\widehat{{}^*\mathcal{E}(\Omega)}$.*

(ii) *$(\iota_\Omega \circ S_\Omega)[\mathcal{D}(\Omega)]$, $(\iota_\Omega \circ S_\Omega)[\mathcal{E}(\Omega)]$, $(\iota_\Omega \circ S_\Omega)[\mathcal{C}(\Omega)]$ and $(\iota_\Omega \circ S_\Omega)[\mathcal{L}_{loc}(\Omega)]$ are $\mathbb{C}$-vector subspaces of $\widehat{{}^*\mathcal{E}(\Omega)}$. Moreover, $(\iota_\Omega \circ S_\Omega)[\mathcal{D}(\Omega)]$ and $(\iota_\Omega \circ S_\Omega)[\mathcal{E}(\Omega)]$ are differential $\mathbb{C}$-vector subspaces of $\widehat{{}^*\mathcal{E}(\Omega)}$. Also, we have $(\iota_\Omega \circ S_\Omega)(f) \cong \sigma(f)$ for all $f \in \mathcal{E}(\Omega)$ (Definition 1).*

(iii) *$(\iota_\Omega \circ S_\Omega)(P) = \sigma(P)$ for all polynomials $P \in \mathbb{C}[\Omega]$. Consequently, $(\iota_\Omega \circ S_\Omega)\big[\mathbb{C}[\Omega]\big]$ is a differential subring (a differential $\mathbb{C}$-subalgebra) of $\widehat{{}^*\mathcal{E}(\Omega)}$, which is isomorphic to $\mathbb{C}[\Omega]$. We summarize these in the chain of embeddings: $\mathbb{C}[\Omega] \subset \mathcal{D}'(\Omega) \subset \widehat{{}^*\mathcal{E}(\Omega)}$, after dropping ι_Ω.*

(iv) *The family $\big\{\iota_\Omega[\mathcal{D}'(\Omega)]\big\}_{\Omega\in\mathcal{T}^d}$ is subsheaf of $\{\widehat{{}^*\mathcal{E}(\Omega)}\}_{\Omega\in\mathcal{T}^d}$ on $\mathbb{R}^d$. Consequently,* $\operatorname{supp}(T) = \operatorname{supp}(\iota_\Omega(T))$ *for all $T \in \mathcal{D}'(\Omega)$.*

(v) *Let $f \in \mathcal{C}(\Omega)$ be continuous function. Then $(\iota_\Omega \circ S_\Omega)(f)$ is an extension of f (from Ω to $\mu(\Omega)$, i.e. $(\iota_\Omega \circ S_\Omega)(f)(x) = f(x)$ for all $x \in \Omega$. In particular, $\partial^\alpha\, (\iota_\Omega \circ S_\Omega)(f)(x) = \partial^\alpha f(x)$ for all $f \in \mathcal{E}(\Omega)$, all $\alpha \in \mathbb{N}_0^d$ and all $x \in \Omega$.*

(vi) *Let X and Y be two open subsets of $\mathbb{R}^d$ and $\theta \in$ Diff(X, Y). Let the mapping $T \to T(\theta)$, from $\mathcal{D}'(X)$ to $\mathcal{D}'(Y)$, stands for the change of variables in the sense of distribution theory (Hörmander [3], §6.3–§6.4) and (Vladimirov [16], p.26). We define $\theta_* : \widehat{{}^*\mathcal{E}(X)} \to \widehat{{}^*\mathcal{E}(Y)}$ by $\theta_*(\widehat{f}) = (f \circ {}^*\theta^{-1}) \upharpoonright \mu(Y)$, where ${}^*\theta^{-1}$ stands for the non-standard extension of θ^{-1}. Then $\theta_*(\iota_X(T)) \cong \iota_Y(T(\theta))$ for all $T \in \mathcal{D}'(X)$.*

Proof Relatively straightforward consequences from Theorem 4. □

Examples 7 (Some Particular Generalized Functions in $\widehat{{}^*\mathcal{E}}(\Omega)$)

(i) $\iota_{\mathbb{R}^d}(\delta) = \widehat{\Delta}$ and more generally, $\iota_{\mathbb{R}^d}(\partial^\alpha\delta) = \widehat{\partial^\alpha\Delta}$ for all $\alpha \in \mathbb{N}_0^d$. We write these more casually as $\partial^\alpha\delta \in \widehat{{}^*\mathcal{E}}(\mathbb{R}^d)$.

(ii) $(\widehat{\Delta})^n \in \widehat{{}^*\mathcal{E}}(\mathbb{R}^d)$ for all $n \in \mathbb{N}$, since $\widehat{{}^*\mathcal{E}}(\mathbb{R}^d)$ is an algebra. We write this more casually as $\delta^n \in \widehat{{}^*\mathcal{E}}(\mathbb{R}^d)$.

(iii) Let $f : \mathbb{C} \mapsto \mathbb{C}$ stand for $f(z) = e^z$. Clearly, $e^{x+iy} \in \mathcal{E}(\mathbb{R}^2)$. Thus $e^\Delta \in {}^*\mathcal{E}(\mathbb{R}^2)$ (we skip the asterisk in front of ${}^*e^z$) and $\widehat{e^\Delta} \in \widehat{{}^*\mathcal{E}}(\mathbb{R}^2)$. We write more causally, $e^\delta \in \widehat{{}^*\mathcal{E}}(\mathbb{R}^2)$. Notice that e^δ makes sense as well in (Egorov [2]), but not in Colombeau algebra, since e^{x+iy} is *non-moderate* in the variable x (Colombeau [1]).

(iv) Let temporarily write Δ_d instead of Δ indicating that $\Delta_d \in {}^*\mathcal{D}(\mathbb{R}^d)$. Let $(e_1, e_2, \dots, e_d)$ be the standard basis for $\mathbb{R}^d$. Then $\widehat{\Delta}_d \restriction \operatorname{span}(e_n) \cong \widehat{\Delta}_1$, where $\widehat{\Delta}_d \restriction \operatorname{span}(e_n) := \Delta_d \restriction \mu(\operatorname{span}(e_n))$. Notice that $\operatorname{span}(e_n)$ is a *smooth submanifold* (not open subset) of $\mathbb{R}^d$.

Very much like in Colombeau and Egorov theories, Schwartz distribution can be multiplied within $\widehat{{}^*\mathcal{E}}(\Omega)$, since the latter is a differential (commutative and associative) algebra. How good (or bad) is this product? In a lack of compelling applications to other branches of mathematics or physics, we making our judgement mostly by applying this product to the ι_Ω-images in $\widehat{{}^*\mathcal{E}}(\Omega)$ of the classical functions. Here is our test:

Corollary 8 (Multiplication of Classical Functions)

(i) *The product in the algebra $\widehat{{}^*\mathcal{E}}(\Omega)$, if restricted to $\mathbb{C}[\Omega]$ (more precisely, on $(\iota_\Omega \circ S_\Omega)\big[\mathbb{C}[\Omega]\big]$), coincides with the usual product between polynomials, i.e. for every $P, Q \in \mathbb{C}[\Omega]$ we have*

$$(\iota_\Omega \circ S_\Omega)(PQ) = (\iota_\Omega \circ S_\Omega)(P) \cdot (\iota_\Omega \circ S_\Omega)(Q).$$

(ii) $(\iota_\Omega \circ S_\Omega)\big[\mathbb{C}[\Omega]\big] = \sigma\big[\mathbb{C}[\Omega]\big] = \sigma[\mathcal{E}(\Omega)] \cap (\iota_\Omega \circ S_\Omega)[\mathcal{E}(\Omega)]$.

(iii) *For every two continuous functions $f, g \in \mathcal{C}(\Omega)$ and for all (standard) $x \in \Omega$ we have $(\iota_\Omega \circ S_\Omega)(fg)(x) = (\iota_\Omega \circ S_\Omega)(f)(x) \cdot (\iota_\Omega \circ S_\Omega)(g)(x) = f(x)g(x)$.*

(iv) *$\iota_\Omega(fT) \cong \sigma(f)\iota_\Omega(T)$ (Definition 1) for all $f \in \mathcal{E}(\Omega)$ and all $T \in \mathcal{D}'(\Omega)$, where the product fT is in the sense of distribution theory (Vladimirov [16], §1.10). In particular, $\iota_\Omega(PT) \cong (\iota_\Omega \circ S_\Omega)(P)\ \iota_\Omega(T)$ for all polynomials $P \in \mathbb{C}[\Omega]$ and all $T \in \mathcal{D}'(\Omega)$.*

6 Regular Algebra

The algebra $\widehat{\mathcal{R}}_\rho(\Omega)$ defined below is similar, but different from the algbra $\mathcal{G}^\infty(\Omega)$ introduced and study in (Oberguggenberger [6], [7]). We should mention that $\widehat{\mathcal{R}}_\rho$ does not contain the counterexample in Vernaeve [15]).

Definition 9 (Regular Algebra) Let Ω and Δ be chosen (and fixed) as in Lemma 3. Let ${}^{\sigma}\mathcal{E}(\Omega) = \{{}^*f : f \in \mathcal{E}(\Omega)\}$ and $\mathcal{M}_\rho = \{\xi \in {}^*\mathbb{C} : |\xi| \leq \rho^{-n}$ for some $n \in \mathbb{N}\}$. Let $\mathcal{R}_\rho(\Omega)$ denote the subring of ${}^*\mathcal{E}(\Omega)$ generated by ${}^{\sigma}\mathcal{E}(\Omega) \cup \mathcal{M}_\rho$, in symbol, $\mathcal{R}_\rho(\Omega) = {}^{\sigma}\mathcal{E}(\Omega)(\mathcal{M}_\rho)$. The *algebra of ρ-regular functions* is defined by $\widehat{\mathcal{R}}_\rho(\Omega) = \{\widehat{f} : f \in \mathcal{R}_\rho(\Omega)\}$ or equivalently, by $\widehat{\mathcal{R}}_\rho(\Omega) = \{f \restriction \mu(\Omega) : f \in \mathcal{R}_\rho(\Omega)\}$.

Theorem 10 *Under the assumption of the above definition we have:*

(i) $\widehat{\mathcal{R}}_\rho$ *is a differential* $\mathbb{C}$*-subalgebra of* $\widehat{{}^*\mathcal{E}}(\Omega)$.
(ii) *If* $\widehat{f} \in \widehat{\mathcal{R}}_\rho(\Omega)$*, then* $(\forall \xi \in \mu(\Omega))(\exists n \in \mathbb{N})(\forall \alpha \in \mathbb{N}_0^d)(|\partial^\alpha \widehat{f}(\xi)| \leq \rho^{-n})$.
(iii) $\widehat{\mathcal{R}}_\rho(\Omega) \cap \iota_\Omega[\mathcal{D}'(\Omega)] = (\iota_\Omega \circ S_\Omega)[\mathbb{C}[\Omega]]$.

Proof We leave the verification to the reader. □

References

1. J.F. Colombeau, *New Generalized Functions and Multiplication of Distributions*. Mathematical Studies, vol. 84 (North-Holland, Amsterdam/New York/Oxford, 1984)
2. Y.V. Egorov, A contribution to the theory of generalized functions (Russian). Uspekhi Mat Nauk (Moskva) **45**(5), 3–40 (1990) . English translation in Russian Math. Surveys **45**(5), 1–49 (1990)
3. L. Hörmander, *The Analysis of Linear Partial Differential Operators I*. Grundlehren der mathematischen Wissenschaften, vol. 256 (Springer, Berlin, 1990)
4. T. Lindstrøm, *An Invitation to Nonstandard Analysis*. Nonstandard Analysis and Its Applications, ed. by N. Cutland (Cambridge University Press, Cambridge, 1988), pp. 1–105
5. P.A. Loeb, M. Wolff (eds.), *Nonstandard Analysis for the Working Mathematician* (Kluwer Academic Publishers, Dordrecht/Boston/London, 2000)
6. M. Oberguggenberger, *Multiplication of Distributions and Applications to Partial Differential Equations*, Pitman Research Notes Mathematics Series, vol. 259 (Longman, Harlow, 1992)
7. M. Oberguggenberger, Regularity Theory in Colombeau Algebras, in *Bulletin T. CXXXIII de l'Academie serbe des sciences et des arts, Classe des sciences mathématiques et naturelles, Sciences mathématique*, vol. 31 (2006), pp. 147–162
8. M. Oberguggenberger, T. Todorov, An embedding of Schwartz distributions in the algebra of asymptotic functions. Int. J. Math. Math. Sci. **21**(3), 417–428 (1998)
9. T. Todorov, A nonstandard delta function. Proc. Am. Math. Soc. **110**(4), 1143–1144 (1990)
10. T. Todorov, Pointwise kernels of Schwartz distributions. Proc. Am. Math. Soc. **114**(3), 817–819 (1992)
11. T. Todorov, An existence of solutions for linear PDE with C^∞-coefficients in an algebra of generalized functions. Trans. Am. Math. Soc. **348**(2), 673–689 (1996). http://www.ams.org/journals/tran/1996-348-02
12. T. Todorov, H. Vernaeve, Full algebra of generalized functions and non-standard asymptotic analysis. Logic Anal **1**(3), 205–234 (2008). Available at: http://www.logicandanalysis.com/index.php/jla/article/view/193/79 and/or at: http://arxiv.org/abs/0712.2603.
13. H. Vernaeve, *Nonstandard Contributions to the Theory of Generalized Functions*. Ph.D. Thesis, University of Gent, 2002

14. H. Vernaeve, Optimal embeddings of distributions into algebras. Proc. Edinb. Math. Soc. **46**, 373–378 (2003)
15. H. Vernaeve, Regularity of nonlinear generalized functions: a counterexample in the nonstandard setting. Examples Counterexamples **1**, 100021 (2021)
16. V.S. Vladimirov, *Generalized Functions in Mathematical Physics* (Mir-Publisher, Moscow, 1979)

Density Conditions for Coherent State Subsystems of Nilpotent Lie Groups

Jordy Timo van Velthoven

Abstract The aim of this note is to present recent work on density conditions for spanning properties of coherent state subsystems for nilpotent Lie groups and provide context.

Keywords Coherent state · Density condition · Nilpotent Lie group

Mathematics Subject Classification (2010) 42C30, 42C40

1 Introduction

A common procedure for the construction of coherent states is through a group-theoretic method. Constructed in this manner, a coherent state system is a subset of the orbit of a vector under an irreducible unitary representation. The most classical example of such a coherent state system is the nowadays called *Gabor* or *Weyl-Heisenberg system* in $L^2(\mathbb{R})$,

$$\{e^{2\pi i \xi \cdot} g(\cdot - x) : (x, \xi) \in \mathbb{R}^2\},$$

which arises from a function $g \in L^2(\mathbb{R})$ through the action of the Schrödinger representation of the Heisenberg group. Gabor systems form a classical object of study in several areas of mathematics, ranging from complex and harmonic analysis to mathematical physics, see, e.g., the books [16, 20, 32, 35, 39].

Recently, there has also been a considerable interest in various aspects of coherent state systems arising from other nilpotent Lie groups, see, e.g., [5, 6, 12, 18, 19, 21, 23, 29, 30]. The aim of this note is to survey recent results on density conditions for spanning properties of such coherent states and to discuss

J. T. van Velthoven (✉)
University of Vienna, Vienna, Austria
e-mail: jordy.timo.van.velthoven@univie.ac.at

D. Cardona et al. (eds.), *Extended Abstracts 2021/2022*, Research Perspectives
Ghent Analysis and PDE Center 3, https://doi.org/10.1007/978-3-031-48579-4_22

the interrelation of these various conditions. Most of the presented results are well-known for Gabor systems (cf. the survey [24]), but require other proof methods for general (classes of) nilpotent Lie groups. For example, a powerful technique for studying Gabor systems is the comparison with a "reference system" (cf. [2, 3, 34]), which are unknown to exist for other nilpotent Lie groups. Some related open problems are stated in Sect. 5.

2 Coherent State Systems for Nilpotent Lie Groups

This section provides the basic notions on representations of nilpotent Lie groups and associated coherent states; see the books [10, 32] for more details.

Let N be a connected, simply connected nilpotent Lie group and let $(\pi, \mathcal{H})$ be an irreducible unitary representation of N. As π is irreducible, there exists a unitary character χ_π of the center $Z = Z(N)$ of N such that $\pi(x) = \chi_\pi(x) \cdot I_{\mathcal{H}}$ for all $x \in Z$. In particular, this implies that, given $f, g \in \mathcal{H}$, the map $x \mapsto |\langle f, \pi(x)g\rangle|^2$ is a well-defined function on the quotient group N/Z. Additionally, it will be assumed that $(\pi, \mathcal{H})$ is *square-integrable* modulo the center $Z = Z(N)$ of N, i.e., there exists nonzero $g \in \mathcal{H}$ such that

$$\int_{N/Z} |\langle g, \pi(x)g\rangle|^2 \, d\mu_{N/Z}(x) < \infty,$$

where $\mu_{N/Z}$ denotes Haar measure on N/Z.

For the construction of coherent states, let $s : N/Z \to N$ be a smooth cross-section of the canonical projection $q : N \to N/Z$, i.e., $q \circ s = \mathrm{id}_{N/Z}$, and define

$$\rho := \pi \circ s : N/Z \to \mathcal{U}(\mathcal{H}).$$

Then the pair $(\rho, \mathcal{H})$ is an irreducible, square-integrable projective unitary representation of $G := N/Z$; in particular, it satisfies

$$\rho(xy) = \sigma(x, y)\rho(x)\rho(y), \quad x, y \in G, \tag{1}$$

for a smooth function $\sigma : G \times G \to \mathbb{T}$, called the *cocycle* of ρ.

Following [28, 31], a *coherent state system* based on $G = N/Z$ is an orbit

$$\rho(G)g = \{\rho(x)g : x \in G\}$$

of a vector $g \in \mathcal{H}$.

By the orthogonality relations for square-integrable representations [27], there exists a unique constant $d_\rho > 0$, called the *formal degree*, such that

$$\int_G |\langle f, \rho(x)g\rangle|^2 \, d\mu_G(x) = d_\rho^{-1} \|f\|_{\mathcal{H}}^2 \|g\|_{\mathcal{H}}^2 \tag{2}$$

for all $f, g \in \mathcal{H}$.

The relation (2) implies, in particular, that the coherent state system $\rho(G)g$ is *overcomplete*, in the sense that it remains complete even after the removal of an arbitrary element. The aim of this note is to provide an overview of quantitative sufficient and necessary conditions on a discrete subset $\Lambda \subseteq G$ such that the associated subsystem of coherent states,

$$\rho(\Lambda)g = \{\rho(\lambda)g : \lambda \in \Lambda\},$$

remains complete in $\mathcal{H}$ (cf. [31]) or even forms a *frame* for $\mathcal{H}$ (cf. [11]), i.e., there exist $A, B > 0$, called *frame bounds*, such that

$$A\|f\|_{\mathcal{H}}^2 \leq \sum_{\lambda \in \Lambda} |\langle f, \rho(\lambda)g\rangle|^2 \leq B\|f\|_{\mathcal{H}}^2 \quad \text{for all} \quad f \in \mathcal{H}.$$

Most of the presented results also admit dual statements for minimal systems and Riesz sequences. These statement can be found in the same references.

3 Density Conditions for Lattice Subgroups

A *lattice* in G is a discrete subgroup $\Gamma \subseteq G$ whose quotient G/Γ is compact. Equivalently, a discrete subgroup $\Gamma \subseteq G$ is a lattice if there exists a relatively compact fundamental domain, i.e., a relatively compact set $\Omega \subseteq G$ such that $\gamma\Omega \cap \gamma'\Omega = \emptyset$ for all $\gamma \neq \gamma'$. The *covolume* of a lattice Γ is defined by $\mathrm{vol}(G/\Gamma) := \mu_G(\Omega)$ and independent of the choice of fundamental domain Ω. The reciprocal of the covolume is often referred to as the *density* of a lattice.

The following fundamental result provides a necessary condition on the density of a lattice admitting a complete system in its orbit. The theorem was first shown in [6, Theorem 3].

Theorem 1 *Let* $\Gamma \subseteq G$ *be a lattice. If there exists* $g \in \mathcal{H}$ *such that* $\rho(\Gamma)g$ *is complete, then* $\mathrm{vol}(G/\Gamma)d_\rho \leq 1$.

Remark 2 *The value of the covolume* $\mathrm{vol}(G/\Gamma)$ *and the formal degree* d_ρ *depend on the choice of Haar measure on* G*, but their product does not.*

The original proof of Theorem 1 given in [6] relies on advanced techniques on group von Neumann algebras. Recently, more elementary proofs were given in [14, 37].

Theorem 1 is sharp, in the sense that for a general vector $g \in \mathcal{H}$ the necessary condition $\mathrm{vol}(G/\Gamma)d_\rho \leq 1$ cannot be improved. This is shown by the following theorem [6, Theorem 3]; see also [15].

Theorem 3 *Let $\Gamma \subseteq G$ be a lattice. If $\mathrm{vol}(G/\Gamma)d_\rho \leq 1$, then there exists $g \in \mathcal{H}$ such that $\rho(\Gamma)g$ forms a frame for $\mathcal{H}$.*

The existence claim in Theorem 3 is not accompanied by constructions of explicit vectors nor does it guarantee additional regularity properties of the vector than membership in $\mathcal{H}$. However, under additional assumptions on the lattice, it was shown in [5] that a generating vector can be chosen to be in the space $\mathcal{H}^\infty$ consisting of smooth vectors of ρ, i.e., vectors $g \in \mathcal{H}$ such that the orbit map $x \mapsto \rho(x)g$ is smooth. For stating this result, a pair (Γ, σ) consisting of a lattice Γ and a cocycle σ as in (1) is said to satisfy *Kleppner's condition* if the conjugacy class $\{\gamma\gamma_0\gamma^{-1} : \gamma \in \Gamma\}$ of an element $\gamma_0 \in \Gamma \setminus \{e\}$ is infinite whenever $\sigma(\gamma_0, \gamma) = \sigma(\gamma, \gamma_0)$ holds for all $\gamma \in \Gamma$ commuting with γ_0.

Theorem 4 *Let $\Gamma \subseteq G$ be a lattice such that (Γ, σ) satisfies Kleppner's condition. If $\mathrm{vol}(G/\Gamma)d_\rho < 1$, then there exists $g \in \mathcal{H}^\infty$ such that $\rho(\Gamma)g$ forms a frame for $\mathcal{H}$.*

Note that, in contrast to Theorem 3, the sufficient density condition in Theorem 4 is a *strict* inequality. For the nonexistence of smooth frames at the *critical density* $\mathrm{vol}(G/\Gamma)d_\rho = 1$, see Theorem 8.

4 Necessary Density Conditions for Irregular Point Sets

This section contains various density theorems for coherent state subsystems associated to nonlattice index sets. The notion of density appearing in these theorems is the so-called *Beurling density*, which will be described next.

Following [33], a sequence $(K_n)_{n\in\mathbb{N}}$ of nonnull compact subsets $K_n \subseteq G$ is said to be a *strong Følner sequence* if it satisfies

$$\lim_{n\to\infty} \frac{\mu_G(K_n K \cap K_n^c K)}{\mu_G(K_n)} = 0$$

for all compact sets $K \subseteq G$. Examples of strong Følner sequences include sequences of balls $(B_{r_n}(e))_{n\in\mathbb{N}}$ with $r_n \to \infty$ associated to a word metric, Riemannian metric or Carnot-Carathéodory metric on G, cf. [8]. Given any strong Følner sequence $(K_n)_{n\in\mathbb{N}}$, the associated (lower) *Beurling density* of a set $\Lambda \subseteq G$ is given by

$$D^-(\Lambda) = \liminf_{n\to\infty} \inf_{x\in G} \frac{\#(\Lambda \cap xK_n)}{\mu_G(K_n)}.$$

The Beurling density of a set is independent of the choice of strong Følner sequence, see [33, Proposition 5.14]. In particular, if $\Gamma \subseteq G$ is a lattice, then

$$D^-(\Gamma) = \mathrm{vol}(G/\Gamma)^{-1},$$

see, e.g., [33, Section 6].

4.1 Approximate Lattices

For a general discrete set $\Lambda \subseteq G$ and a vector $g \in \mathcal{H}$, there are no necessary density conditions (involving Beurling densities) for $\rho(\Lambda)g$ to be complete. Namely, as shown in [36, 38], there are complete Gabor systems for which the Beurling density of its index set is zero. However, an extension of Theorem 1 to so-called *approximate lattices* [7] was recently obtained in [14].

A set $\Lambda \subseteq G$ is called a *Delone set* if there exist a compact set $K \subseteq G$ such that $\Lambda K = G$ and an open set $U \subseteq G$ such that $\#(\Lambda \cap xU) \leq 1$ for all $x \in G$. A Delone set Λ is called a *k-approximate lattice* ($k \in \mathbb{N}$) if it additionally satisfies

(a1) The identity $e \in G$ is contained in Λ;
(a2) $\Lambda^{-1} = \Lambda$;
(a3) there exists a finite $F \subseteq G$ of cardinality $\#F \leq k$ such that $\Lambda^2 \subseteq F\Lambda$.

Note that a lattice is precisely a 1-approximate lattice.

The following theorem is the main result of [14].

Theorem 5 *Let $\Lambda \subseteq G$ be a k-approximate lattice. If there exists $g \in \mathcal{H}$ such that $\rho(\Lambda)g$ is complete in $\mathcal{H}$, then*

$$D^-(\Lambda) \geq d_\rho / k.$$

Theorem 5 is optimal in the sense that $D^-(\Lambda)$ could be arbitrary small while $d_\rho = 1$, cf. [14, Proposition 3.7].

4.2 General Point Sets

This subsection contains various results on the density of frames $\rho(\Lambda)g$ for general point sets $\Lambda \subseteq G$. In order to state these results, an additional condition on the generating vector $g \in \mathcal{H}$ is required.

Given $p \in \{1, 2\}$ and a symmetric compact unit neighborhood $Q \subseteq G$, define

$$\mathcal{B}^p := \left\{ g \in \mathcal{H} : \int_G \sup_{y \in Q} |\langle g, \pi(xy)g \rangle|^p \, d\mu_G(x) < \infty \right\}.$$

Then $\mathcal{B}^p$ is independent of the choice of defining neighbourhood, and $\mathcal{B}^1 \subseteq \mathcal{B}^2$. In particular, the space of smooth vectors $\mathcal{H}^\infty$ is contained in $\mathcal{B}^1$, and hence $\mathcal{B}^p$ is norm dense in $\mathcal{H}$.

The following theorem was first shown in [18]; see also [9, 13, 14].

Theorem 6 *Let $\Lambda \subseteq G$. If there exists $g \in \mathcal{B}^2$ such that $\rho(\Lambda)g$ is a frame for $\mathcal{H}$, then $D^-(\Lambda) \geq d_\rho$.*

Theorem 6 provides a strong necessary condition on a discrete set admitting a frame. If, under the same hypotheses as Theorem 6, the system $\rho(\Lambda)g$ is even an orthonormal (or Riesz) basis, then necessarily $D^-(\Lambda) = d_\rho$, cf. [9, 13, 14, 18]. In particular, a frame $\rho(\Lambda)g$ with $g \in \mathcal{B}^2$ and $D^-(\Lambda) > d_\rho$ is overcomplete.

A criterion for a frame under which a set of positive density can be removed yet leave a frame is given by the following theorem from [9].

Theorem 7 *Suppose $\rho(\Lambda)g$ is a frame for $\mathcal{H}$ with $g \in \mathcal{B}^1$ and $D^-(\Lambda) > d_\rho$. Then there exists $\Lambda' \subseteq \Lambda$ with $D^-(\Lambda') > 0$ such that $\{\rho(\lambda)g\}_{\lambda \in \Lambda \setminus \Lambda'}$ is a frame for $\mathcal{H}$.*

The last presented result is a theorem asserting that the *strict* density inequality appearing in Theorem 7 is automatic for frames $\rho(\Lambda)g$ with a vector $g \in \mathcal{B}^1$. See [17] for the definition of a homogeneous (nilpotent) group.

Theorem 8 *Let G be a homogeneous group and let $\Lambda \subseteq G$. If $\rho(\Lambda)g$ is a frame for $\mathcal{H}$ with $g \in \mathcal{B}^1$, then $D^-(\Lambda) > d_\rho$.*

In particular, if $\rho(\Lambda)g$ is an orthonormal basis for $\mathcal{H}$, then $g \notin \mathcal{B}^1$.

Theorem 8 is proved in [23], and forms an extension of a corresponding result for Gabor systems [1, 22].

5 Open Problems and Questions

This section contains four open problems on density conditions for coherent state subsystems of nilpotent Lie groups.

The first question concerns the validity of Theorem 3 without Kleppner's condition.

Question 9 Given a lattice $\Gamma \subseteq G$ satisfying $\mathrm{vol}(G/\Gamma)d_\rho < 1$, does there exist $g \in \mathcal{H}^\infty$ such that $\rho(\Gamma)g$ is a frame for $\mathcal{H}$?

Question 9 is still open in the special case of Gabor systems. See [25, 26] for related sufficient density conditions in this setting.

Second, the question whether Theorem 6 is still valid for frames $\rho(\Lambda)g$ for $\mathcal{H}$ with generating vector $g \notin \mathcal{B}^2$.

Question 10 Given a frame $\rho(\Lambda)g$ for $\mathcal{H}$ with $g \notin \mathcal{B}^2$, is it necessary that $D^-(\Lambda) \geq d_\rho$?

The additional condition $g \in \mathcal{B}^2$ in Theorem 6 is known to be unnecessary for Gabor systems [34] and more generally for groups G with an invariant neighbourhood [13], so for these cases Question 10 is true.

The third question concerns a quantitative version of Theorem 7.

Question 11 Let $\varepsilon > 0$. Given a frame $\rho(\Lambda)g$ for $\mathcal{H}$ with $g \in \mathcal{B}^1$ and $D^-(\Lambda) > d_\rho$, does there exist $\Lambda' \subseteq \Lambda$ such that

$$(1+\varepsilon)d_\rho \geq D^-(\Lambda \setminus \Lambda') \geq d_\rho$$

and $(\rho(\lambda)g)_{\lambda \in \Lambda \setminus \Lambda'}$ is still a frame for $\mathcal{H}$?

Question 11 **has an affirmative answer for Gabor systems, cf.** [4]. A positive answer to Question 11 would show, in particular, that the strict density conditions of Theorem 8 are optimal.

Lastly, the question whether Theorem 8 remains valid for general (nonhomogeneous) nilpotent Lie groups.

Question 12 Given a frame $\rho(\Lambda)g$ for $\mathcal{H}$ with $g \in \mathcal{B}^1$, is it necessary that $D^-(\Lambda) > d_\rho$?

Acknowledgments The author gratefully acknowledges support from the Austrian Science Fund (FWF) project J4555. Thanks are due to Ulrik Enstad for helpful comments on a preliminary version of this note.

References

1. G. Ascensi, H.G. Feichtinger, N. Kaiblinger, Dilation of the Weyl symbol and Balian-low theorem. Trans. Am. Math. Soc. **366**(7), 3865–3880 (2014)
2. R. Balan, P.G. Casazza, C. Heil, Z. Landau, Density, overcompleteness, and localization of frames. I: theory. J. Fourier Anal. Appl. **12**(2), 105–143 (2006)
3. R. Balan, P.G. Casazza, C. Heil, Z. Landau, Density, overcompleteness, and localization of frames. II: Gabor systems. J. Fourier Anal. Appl. **12**(3), 307–344 (2006)
4. R. Balan, P. Casazza, Z. Landau, Redundancy for localized frames. Isr. J. Math. **185**, 445–476 (2011)
5. E. Bédos, U. Enstad, J.T. van Velthoven, Smooth lattice orbits of nilpotent groups and strict comparison of projections. J. Funct. Anal. **283**(6), 48 (2022). Id/No 109572.
6. B. Bekka, Square integrable representations, von Neumann algebras and an application to Gabor analysis. J. Fourier Anal. Appl. **10**(4), 325–349 (2004)
7. M. Björklund, T. Hartnick, Approximate lattices. Duke Math. J. **167**(15), 2903–2964 (2018)
8. E. Breuillard, Geometry of locally compact groups of polynomial growth and shape of large balls. Groups Geom. Dyn. **8**(3), 669–732 (2014)
9. M. Caspers, J.T. van Velthoven, Overcompleteness of coherent frames for unimodular amenable groups. Ark. Math. **61**(2), 277–299 (2023)
10. L.J. Corwin, F.P. Greenleaf, *Representations of Nilpotent Lie Groups and Their Applications. Part 1: Basic Theory and Examples*. Cambridge Studies in Advanced Mathematics, vol. 18 (Cambridge University Press, Cambridge, 1990)

11. I. Daubechies, A. Grossmann, Y. Meyer, Painless nonorthogonal expansions. J. Math. Phys. **27**, 1271–1283 (1986)
12. U. Enstad, The density theorem for projective representations via twisted group von Neumann algebras. J. Math. Anal. Appl. **511**(2), 25 (2022) Id/No 126072
13. U. Enstad, S. Raum, A dynamical approach to non-uniform density theorems for coherent systems. Preprint (2022). arXiv:2207.05125
14. U. Enstad, J.T. Van Velthoven, Coherent systems over approximate lattices in amenable groups. Preprint (2022). Ann. Inst. Fourier (To Appear)
15. U. Enstad, J.T. van Velthoven, On sufficient density conditions for lattice orbits of relative discrete series. Arch. Math. **119**(3), 279–291 (2022)
16. G.B. Folland, *Harmonic Analysis in Phase Space*. Annals of Mathematics Studies, vol. 122 (Princeton University Press, Princeton, 1989)
17. G.B. Folland, E.M. Stein, *Hardy Spaces on Homogeneous Groups*. Mathematical Notes (Princeton), vol. 28 (Princeton University Press, Princeton, 1982)
18. H. Führ, K. Gröchenig, A. Haimi, A. Klotz, J.L. Romero, Density of sampling and interpolation in reproducing kernel Hilbert spaces. J. Lond. Math. Soc., II. Ser. **96**(3), 663–686 (2017)
19. J.-P. Gabardo, C.-K. Lai, V. Oussa, On exponential bases and frames with non-linear phase functions and some applications. J. Fourier Anal. Appl. **27**(2), 24 (2021) Id/No 9
20. K. Gröchenig, *Foundations of Time-Frequency Analysis*. Applied and Numerical Harmonic Analysis (Birkhäuser, Boston, 2001)
21. K. Gröchenig, D. Rottensteiner, Orthonormal bases in the orbit of square-integrable representations of nilpotent Lie groups. J. Funct. Anal. **275**(12), 3338–3379 (2018)
22. K. Gröchenig, J. Ortega-Cerdà, J.L. Romero, Deformation of Gabor systems. Adv. Math. **277**, 388–425 (2015)
23. K. Gröchenig, J.L. Romero, D. Rottensteiner, J.T. Van Velthoven, Balian-Low type theorems on homogeneous groups. Anal. Math. **46**(3), 483–515 (2020)
24. C. Heil, History and evolution of the density theorem for Gabor frames. J. Fourier Anal. Appl. **13**(2), 113–166 (2007)
25. M.S. Jakobsen, F. Luef, Duality of Gabor frames and Heisenberg modules. J. Noncommut. Geom. **14**(4), 1445–1500 (2020)
26. F. Luef, X. Wang, Gaussian gabor frames, seshadri constants and generalized buser–sarnak invariants. Geom. Funct. Anal. **33**(3), 778–823 (2023)
27. C.C. Moore, J.A. Wolf, Square integrable representations of nilpotent groups. Trans. Am. Math. Soc. **185**, 445–462 (1974)
28. H. Moscovici, Coherent state representations of nilpotent Lie groups. Commun. Math. Phys. **54**, 63–68 (1977)
29. V. Oussa, Compactly supported bounded frames on Lie groups. J. Funct. Anal. **277**(6), 1718–1762 (2019)
30. V. Oussa, Orthonormal bases arising from nilpotent actions. Trans. Am. Math. Soc. (to Appear)
31. A.M. Perelomov, Coherent states for arbitrary Lie group. Commun. Math. Phys. **26**, 222–236 (1972)
32. A. Perelomov, *Generalized Coherent States and Their Applications*. Texts and Monographs in Physics (Springer, New York, 1986)
33. F. Pogorzelski, C. Richard, N. Strungaru, Leptin densities in amenable groups. J. Fourier Anal. Appl. **28**(6), 36 (2022). Id/No 85
34. J. Ramanathan, T. Steger, Incompleteness of sparse coherent states. Appl. Comput. Harmon. Anal. **2**(2), 148–153 (1995)
35. D. Robert, M. Combescure, *Coherent States and Applications in Mathematical Physics*. Theoretical and Mathematical Physics, 2nd revised and enlarged edition (Springer, Cham, 2021)
36. E. Romero, A complete Gabor system of zero Beurling density. Sampl. Theory Signal Image Process. **3**(3), 299–304 (2004)
37. J.L. Romero, J.T. van Velthoven, The density theorem for discrete series representations restricted to lattices. Expo. Math. **40**(2), 265–301 (2022)

38. Y. Wang, Sparse complete Gabor systems on a lattice. Appl. Comput. Harmon. Anal. **16**(1), 60–67 (2004)
39. K. Zhu, *Analysis on Fock Spaces*. Graduate Texts in Mathematics, vol. 263 (Springer, New York, 2012)

Space-Time Mixed Norm Estimates in Riemannian Symmetric Spaces of Non-Compact Type

Hong-Wei Zhang

Abstract We present a summary of recent advances in the Strichartz inequality and the smoothing property on non-compact type and general rank Riemannian symmetric spaces.

Keywords Dispersive equation · Non-compact symmetric space · Strichartz inequality · Smoothing property

2020 Mathematics Subject Classification 22E30, 35J10, 35P25, 43A85, 43A90

1 Introduction

Two primary instruments for addressing non-linear dispersive equations are the Strichartz inequality and the smoothing property. Both these estimates necessitate an exploration of the solution for the corresponding linearized equation within the space-time mixed Lebesgue norm. In the Euclidean setting, the estimates discussed have been extensively covered in the existing literature. We recommend referring to [8, 17] for a more comprehensive review. In this brief note, we aim to highlight recent progress concerning these estimates in non-compact Riemannian symmetric spaces. These spaces are Riemannian manifolds with non-positive sectional curvature and grow exponentially fast to infinity. In particular, the techniques of the Fourier analysis are available in such a context.

Notation We utilize standard notation, which can be found in [9] for more explanation. Let $\mathbb{X}$ be a symmetric space of rank $\ell \geq 1$, dimension $n \geq 2$, and pseudo-dimension $\nu \geq 3$. We denote by Δ the Laplace-Beltrami operator on $\mathbb{X}$ and $|\rho|^2$ the bottom of its L^2-spectrum. Let $\mathfrak{a}$ be the Cartan subspace which is a

H.-W. Zhang (✉)
Department of Mathematics: Analysis, Logic and Discrete Mathematics, Ghent University, Ghent, Belgium
e-mail: hongwei.zhang@ugent.be

D. Cardona et al. (eds.), *Extended Abstracts 2021/2022*, Research Perspectives Ghent Analysis and PDE Center 3, https://doi.org/10.1007/978-3-031-48579-4_23

flat submanifold of $\mathbb{X}$. Recall that the dimension of $\mathfrak{a}$ is ℓ, that is the rank of $\mathbb{X}$. Throughout this note, we denote by $C > 0$ a constant independent of variables.

2 Strichartz Inequality

To provide a simplified perspective, we consider the free Schrödinger equation as an elementary model. Consider the homogeneous Cauchy problem

$$(i\partial_t + \Delta_x)\, u(t,x) = 0, \qquad u(0,x) = f(x). \tag{1}$$

The Strichartz inequality aims to establish a relationship between the initial data $f(x)$ and the solution $u(t,x) = e^{it\Delta} f(x)$ to the Eq. (1). It seeks to identify suitable pairs (p,q) for which the following inequality holds:

$$\|u\|_{L_t^p(\mathbb{R};\, L_x^q(\mathbb{X}))} = \left\{ \int_{\mathbb{R}} \mathrm{d}t\, \|e^{it\Delta} f\|_{L^q(\mathbb{X})}^p \right\}^{1/p} \le C\, \|f\|_{L^2(\mathbb{X})}. \tag{2}$$

In the Euclidean setting, the explicit expression of the convolution kernel of the Schrödinger propagator directly implies the L^1-L^∞ dispersive property. This property plays a central role in establishing the Strichartz inequality through the use of the TT^* duality argument. However, it is not always possible to expect such an explicit expression on more general manifolds, even in symmetric spaces where the Fourier analysis techniques are available. In fact, by using the inverse formula of the spherical Fourier-Helgason transform, one can write the Schrödinger kernel as

$$s_t(x) = C\, e^{-it|\rho|^2} \int_{\mathfrak{a}} \mathrm{d}\lambda\, |\mathbf{c}(\lambda)|^{-2}\, \varphi_\lambda(x)\, e^{-it|\lambda|^2},$$

where $\mathbf{c}(\lambda)$ is the so-called Harish-Chandra function and $\varphi_\lambda(x)$ is the spherical function which plays a similar role as the exponential factor in the Euclidean Fourier transform. In harmonic analysis of higher rank symmetric spaces, one of the well-known difficulties arises from the fact that the Plancherel density $|\mathbf{c}(\lambda)|^{-2}$ is not always a differential symbol. A recent breakthrough [3] has successfully tackled this difficulty. The authors introduced a novel spectral decomposition method that divides the Cartan subspace $\mathfrak{a}$ into distinct subcones. Within each subcone, the Plancherel density behaves as if it were a symbol along a well-chosen direction. Together with an explicit Hadamard parametrix and the stationary phase method, the authors in [5] have established the following pointwise estimate for the Schrödinger kernel:

$$|s_t(x)| \le C\,(1+|x|)^M\, \varphi_0(x) \begin{cases} |t|^{-\frac{n}{2}} & \text{if} \quad 0 < |t| < 1, \\ |t|^{-\frac{\nu}{2}} & \text{if} \quad |t| \ge 1, \end{cases} \tag{3}$$

for some constants $C > 0$ and $M > 0$. Unlike the kernel estimate in the Euclidean setting, the decay at the large time in estimate (3) is solely determined by the pseudo-dimension ν, rather than the dimension of the manifold n. By leveraging the Kunze-Stein phenomenon in conjunction with estimate (3), one derives the following stronger dispersive property:

$$\|e^{it\Delta}\|_{L^{q'}(\mathbb{X})\to L^{q}(\mathbb{X})} \leq C \begin{cases} |t|^{-(\frac{1}{2}-\frac{1}{q})n} & \text{if} \quad 0 < |t| < 1, \\ |t|^{-\frac{\nu}{2}} & \text{if} \quad |t| \geq 1, \end{cases} \tag{4}$$

where the dispersion in large time is independent of the index $q \geq 2$. This phenomenon was initially discovered in the context of real hyperbolic spaces, which are non-compact symmetric spaces of rank one, see [1, 6, 10]. In these spaces, the pseudo-dimension is consistently equal to 3, and the challenge mentioned earlier regarding the behavior of the Plancherel density does not arise. See also [4] for results on Damek-Ricci spaces, which include all the symmetric spaces of rank one, and [7] where the author have obtained an improved Strichartz inequality without using the spherical Fourier analysis in a class of manifolds including the hyperbolic space and the Damek-Ricci space.

Once the dispersive property is established, the Strichartz inequality can be deduced through the standard duality argument. Thanks to the stronger dispersive estimate (4), a wide range of pairs (p, q) can satisfy the Strichartz inequality (2).

Theorem 1 *The Strichartz inequality* (2) *holds for all* (p, q) *belongs to the admissible triangle*

$$\Big\{\Big(\frac{1}{p}, \frac{1}{q}\Big) \in \Big(0, \frac{1}{2}\Big] \times \Big(0, \frac{1}{2}\Big) \,\Big|\, \frac{2}{p} + \frac{n}{q} \geq \frac{n}{2}\Big\} \bigcup \Big\{\Big(0, \frac{1}{2}\Big)\Big\}. \tag{5}$$

Note that in symmetric spaces, the admissible set is significantly larger than in the Euclidean setting. The latter corresponds only to the lower edge of the triangle shown in Fig. 1. The enlarged admissible set in symmetric spaces can be attributed to the stronger dispersive property arising from the large-scale geometry inherent in these spaces.

By using a similar argument, it is possible to derive an improved Strichartz inequality for the non-homogeneous linear Schrödinger equation, that is Eq. (1) with a linear forcing term on the right-hand side. This improved inequality, when combined with the fixed-point argument, enables one to prove the stronger global well-posedness and scattering results in the analysis of the corresponding non-linear Schrödinger equation, see [1, 5, 10]. Similar results also hold for the wave equation, see, for instance, [2, 3, 14, 18, 19].

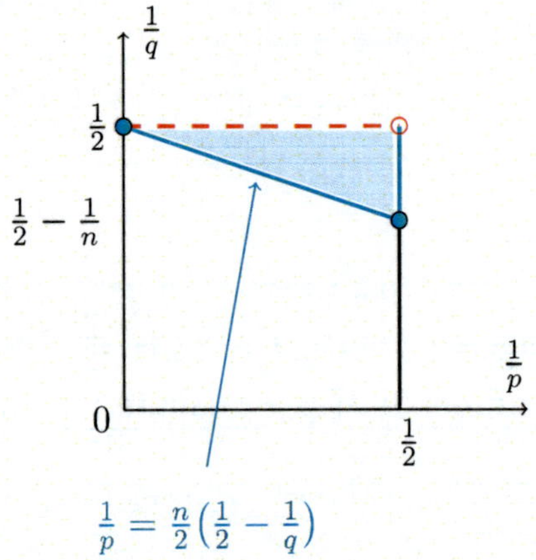

Fig. 1 Admissible triangle in dimension $n \geq 3$

3 Smoothing Property

It is known that the Schrödinger propagator $e^{it\Delta}$ preserves the L^2 norm for each fixed time $t \in \mathbb{R}$. Hence, the inequality (2) cannot hold for any $2 \leq p < \infty$ when $q = 2$ (see the red dashed edge in Fig. 1). The smoothing property refers to the L^2-L^2 space-time estimate, which enables one to gain extra regularity in comparison to the initial data. Recently, in [13], various types of smoothing properties have been established for the Cauchy problem involving the more general m-order operator $D^m = (-\Delta - |\rho|)^{m/2}$, namely,

$$(i\partial_t - D^m)\, u(t,x) = 0, \tag{6}$$

Theorem 2 *Suppose that $m > 0$ and $A(x,D)$ is defined as one of the following*

(I) $A(x,D) = |x|^{\alpha - \frac{m}{2}} D^{\alpha}$; $\frac{m-3}{2} < \alpha < \frac{m-1}{2}$ *if* $\ell = 1$ *or* $\frac{m-\min\{n,\nu\}}{2} < \alpha < \frac{m-1}{2}$ *if* $\ell \geq 2$,

(II) $A(x,D) = \langle x\rangle^{-s} D^{\frac{m-1}{2}}$; $s > \frac{1}{2}$ *and* $m > 0$,

(III) $A(x,D) = \langle x\rangle^{-s} \langle D\rangle^{\frac{m-1}{2}}$; $s \geq \frac{m}{2}$ *and* $1 < m < \nu$.

Then, the solution to the Cauchy problem (6) satisfies the smoothing property:

$$\|A(x,D_x)\, u\|_{L^2(\mathbb{R}_t \times \mathbb{X})} \lesssim \|u_0\|_{L^2(\mathbb{X})}.$$

The type (II) and (III) estimates have been partially derived in [11]. Notably, the regularity condition of the type (III) estimate relies solely on the pseudo-dimension ν, in contrast to the Euclidean setting where it depends on the manifold dimension. It is worth noting that this range of regularity is optimal, as the type (III) estimate fails for all $m \geq \nu$. The type (I) estimate is also optimal in the case where the homogeneous space $\mathbb{X} = G/K$ involves a complex G. In such a case, the manifold dimension and the pseudo-dimension coincide, the type (I) estimate fails for any $\alpha \leq (m-\nu)/2$ or $\alpha \geq (m-1)/2$.

An interesting observation arises in the case where $\mathbb{X}$ has rank one (then the pseudo-dimension $\nu = 3$). For instance, consider the Schrödinger equation ($m = 2$) on a 2-dimensional hyperbolic plane, the above theorem shows that $A(x, D) = |x|^{-1}$ satisfies the smoothing property. It is known that the weight $|x|^{-1}$ does not enjoy such a property on a Euclidean plane, see for instance [12, 15, 16]. Such differences highlight the particular geometry at infinity of the hyperbolic plane in comparison to the Euclidean setting.

4 Conclusion

There are several open questions that require further investigation. Firstly, it remains unclear whether the regularity condition of the type (I) estimate is sharp for higher ranks. It is expected that it depends solely on the pseudo-dimension, similar to the type (III) estimate. Additionally, in the non-linear PDE studies, it is crucial to establish smoothing properties for (6) with a linear forcing term, which has been partially accomplished in [11]. By utilizing the arguments presented therein, in conjunction with the improved Stein-Weiss inequality derived in [13], one can obtain the necessary estimates to analyze the corresponding non-linear equations. Furthermore, it is also natural to look for the geometric conditions under which the discrete group should satisfy in order to obtain similar results on corresponding locally symmetric spaces with regard to the Strichartz inequality.

Acknowledgments The authors are supported by the FWO Odysseus 1 grant G.0H94.18N: Analysis and Partial Differential Equations and by the Methusalem programme of the Ghent University Special Research Fund (BOF) (Grant number 01M01021).

References

1. J.-P. Anker, V. Pierfelice, Nonlinear Schrödinger equation on real hyperbolic spaces. Ann. Inst. H. Poincaré Anal. Non Linéaire **26**, 1853–1869 (2009)
2. J.-P. Anker, V. Pierfelice, Wave and Klein-Gordon equations on hyperbolic spaces. Anal. PDE **7**, 953–995 (2014)
3. J.-P. Anker, H.-W. Zhang, Wave equation on general noncompact symmetric spaces. Am. J. Math. (to appear). arXiv: 2010.08467
4. J.-P. Anker, V. Pierfelice, M. Vallarino, Schrödinger equations on Damek-Ricci spaces. Commun. Partial Differ. Equ. **36**, 976–997 (2011)
5. J.-P. Anker, S. Meda, V. Pierfelice, M. Vallarino, H.-W. Zhang, Schrödinger equation on noncompact symmetric spaces. J. Differ. Equ. **356**, 163–187 (2023)
6. V. Banica, The nonlinear Schrödinger equation on hyperbolic space. Commun. Partial Differ. Equ. **32**, 1643–1677 (2007)

7. V. Banica, T. Duyckaerts, Weighted Strichartz estimates for radial Schrödinger equation on noncompact manifolds. Dyn. PDE **4**, 337–359 (2007)
8. T. Cazenave, *Semilinear Schrödinger Equations*. Courant Lecture Notes in Mathematics, vol. 10 (New York University, Courant Institute of Mathematical Sciences/American Mathematical Society, New York/Providence, 2003)
9. S. Helgason, *Groups and Geometric Analysis: Integral Geometry, Invariant Differential Operators, and Spherical Functions* (Corrected reprint of the 1984 original). Mathematical Surveys and Monographs, vol. 83 (American Mathematical Society, Providence, 2000)
10. A.D. Ionescu, G. Staffilani, Semilinear Schrödinger flows on hyperbolic spaces: scattering H1. Math. Ann. **345**, 133–158 (2009)
11. K. Kaizuka, Resolvent estimates on symmetric spaces of noncompact type. J. Math. Soc. Jpn. **66**, 895–926 (2014)
12. T. Kato, K. Yajima, Some examples of smooth operators and the associated smoothing effect. Rev. Math. Phys. **1**, 481–496 (1989)
13. V. Kumar, M. Ruzhansky, H.-W. Zhang, Smoothing properties of dispersive equations on non-compact symmetric spaces. Preprint. arXiv: 2302.03961
14. J. Metcalfe, M. Taylor, Nonlinear waves on 3D hyperbolic space. Trans. Am. Math. Soc. **363**, 3489–3529 (2011)
15. M. Ruzhansky, M. Sugimoto, Smoothing properties of evolution equations via canonical transforms and comparison principle. Proc. Lond. Math. Soc. (3) **105**, 393–423 (2012)
16. M. Sugimoto, Global smoothing properties of generalized Schrödinger equations. J. Anal. Math. **76**, 191–204 (1998)
17. T. Tao, Nonlinear dispersive equations: local and global analysis, in *CBMS Regional Conference Series in Mathematics*, vol. 106 (American Mathematical Society, Providence, 2006)
18. D. Tataru, Strichartz estimates in the hyperbolic space and global existence for the semilinear wave equation. Trans. Am. Math. Soc. **353**, 795–807 (2001)
19. H.-W. Zhang, Wave equation on certain non-compact symmetric spaces. Pure Appl. Anal. **3**, 363–386 (2021)

Analysis on Compact Symmetric Spaces: Eigenfunctions and Nonlinear Schrödinger Equations

Yunfeng Zhang

Abstract We discuss several open problems on harmonic analysis on compact globally symmetric spaces, and their applications towards nonlinear Schrödinger equations.

Keywords Compact globally symmetric space · Joint eigenfunction · Spherical function · Laplacian · Nonlinear Schrödinger equation · Strichartz estimate · Multilinear estimate

2000 Mathematics Subject Classification 22E30, 35Q55, 43A85, 43A90, 58J05

1 Compact Globally Symmetric Spaces

Compact globally symmetric spaces are compact Riemannian manifolds whose geodesic symmetries are global isometries. Despite this seemlingly simple description of them, they are very rigid structures. Through the fundamental theory of Cartan-Weyl on classification and representations of compact Lie groups, (simply connected) compact globally symmetric spaces can be classified in terms of certain combinatorial structures called the Satake diagrams, and a basic spectral decomposition of the space of square-integrable functions as acted upon by the isometry group can be obtained. We briefly describe this spectral decomposition, based on which we unfold this paper.

Let $X = U/K$ be a compact globally symmetric space (CGSS), realized as the quotient of the compact isometry group U by a certain subgroup K of U. X is canonically equipped with a U-invariant measure and so one may consider the space $L^2(X)$ of square integrable functions with respect to this measure. Under the action

Y. Zhang (✉)
School of Mathematical Sciences, Peking University, Beijing, China

D. Cardona et al. (eds.), *Extended Abstracts 2021/2022*, Research Perspectives Ghent Analysis and PDE Center 3, https://doi.org/10.1007/978-3-031-48579-4_24

of U by translation, $L^2(X)$ disintegrate into a direct sum of spherical representations of U so that one may write

$$f = \sum_{\lambda \in \Lambda^+} f_\lambda, \ f \in L^2(X), \tag{1}$$

where Λ^+ is the set of isomorphism classes of spherical representations of U with respect to K, and f_λ lies in the representation space $L^2_\lambda(X)$ (consisting of smooth functions) of finite dimension d_λ corresponding to $\lambda \in \Lambda^+$. Λ^+ has the structure of being the Weyl chamber part of a lattice $\Lambda \cong \mathbb{Z}^r$ of a Euclidean space $\mathbb{R}^r$ acted upon by a Weyl group, with r being the rank of the X, i.e., the maximal dimension of a flat totally geodesic submanifold of X. We let d denote the dimension of X throughout the paper.

2 Eigenfunction Bounds and Spherical Functions

It turns out there is another description of the above spectral decomposition of $L^2(X)$ using differential calculus. As a Riemannian manifold, X is equipped with the Laplace-Beltrami differential operator (or simply the Laplacian) Δ_X which plays important roles in the analysis of X. But Δ_X is only a special element of the ring $\mathbf{D}(X)$ of U-invariant differential operators on X. A fundamental theorem is that $\mathbf{D}(X)$ as an algebra is commutative and isomorphic to a polynomial ring $\mathbb{C}[Z_1, \ldots, Z_r]$. Simultaneously diagonalizing all elements of $\mathbf{D}(X)$ as unbounded operators on $L^2(X)$ yields exactly (1). In other words, each f_λ in (1) is a joint eigenfunction of $\mathbf{D}(X)$, so that

$$Df_\lambda = \Gamma(D)(\lambda) f_\lambda, \ D \in \mathbf{D}(X).$$

Here $\Gamma : \mathbf{D}(X) \to \mathbb{C}[Z_1, \ldots, Z_r]$ is an appropriately chosen isomorphism of $\mathbb{C}$-algebras, and $\Gamma(D)(\lambda)$ is the evaluation of the polynomial $\Gamma(D)$ at $\lambda \in \Lambda^+ \subset \mathbb{Z}^r$. In particular, f is an eigenfunction of the Laplacian Δ_X of eigenvalue $-N^2$ if and only if there exists $\lambda \in \Lambda^+$ such that $\Gamma(\Delta_X)(\lambda) = -N^2$, and that

$$f = \sum_{\Gamma(\Delta_X)(\lambda) = -N^2} f_\lambda. \tag{2}$$

A major problem of analysis on manifolds is estimating L^p norms of Laplacian eigenfunctions in terms of their Laplacian eigenvalues. Equation (2) tells that this problem posed on CGSSs is closely related to the problem of estimating L^p norms of joint eigenfunctions, for which we have a precise conjecture due to Marshall [6].

Conjecture 1 Let ψ be a joint eigenfunction on an irreducible CGSS of Laplacian eigenvalue $-N^2 \leq -1$. Then

$$\|\psi\|_{L^p(X)} \leq \begin{cases} CN^{\frac{d-r}{2}-\frac{d}{p}}\|\psi\|_{L^2(X)}, & \text{if } p > \frac{2(d+r)}{d-r}, \\ CN^{\frac{d-r}{2}\left(\frac{1}{2}-\frac{1}{p}\right)}\|\psi\|_{L^2(X)}, & \text{if } 2 \leq p < \frac{2(d+r)}{d-r}. \end{cases} \tag{3}$$

Marshall in [6] established this conjecture under a regularity condition on the spectral parameter λ of ψ, that λ stays a certain distance away from the walls of the Weyl chamber. He also showed that either of the two pieces of the bound (3) is saturated on any CGSS. The piece for $p > \frac{2(d+r)}{d-r}$ is saturated by spherical functions, which are K-invariant joint eigenfunctions. With the results of [11] on compact Lie groups in mind, we also conjecture the following bounds on general spherical functions.

Conjecture 2 Let ψ be a spherical function of Laplacian eigenvalue $-N^2 \leq -1$ on an irreducible CGSS. Then

$$\|\psi\|_{L^p(X)} \leq \begin{cases} CN^{\frac{d-r}{2}-\frac{d}{p}}\|\psi\|_{L^2(X)}, & \text{if } p > \frac{2d}{d-r}, \\ C, & \text{if } 2 \leq p < \frac{2d}{d-r}. \end{cases} \tag{4}$$

Spherical functions are important, as they also appear as convolution kernels for the projection operators $L^2(X) \to L^2_\lambda(X)$, $f \mapsto f_\lambda$. There is a unique spherical function ψ_λ of spectral parameter λ which evaluates to be one at the identity coset. Then

$$f_\lambda = f * (d_\lambda \psi_\lambda).$$

On the other hand, the other piece of (3) for $2 \leq p < \frac{2(d+r)}{d-r}$ is saturated by the higher-rank Gaussian beam functions e_λ, which corresponds to the highest weight vector for the spherical representation of parameter λ. The reader may wonder if there are any relations between spherical functions ψ_λ and Gaussian beams e_λ, and indeed there are. First, the integral formula holds

$$\psi_\lambda(x) = \int_K e_\lambda(k^{-1}x)\, dk. \tag{5}$$

There is a way to rewrite this formula into a more complicated one, replacing the integrand $e_\lambda(kx)$ by a globally defined exponential function with explicit phrase [3], which can then be applied nicely to establish certain Laplacian eigenfunction bounds on CGSSs [9]. Furthermore, one can develop a finer Fourier theory than (1) at least locally. For smooth functions f supported in a small neighborhood X_o of the identity coset, a Helgason-Fourier transform $\tilde{f}(\lambda, k)$ of f may be defined [4] so that the following Poisson integral formula holds

$$f_\lambda(x) = d_\lambda \int_K \tilde{f}(\lambda, k) e_\lambda(k^{-1}x)\, dk, \ x \in X_o. \tag{6}$$

Question 3 Can (6) be globalized?

For rank-one spaces, this question was answered in the positive by Sherman [7]. A answer in the positive for general higher rank spaces could be useful to bound joint eigenfunctions and establish unconditionally Conjecture 1.

Straightforwardly combing (2) with standard estimate of representations of an integer by a positive definite integral quadratic form, one could get certain estimates on Laplacian eigenfunctions, which are however never sharp. A major open question is as follows.

Question 4 Let f be a Laplacian eigenfunction of eigenvalue $-N^2 \leq -1$ on an (irreducible) CGSS. Consider bound of the form

$$\|f\|_{L^p(X)} \leq C N^{s(p)} \|f\|_{L^2(X)}.$$

What is the sharp exponent $s(p)$ for all $p \geq 2$?

For partial progress to this question, we refer the reader to our works [9, 11].

3 NLS on CGSSs

The eigenfunction bounds as explained in the previous section are closedly related to solving the following initial value problem for the nonlinear Schrödinger equation

$$\begin{cases} i\partial_t u + \Delta u = \pm |u|^{\beta-1} u, \ u = u(t, x), \ t \in \mathbb{R}, \ x \in X, \\ u(0, x) = u_0(x) \in H^s(X), \ x \in X. \end{cases}$$

Pretending X to be $\mathbb{R}^d$, one has the scaling symmetry $u(t, x) \mapsto \lambda^{\frac{2}{\beta-1}} u(\lambda^2 t, \lambda x)$ for the solutions, and the H^s-norm of the initial data is invariant under this scaling symmetry if and only if

$$s = s_c := \frac{d}{2} - \frac{2}{\beta - 1},$$

and in this case the IVP is called of critical regularity. Solving the IVP of critical regularity ($s = s_c$) or of subcritical regularity for the full range $s > s_c$ of Sobolev exponent usually requires establishing scale invariant Strichartz estimate for the linear Schödinger propagator, for which we formulate the following open question. Let I be a finite time interval.

Question 5 Consider Strichartz estimate of the form

$$\|e^{it\Delta}f\|_{L^p(I\times X)} \leq C\|f\|_{H^{s(p)}(X)} \tag{7}$$

on an (irreducible) CGSS. What is the optimal range of p for which the above estimate holds with the scale invariant exponent $s(p) = \frac{d}{2} - \frac{d+2}{p}$? Futhermore, what is the optimal Sobolev exponent $s(p)$ for all $p \geq 2$?

We refer the reader to our works [8–11] for progress towards this question. Through the works of Burq-Gérard-Tzvetkov [1, 2], in addition to the linear Strichartz estimate (7), bilinear or multilinear Strichartz versions would also be need to treat the IVPs of critical regularity. We now focus on bilinear estimates tailored for the case $\beta = 3$. For this we formulate the following conjecture, inspired by the works of Burq-Gérard-Tzvetkov [1, 2] as well as Herr-Strunk [5].

Conjecture 6 Suppose the rank of X is at least two. Let u_i be functions on X with localized Laplacian spectrum so that $\mathbb{1}_{[N_i,2N_i]}(\sqrt{-\Delta})u_i = u_i$, $i = 1, 2$. Suppose $N_1 \geq N_2 \geq 1$.

(i) Suppose the dimension of each irreducible factor of X is at least 3. Then

$$\|e^{it\Delta}u_1\, e^{it\Delta}u_2\|_{L^2(I\times X)} \leq C_\varepsilon N_2^{\frac{d}{2}-1+\varepsilon}\|u_1\|_{L^2(X)}\|u_2\|_{L^2(X)}.$$

This would imply local well-posedness of the IVP for $s > s_c = \frac{d}{2} - 1$ ($\beta = 3$).

(ii) Suppose the dimension of each irreducible factor of X is at least 4. Then there exists $\delta > 0$ such that

$$\|e^{it\Delta}u_1\, e^{it\Delta}u_2\|_{L^2(I\times X)} \leq C\left(\frac{N_2}{N_1} + \frac{1}{N_2}\right)^{\delta} N_2^{\frac{d}{2}-1}\|u_1\|_{L^2(X)}\|u_2\|_{L^2(X)}.$$

This would imply local well-posedness for $s = s_c = \frac{d}{2} - 1$ ($\beta = 3$).

In [9] we established this conjecture when X is a product of rank-one CGSSs. We reduced the above conjecture to bilinear estimates for joint eigenfunctions, for which we conjecture the following.

Conjecture 7 Suppose the rank of X is at least two. Let f_i be joint eigenfunctions of the ring of invariant differential operators of Laplacian eigenvalue $-N_i^2 \leq -1$, $i = 1, 2$.

(i) Suppose the dimension of each irreducible factor of X is at least 3. Then

$$\|f_1 \cdot f_2\|_{L^2(X)} \leq C_\varepsilon(\min(N_1, N_2))^{\frac{d}{2}-r+\varepsilon}\|f_1\|_{L^2(X)}\|f_2\|_{L^2(X)}.$$

(ii) Suppose the dimension of each irreducible factor of X is at least 4.Then

$$\|f_1 \cdot f_2\|_{L^2(X)} \le C(\min(N_1, N_2))^{\frac{d}{2}-r}\|f_1\|_{L^2(X)}\|f_2\|_{L^2(X)}.$$

Note that when $f_1 = f_2$, the estimates in the above conjecture reduce to the L^4 case of the bounds (3) of Marshall.

4 Conclusion

Compact globally symmetric spaces, as classical objects in differential geomety, still carry a lot of fundamental open questions in their analysis, such as bound of spherical functions, bound of joint eigenfunctions of invariant differential operators, bound of Laplacian eigenfunctions, Strichartz estimate, and globalization of Poisson integrals. These questions are closedly related, and resolution of them will for sure deepen our understanding of CGSSs and open new routes for analysis on general compact manifolds.

References

1. N. Burq, P. Gérard, N. Tzvetkov, Bilinear eigenfunction estimates and the nonlinear Schrödinger equation on surfaces. Invent. Math. **159**(1), 187–223 (2005)
2. N. Burq, P. Gérard, N. Tzvetkov, Multilinear eigenfunction estimates and global existence for the three dimensional nonlinear Schrödinger equations. Ann. Sci. École Norm. Sup. (4) **38**(2), 255–301 (2005)
3. J.-L. Clerc, Fonctions sphériques des espaces symétriques compacts. Trans. Am. Math. Soc. **306**(1), 421–431 (1988)
4. S. Helgason, *Geometric Analysis on Symmetric Spaces*. Mathematical Surveys and Monographs, vol. 39, 2nd edn. (American Mathematical Society, Providence, 2008)
5. S. Herr, N. Strunk, The energy-critical nonlinear Schrödinger equation on a product of spheres. Math. Res. Lett. **22**(3), 741–761 (2015)
6. S. Marshall, L^p norms of higher rank eigenfunctions and bounds for spherical functions. J. Eur. Math. Soc. **18**(7), 1437–1493 (2016)
7. T. Sherman, The Helgason Fourier transform for compact Riemannian symmetric spaces of rank one. Acta Math. **164**, 73–144 (1990)
8. **Y. Zhang, Strichartz estimates for the Schrödinger flow on compact Lie groups. Anal. PDE 13**(4), 1173–1219 (2020)
9. Y. Zhang, Schrödinger equations on compact globally symmetric spaces. J. Geom. Anal. **31**(11), 10778–10819 (2021)
10. Y. Zhang, Strichartz estimates for the Schrödinger equation on products of odd-dimensional spheres. Nonlinear Anal. 199, 112052, 21 pp. (2020). arXiv:2301.02823
11. Y. Zhang, On Fourier restriction type problems on compact Lie groups. Indiana Univ. Math. J., to appear (2023). arXiv:2005.11451

Part III
Applied Mathematics

On Empirical Bayes Approach to Inverse Problems

E. Belitser

Abstract Inverse problems arise naturally in many scientific disciplines, such as physics, imaging, tomography, medicine, material sciences and engineering, when one wants to extract information from indirect and noisy measurements. Observed are noisy results of certain (forward) operator evaluated at an element from certain Hilbert space. The general objective is to recover that element using the observed data. Some important inverse problems arise in the area of partial differential equations (PDEs) which describe some physical systems. We consider the inverse problem in a statistical setting and apply an empirical Bayes approach. We address the issue of (local) optimality in the framework of oracle inequalities which is stronger than the traditional minimax (global) optimality. The Bayesian modeling allows to solve a local version of the problem in the oracle formulation, leading to intrinsically adaptive results, i.e., we establish the optimality of the proposed procedure without knowledge of the smoothness/sparsity structure of the underlying function of interest.

Keywords Empirical Bayes · Inverse problem · Oracle rate · PDE · Posterior contraction

2000 Mathematics Subject Classification 35Q62, 62C05, 62C10

1 Introduction

Inverse problems arise naturally in many scientific disciplines, such as physics, imaging, tomography, medicine, material sciences and engineering, when one wants to extract information from indirect measurements. The general inverse problem consists of the solution f of the equation $\mathcal{G}(f) = y$, where operator $\mathcal{G}$ maps a subset

E. Belitser (✉)
VU Amsterdam, Amsterdam, The Netherlands
e-mail: e.n.belitser@vu.nl

D. Cardona et al. (eds.), *Extended Abstracts 2021/2022*, Research Perspectives
Ghent Analysis and PDE Center 3, https://doi.org/10.1007/978-3-031-48579-4_25

of one Banach space to another Banach space. Here we assume more structure by letting the involved spaces to be Hilbert. Inverse problems can be ill-posed: there may be no solution, or the solution may not be unique and may depend sensitively on y; cf. [6, 18]. We study the inverse problem from the statistical perspective, namely, we observe typically noisy results of certain (forward) operator evaluated at an element from certain Hilbert space; cf. [4, 5, 8, 9, 15, 17]. The measurement errors are modeled as independent random elements of an appropriate Hilbert space. Each error being itself a superposition of many independent random effects, a Gaussian model for the errors would be reasonable in view of the central limit theorem.

Formally, let $\mathcal{G} : \mathbb{H}_X \mapsto \mathbb{H}_Y$, where $\mathcal{G}$ is an operator mapping separable Hilbert space $\mathbb{H}_X$ to another separable Hilbert space $\mathbb{H}_Y$. In practice observations are typically discrete, as a sample at a number of 'design points', but we adopt the paradigm 'first regularization, then discretization', so we formulate the problem in general functional form. For $f \in \mathbb{H}_X$, we observe

$$Y = \mathcal{G}(f) + \varepsilon\xi, \tag{1}$$

in the sense that $Y(g) = \langle \mathcal{G}(f), g\rangle + \varepsilon\xi(g)$, $g \in \mathbb{H}_Y$, where $\varepsilon > 0$ is the noise level, ξ is Gaussian white noise on $\mathbb{H}_Y$, i.e., $\xi : \mathbb{H}_Y \to \mathbb{R}$ is a linear mapping on $\mathbb{H}_Y$ such that $\xi(g) \sim N(0, \|g\|^2)$ and $\mathrm{Cov}(\xi(g), \xi(g)) = \langle g, g'\rangle$ for any $g, g' \in \mathbb{H}_Y$, $\|\cdot\|$ and $\langle\cdot,\cdot\rangle$ denote the norm and scalar product in $\mathbb{H}_Y$ (or in $\mathbb{H}_X$, this should always be clear from the context). The sequence version of (1) is obtained by taking an orthonormal basis $(\phi_k)_{k\in\mathbb{N}}$ in $\mathbb{H}_Y$:

$$Y_k = g_k(f) + \varepsilon\xi_k, \quad k \in \mathbb{N}, \tag{2}$$

where $Y_k = Y(\phi_k)$, $g_k(f) = \langle \mathcal{G}(f), \phi_k\rangle$, $\xi_k \overset{\text{ind}}{\sim} N(0, 1)$. We refer to Y as data and the goal is (for now loosely formulated) to recover $f \in \mathbb{H}_X$ using the data. Typically, the problem is studied in asymptotic regime $\varepsilon \to 0$, but, whenever possible, we prefer non-asymptotic characterizations as asymptotic claims would follow from it. Noise level ε in model (2) can be associated to sample size n in some other related statistical models as $\varepsilon^2 = n^{-1}$.

In this note we focus on the case when $\mathcal{G}$ is a continuous linear operator and $\mathcal{G}^*\mathcal{G}$ is compact ($\mathcal{G}^*$ stands for the adjoint of $\mathcal{G}$). Examples of the operator $\mathcal{G}$ (for appropriate functional spaces $\mathbb{H}_X, \mathbb{H}_Y$) include various integral operators (in particular, convolution operator, Radon transform, etc.), fractional derivatives (including negative ones). Some important inverse problems arise in the area of PDEs and ODEs. The observations in such systems are the noisy solutions of a PDE and the object of interest is, for example, a functional parameter of the equation (say, functional coefficient). A couple interesting examples motivated by the PDEs are provided in Sect. 3. The case of non-linear $\mathcal{G}$ is briefly discussed also in Sect. 3.

The case of linear compact $\mathcal{G}$ is relatively well investigated in the literature; cf. [1, 2, 4, 5, 12]. The main feature in this case is that the effect of operator $\mathcal{G}$ (how ill-posed it is) can be explicitly expressed in terms of the behavior of the eigenvalues of $\mathcal{G}^*\mathcal{G}$. However, one problem is that one typically needs to impose the smoothness assumptions of the underlying functional parameter $f \in \mathbb{H}_X$ in terms of the Fourier coefficients $\{f_k\}_{k\in\mathbb{N}}$ with respect to the orthonormal basis of the eigenfunctions of $\mathcal{G}^*\mathcal{G}$, e.g., Sobolev ellipsoid of smoothness $\alpha > 0$: $f \in \mathcal{E}_\alpha = \{f \in \mathbb{H}_X : \sum_{k\in\mathbb{N}} k^{2\alpha} f_k^2 \leq Q\}$. Precisely, if the eigenvalues of $\mathcal{G}^*\mathcal{G}$ behave like $\lambda_k \asymp k^{-2\beta}$, then the minimax risk for estimating f is of the order $R(\mathcal{E}_\alpha) = \inf_{\hat{f}} \sup_{f\in\mathcal{E}_\alpha} E_\theta\|\hat{f} - f\|^2 \asymp \varepsilon^{4\alpha/(2\alpha+2\beta+1)}$ (in asymptotic regime $\varepsilon \to 0$), the infimum is taken over all estimators $\hat{f} = \hat{f}(Y)$, measurable functions of the observed data. The minimax risk $R(\mathcal{E}_\alpha)$ measures the statistical difficulty of estimating uniformly over the class $\mathcal{E}_\alpha$. The rate exhibits explicitly the effect of smoothness α of f and ill-posedness β of the operator $\mathcal{G}$.

We tackle the above mentioned problem of smoothness assumption on f by providing local results of oracle type in this note. A local result basically asserts that the method is locally adaptive in the sense that no specific smoothness (or sparsity) assumption on f is imposed, instead, the procedure is shown to utilise as much of the smoothness as there is in f: if the true f is 'well approximated' by a smooth function f_{appr}, the procedure provides the estimation quality as good as one would have for that smooth function f_{appr}. The idea is to slice the 'big' space $\mathbb{H}_X \ni f$ in layers $\mathbb{H}_X = \cup_{S\in\mathbb{S}}\mathcal{F}_S$, where $\mathbb{S}$ is a family of *structures* S, and then show that the procedure 'mimics the oracle' over family $\mathbb{S}$. This means roughly that our procedure performs as good as for the best (oracle) choice of structure $S_0 \in \mathbb{S}$, which the oracle would take by using the knowledge of the true f. Another interesting notion is covering smoothness scales by the oracle over a family of structures, introduced and discussed in [2]. For example, if we knew that f has certain smoothness, say, α, then there exists an $S_0 \in \mathbb{S}$ such that the performance over the layer $\mathcal{F}_{S_0}$ is always not worse than over the set of all functions of smoothness α. This means that the local result in terms of oracle is genuinely adaptive: if f is α-smooth (and α is unknown to the observer), mimicking oracle guarantees that we always provide at least the quality for estimating α-smooth functions.

We apply the Bayesian approach. Bayesian methods are widely used for inverse problems, we refer to [20] for an extensive overview on the topic; see also [1, 2, 7, 10–14, 16, 19, 21]. Basically, when formulated in a Bayesian fashion, a wide range of inverse problems can be treated within a common mathematical framework, with a certain well-posedness that stems from this. The point is that, when applying Bayesian approach, even if the operator may not be invertible, the posterior for the unknown (possibly infinite-dimensional) functional parameter may still be well defined, thus in doing so regularizing the problem even if it was originally ill-posed. This well-posedness implied by the Bayesian method provides the basis for some useful stability and approximation results, including some specific algorithms used when adopting the Bayesian approach to inverse problems, e.g., MCMC methods, filtering and the variational approach.

We propose a hierarchical prior, namely, first, given structure S, we put a prior on f conjugate to the model, next, instead of putting a prior on the family of structures $\mathbb{S}$, we propose a data dependent choice $\hat{S}$ of the structure (which we can use for model selection problem) and use it in the resulting posterior for f and in the construction of an estimator of f. This is a version of the *empirical Bayes* approach. The performance of this resulting empirical Bayes posterior is measured from the frequentists perspective: the rate with which the posterior concentrates around the 'true' $f \in \mathbb{H}_X$ when the data is assumed to come from the model (2) with the true f. The distinctive feature of our approach is that the established concentration posterior rate is local, i.e., it depends on the 'true' f, hence intrinsically adaptive. Global results over suitable scale of smoothness classes follows from our local result: if f happens to be α-smooth (α is not known to the observer), the local rate is always not worse than the optimal rate over the class of α-smooth functions. The posterior can further be used to construct point estimators of f (also of the local rate), e.g., the posterior expectation.

2 Compact $\mathcal{G}^*\mathcal{G}$, Inverse Signal-in-White-Noise Model

Coming back to model (2), suppose that $\mathcal{G}^*\mathcal{G}$ ($\mathcal{G}^*$ stands for the adjoint of $\mathcal{G}$) is a compact operator so that it has a complete orthonormal system of eigenvectors $(\psi_k)_{k\in\mathbb{N}}$ in $\mathbb{H}_X$ with corresponding eigenvalues $\lambda_k > 0$, i.e., $\mathcal{G}^*\mathcal{G}\psi_k = \lambda_i\psi_i$. Then $(\phi_k)_{k\in\mathbb{N}}$, with $\phi_k = \lambda_k^{-1/2}\mathcal{G}\psi_k$, is an orthonormal basis in $\mathbb{H}_Y$, and $\mathcal{G}^*\phi_k = \lambda_k^{-1/2}\mathcal{G}^*\mathcal{G}\psi_k = \lambda^{1/2}\psi_k$, which we now going to use in (2). Denoting by $f_k = \langle f, \psi_k\rangle$ the Fourier coefficient of f with respect to $(\psi_k)_{k\in\mathbb{N}}$, we have $\mathcal{G}^*\mathcal{G}f = \mathcal{G}^*\mathcal{G}\sum_{k\in\mathbb{N}} f_k\psi_k = \sum_{k\in\mathbb{N}}\lambda_k f_k\psi_k$. Finally we obtain (2) with $g_k(f) = \langle\mathcal{G}f, \phi_k\rangle = \lambda_k^{-1/2}\langle\mathcal{G}f, \mathcal{G}\psi_k\rangle = \lambda_k^{-1/2}\langle\mathcal{G}^*\mathcal{G}f, \psi_k\rangle = \lambda_k^{1/2} f_k$: with $\kappa_k = \lambda_k^{-1/2}$,

$$Y_k = f_k/\kappa_k + \varepsilon\xi_k, \quad \text{or, equivalently,} \quad X_k = f_k + \varepsilon\kappa_k\xi_k, \quad k \in \mathbb{N}, \tag{3}$$

where $X_k = \kappa_k Y_k$, the underlying signal $\theta = (f_k)_{k\in\mathbb{N}} \in \ell_2$, the noises $\xi_k \overset{\text{ind}}{\sim} N(0, 1)$. Since the sequence $\lambda_k > 0$ can only have one accumulation point equal to zero, without loss of generality we can assume $\kappa_k \geq 1$.

Model (3) is known to be the sequence version of the *inverse signal-in-white-noise model*. This model is often considered for the estimation inference in the inverse problem literature, cf. [4, 18]. It captures many of the conceptual issues associated with nonparametric estimation.

Introduce the notion of *structure*: the structure $S = (n, I_n)$ is a pair consisting of $n \in \mathbb{N}_0 = \{0\} \cup \mathbb{N}$ and a set $I_n \subseteq [n] \equiv \{1, \ldots, n\}$, with the convention that $I_n = \varnothing$ if $n = 0$ in the pair. The family of all possible structures is denoted by $\mathbb{S} = \{S = (n, I_n), n \in \mathbb{N}_0, I_n \subseteq [n]\}$. We say that the underlying signal f has structure $S = (n, I_n)$ if $f = \sum_{k\in I_n} f_k\psi_k = f_S \in \mathbb{L}_S \subset \mathbb{H}_X$, where $\mathbb{L}_S$ is the $|I_n|$-dimensional linear subspace spanned by $(\psi_k)_{k\in I_n}$ and f_S is the projection of f

onto $\mathbb{L}_S$. Or, we say that f has approximately structure S if $\|f - f_S\|^2$ is 'small'. Introduce the *oracle* rate of structure S at f (i.e., the rate is local) as

$$r(f, S) = \|f - f_S\|^2 + \varepsilon^2 P_{\mathcal{G}}(S) = \sum_{k>n} f_k^2 + \sum_{k \in I_n^c} f_k^2 + \varepsilon^2 \log(\tfrac{en}{|I_n|}) \sum_{k \in I_n} \kappa_k^2,$$

where the first term $\|f - f_S\|^2$ reflects the approximation error of layer $\mathbb{L}_S$ and the second (penalty) term $\varepsilon^2 P_{\mathcal{G}}(S)$ describes its complexity and the level of ill-posedness of operator $\mathcal{G}$. For a fine tuning in further constructions, we could introduce different weights of the two terms by setting a constant $\varkappa'$ in front of the penalty, but not in this note. Basically, we slice the original 'big' space $\mathbb{H}_X = \cup_{S \in \mathbb{S}} \mathbb{L}_S$ in 'layers' $\mathbb{L}_S$ indexed by structures $S \in \mathbb{S}$, and the oracle strucure represent the best trade-off between the ability of the layer $\mathbb{L}_{S_o}$ to approximate (f) and the complexity of $\mathbb{L}_{S_o}$.

Using $r(f, S)$ defined above, for any $f \in \mathbb{H}_X$ define further the *oracle rate* as

$$r(f) = \min_{S \in \mathbb{S}} r(f, S) = r(f, S_o).$$

The structure $S_o = S_o(f) = (n_o, I_{n_o})$ (take any if there are many) where the minimum of $r(\theta, S)$ is achieved is called the *oracle structure* (or just the oracle).

2.1 *Constructing an Empirical Bayes Posterior*

Abusing slightly notation, set $f = (f_k)_{k \in \mathbb{N}}$, then Fourier coefficients of f with respect to the basis $(\psi_k)_{k \in \mathbb{N}}$. Introduce the hierarchical prior Π: for an $S \in \mathbb{S}$, $S = (n, I_n)$, $n \in \mathbb{N}$, $I_n \subseteq [n]$,

$$f|S \sim \Pi_{S,\mu(S)} = \bigotimes_k N\big(\mu_k(S), \tau_k^2(S)\big), \tag{4}$$

where $\mu(S) = (\mu_k(S), k \in [n])$ with $\mu_k(S) = \mu_{S,k} 1\{k \in I_n\}$, $\tau_k^2(S) = \varepsilon^2 \kappa_k^2 1\{k \in I_n\}$. The idea of structure $S = (n, I_n)$ is to model in the prior the truncating level n and the sparsity pattern I_n. Let $\varphi(x, \mu, \sigma^2)$ denote the normal density at point x with mean μ and variance σ^2. Prior (4) is conjugate to model (3) yielding immediately the corresponding posterior

$$\Pi_{S,\mu}(\cdot|X) = \bigotimes_k N\bigg(\frac{\tau_k^2(S) X_k^2 + \varepsilon\kappa_k^2 \mu_k(S)}{\varepsilon\kappa_k^2 + \tau_k^2(S)}, \frac{\varepsilon\kappa_k^2 \tau_k^2(S)}{\varepsilon\kappa_k^2 + \tau_k^2(S)}\bigg).$$

If we would put a prior λ_S on $S \in \mathbb{S}$, then the resulting two-level prior would have led to the corresponding marginal $P_{X,\mu}(X) = \sum_{S \in \mathbb{S}} \lambda_S \prod_k \varphi\big(X_k, \mu_k(S), \tau_k^2(S) +$

$\varepsilon^2\kappa_k^2)$. Even without specifying the prior λ_S, we can easily derive the estimator of the parameter $\mu = (\mu(S), S \in \mathbb{S})$ by maximizing the marginal $\mathrm{P}_{X,\mu}(X)$ with respect to μ: $\hat{\mu} = (\hat{\mu}(S), S \in \mathbb{S})$, $\hat{\mu}_k(S) = X_k 1\{k \in I_n\}$, $k \in \mathbb{N}$. Substituting $\hat{\mu}(S)$ into the expression for posterior $\Pi_{S,\mu}(\cdot|X)$ yields the so called *empirical Bayes posterior* with respect to $\mu(S)$: $\Pi_S(\cdot|X) = \bigotimes_k N\big(X_k 1\{k \in I_n\}, \frac{1}{2}\varepsilon^2\kappa_k^2 1\{k \in I_n\}\big)$.

We continue to use the empirical Bayes strategy also with respect to the choice of structure S. Namely, the data-dependent choice of structure is based on the following criterion: for $\varkappa > 0$,

$$\hat{S} = \hat{S}_\varkappa = \arg\max_{S\in\mathbb{S}} \Big\{ \sum_{k\in I_n} \Big(X_k^2 - \varkappa\varepsilon^2\kappa_k^2 \log(\tfrac{en}{|I_n|}) \Big) \Big\}.$$

A motivation for this criterion can be provided by connecting to model selection problems addressed by the penalization methodology. Here we confine ourselves to mentioning that the above criterion is an extension of the idea of *risk hull method* proposed in [5]. Plugging in $\hat{S} = (\hat{n}, \hat{I}_{\hat{n}})$ in $\Pi_S(\cdot|X)$ results in the empirical Bayes posterior

$$\Pi(\cdot|X) = \Pi_{\hat{S}}(\cdot|X) = \bigotimes_k N\big(X_k 1\{k \in \hat{I}_{\hat{n}}\}, \tfrac{1}{2}\varepsilon^2\kappa_k^2 1\{k \in \hat{I}_{\hat{n}}\}\big). \qquad (5)$$

To estimate f, we take the expectation with respect to the posterior $\Pi(\cdot|X)$:

$$\hat{f}_k = E_{\Pi(\cdot|X)}(f_k) = X_k 1\{k \in \hat{I}_{\hat{n}}\}, \quad k \in \mathbb{N}. \qquad (6)$$

Some constants in the sequel depend on $\varkappa$, but we will omit this dependence.

2.2 Main Result

Here we formulate the main result for the observation model (3). For any $f \in \mathbb{H}_X$, introduce the quantity $v(f) = \varepsilon^2 \log(\frac{en}{|I_o|}) \sum_{k\in I_o} \kappa_k^2$, where $I_o = I_o(f)$ is the oracle sparsity pattern coming from the oracle structure $S_o(f) = (n_o, I_o)$ pertinent to f. This is a part of the complexity term of the oracle rate, note that $v(f) \le r(f)$.

Theorem 1 *Consider model* (3). *Let the empirical Bayes posterior* $\Pi(\cdot|X)$ *and the estimator* $\hat{f}$ *be defined by* (5) *amd* (6) *respectively. Then there exist* $C_0, C_1, c_0, C_3, C_4, c_1 > 0$ *such that for any* $f_0 \in \mathbb{H}_X$ *and any* $M > 0$

$$E_{f_0}\Pi\big(\|f - f_0\|^2 \ge C_0 r(f_0) + M v_o(f_0)|X\big) \le C_1 e^{-c_0 M},$$
$$P_{f_0}\big(\|\hat{f} - f_0\|^2 \ge C_2 r(f_0) + M v_o(f_0)\big) \le C_3 e^{-c_1 M}.$$

Let us consider an example. By C, c denote generic constants varying from line to line. Suppose, f is Sobolev α-smooth in the sense that it belongs to the Sobolev ellipsoid of smoothness $\alpha > 0$: $f \in \mathcal{E}_\alpha = \{f \in \mathbb{H}_X : \sum_{k \in \mathbb{N}} k^{2\alpha} f_k^2 \leq Q\}$ and the eigenvalues of $\mathcal{G}^*\mathcal{G}$ behave like $\lambda_k \asymp k^{-2\beta}$, i.e., $\kappa_k^2 \asymp k^{2\beta}$. Then the minimax risk for estimating f is known to be of the order $R(\mathcal{E}_\alpha) = \inf_{\hat{f}} \sup_{f \in \mathcal{E}_\alpha} E_\theta \|\hat{f} - f\|^2 \asymp \varepsilon^{4\alpha/(2\alpha+2\beta+1)}$ (in the asymptotic regime $\varepsilon \to 0$). Our local result delivers this as a consequence. Indeed, take the structure $S_0 = (N_\varepsilon, [N_\varepsilon])$, where $N_\varepsilon = \lfloor \varepsilon^{2/(2\alpha+2\beta+1)} \rfloor$. Next we evaluate

$$r(f) \leq r(f, S_0) = \sum_{k > N_\varepsilon} f_k^2 + \varepsilon^2 \sum_{k=1}^{N_\varepsilon} \kappa_k^2 \leq \tfrac{Q}{N_\varepsilon^{2\alpha}} + C\varepsilon^2 N_\varepsilon^{2\beta+1} \leq c\varepsilon^{4\alpha/(2\alpha+2\beta+1)}.$$

The last relation, the second claim of the theorem and the fact $v(f_0) \leq r(f_0)$ imply the oracle and global optimal estimation results in expectation

$$\begin{aligned} E_{f_0}\|\hat{f} - f_0\|^2 &= \int_0^\infty P_{f_0}(\|\hat{f} - f_0\|^2 \geq t)dt \leq C_2 r(f_0) + C_3 \int_0^{+\infty} e^{-c_1 u/v(f_0)} du \\ &= C_2 r(f_0) + \tfrac{C_3 v(f_0)}{c_1} \leq C r(f_0) \leq c\varepsilon^{4\alpha/(2\alpha+2\beta+1)}. \end{aligned}$$

We emphasize that the scope of our local results is much broader than the above example might suggest. Typically, in the literature, particular situations in global settings are studied: e.g., the case $f \in \mathcal{E}_\alpha$ and $\kappa_k^2 \asymp k^{2\beta}$ is called mildly ill-posed case for the scale of Sobolev ellipsoids $\{\mathcal{E}_\alpha, \alpha > 0\}$. But then there are separate global results for the so called *severely ill-posed* cases such as $\kappa_k^2 \asymp e^{\beta k^\gamma}$ for some $\beta, \gamma > 0$, Sobolev hyper-rectangles, super-smooth scales of analytic ellipsoids, Besov classes, ℓ_p-bodies, tale classes, general ellipsoids, general hyper-rectangles, etc. Our local result covers all these cases at once, in fact it covers the global results for all scales $\{\mathcal{F}_\alpha, \alpha \in \mathcal{A}\}$ simultaneously for which we establish that the local (oracle) rate is dominated by the minimax rate over classes $\mathcal{F}_\alpha$, i.e., when we have $r(f) \leq cR(\mathcal{F}_\alpha)$ for all $f \in \mathcal{F}_\alpha$ and $\alpha \in \mathcal{A}$.

Another novel aspect of our approach is that the oracle S_o, next to the best smoothness structure (modeled by n_o), also contains the best sparsity structure (modeled by the sparsity pattern I_o). Exactly, assume that $f \in \ell_0[p_\varepsilon, N_\varepsilon] = \{(f_k)_{k \in \mathbb{N}} : \sum_{k=1}^{N_\varepsilon} 1\{f_k \neq 0\} \leq p_\varepsilon, f_k = 0, k > N_\varepsilon\}$, where $N_\varepsilon \in \mathbb{N}$, $p_\varepsilon \in [N_\varepsilon]$, which can be seen as at an extension of the traditional sparsity class with additional 'cut-off' parameter N_ε, introduced to connect an infinite dimensional setting to the standard high-dimensional sparsity setting. Let $I_\kappa = I_\kappa(N_\varepsilon, p_\varepsilon, \mathcal{G})$ be such that $\sum_{k \in I_\kappa} \kappa_k^2 = \max_{I \subseteq [N_\varepsilon], |I| \leq p_\varepsilon} \sum_{k \in I} \kappa_k^2$. If there are multiple minimizers, we take for example the one with the smallest cardinality and the smallest sum of indices. Then Theorem 1 delivers an oracle rate $r(f)$ for which we can show

$$r(f) \leq C\varepsilon^2 \log(\tfrac{eN_\varepsilon}{|I_\kappa|}) \sum_{k \in I_\kappa} \kappa_k^2.$$

Considering the case of $N_\varepsilon = n, \kappa_k = 1, k \in [n]$ (i.e., 'direct problem' instead of inverse problem), leads to the classical high-dimensional framework with a sparse signal $f \in \ell_0[p] = \{(f_k)_{k\in[n]} : \sum_{k\in[n]} 1\{f_k \neq 0\} \leq p\}$. The minimax estimation rate over $\ell_0[p]$ in the direct problem is known to be $\varepsilon^2 p \log(\frac{en}{p})$, which is covered by our local oracle rate: $r(f) \leq C\varepsilon^2 p \log(\frac{en}{p})$ as it follows from the last display.

3 Further Discussion

3.1 Uncertainty Quantification (UQ) Problem

Here we only shortly mention the uncertainty quantification (UQ) problem, this problem will be studied in detail elsewhere. Namely, the Bayesian approach allows to tackle the UQ problem which is in general a much more delicate than the estimation problem. This is because of the so called *deceptiveness phenomenon* which is in detain discussed in [2, 3]. The local nature of our Bayesian approach allows to address the more refined local version of the UQ problem; cf. [2, 3].

3.2 The Case of Non-Linear $\mathcal{G}$

The literature on the non-linear operator $\mathcal{G}$ is much more limited, and we are not aware of any general universal approach to such situation, only some situation-specific methods have been developed. One useful strategy based on the modulus of continuity for the operator $\mathcal{G}$ stems from the idea in [12]. We speculate that the next step would be to develop a local version of this method. Such a method can lead to locally optimal procedure in some situations, but in general would give no guarantee for optimality. On the other hand, at least it provides some claims on the quality of the resulting (Bayesian) procedure.

3.3 Examples of Inverse Problems Originated from PDEs

In case of an inverse problem originating from PDE(s), typically some partial differential operator $\mathcal{L}_f$ acting on $f : \mathcal{D} \mapsto \mathbb{R}$ is given, where f is the unknown functional parameter of interest defined on some regular bounded domain $\mathcal{D} \subseteq \mathbb{R}^d$, $\partial\mathcal{D}$ denotes the boundary of $\mathcal{D}$. Data is given in the form of some solution u of an operator equation $F(\mathcal{L}_f, u) = 0$ subject to some boundary conditions. We adopt the usual notation conventions of partial derivatives, like u_t, u_{xx}, etc; ∇ is the gradient and $\nabla\cdot$ is the divergence operator.

3.3.1 Heat Equation: Recovery of Initial Condition

Consider a problem of recovering the initial condition for the heat equation; cf. [13]. Specifically, assume we observe noisy observations of the solution u to the Dirichlet problem for the heat equation

$$u_t(x,t) = u_{xx}(x,t), \quad u(x,0) = f(x), \quad u(0,t) = u(1,t) = 0,$$

where u is defined on $[0,1] \times [0,T]$ and the function $f \in L_2[0,1]$ satisfies $f(0) = f(1) = 0$. The goal is to recover the initial condition f in the asymptotics $\varepsilon \to 0$.

It is known that the solution of the above boundary problem can be represented as $u(x,t) = \sqrt{2}\sum_{k\in\mathbb{N}} f_k e^{-k^2\pi^2 t}\sin(k\pi x)$, where the f_k's are the Fourier coefficients of f in the basis $(\phi_k)_{k\in\mathbb{N}}$ of eigenfunctions $\phi_k = \sqrt{2}\sin(k\pi x)$ of $\mathcal{G}$ on $L_2[0,1]$, and the corresponding eigenvalues are $\kappa_k = e^{k^2\pi^2 T}$, $k \in \mathbb{N}$. Thus $\mathcal{G}f = u(\cdot, T)$ is linear, and we are in the setting (3). In view of exponential behavior of the κ_k's, one speaks of severely ill-posed inverse problem. Assuming a Sobolev α-smoothness, i.e., $f \in \mathcal{E}_\alpha$, the reader is invited to derive that Theorem 1 yields the oracle rate $r(\theta) \le c\big[\log(\epsilon^{-2})\big]^\alpha$, the right hand side being known the optimal (minimax) rate over Sobolev α-smooth functions for the severely ill-posed inverse problem.

3.3.2 Elliptic PDE: Recovery of a Functional Coefficient

Let the operator $\mathcal{G}$ (non-linear this time) be described as $u_f = \mathcal{G}f$, a solution of elliptic PDE

$$\mathcal{L}_f u = \nabla\cdot(f_1\nabla u) - f_0 u = 0 \quad \text{on } \mathcal{D}, \quad \text{and} \quad u = g \quad \text{on } \partial\mathcal{D}. \tag{7}$$

Here f_1 models the coefficient of the PDE and f_0 is a potential term. We can treat either f_0 or f_1 as the unknown f of interest (assuming the other one known). If a sufficiently smooth 'source' function f_0 is given and $g = 0$, the problem of recovery of the unknown f_1 (under some mild regularity conditions and $f \ge k_0 > 0$) has been considered by Stuart [20], Vollmer [21], and Giordano and Nickl [7]. The case $f_1 = 1/2$ becomes a steady state Schrödinger equation, the problem of recovery of f_0 in this case is studied in [17].

Under suitable conditions the map $f \mapsto u_f$ is injective and we arrive at the problem of recovery of function f given the solution $u_f = \mathcal{G}f$ of the above PDE, corrupted by additive Gaussian white noise as in (1).

3.3.3 Parabolic PDE: Recovery of a Functional Coefficient

Introduce now some time evolution dynamics on a time interval $[0, T]$ to the previous example by considering solutions $u(x, t)$ to the parabolic PDE (the heat equation)

$$u_t(x, t) - \mathcal{L}_f u(x, t) = 0 \quad \text{for all} \quad (x, t) \in \mathcal{D} \times [0, T],$$

subject to an initial condition $u(\cdot, 0) = u_0$ and some boundary condition $u = g$ on $\partial\mathcal{D} \times (0, T)$, where $\mathcal{L}_f$ is given by (7) and g and u_0 are sufficiently smooth boundary and initial value functions respectively. Either f_0 or f_1 can be the unknown functional parameter f of interest, which we wish to recover on the basis of observed noisy solution $u_f = \mathcal{G}f$ of the PDE, as in (1). $\mathcal{G}$ is in general non-linear.

If $f_1 = 1/2$ and the functional parameter of interest is $f = f_0$, we obtain reaction-diffusion type of equation which can be used to describe ecological system with u being density of prey and f describing resources or the effect of predators. Under mild conditions providing $u > 0$, we can write the inverse operator $f = \frac{u_{xx} - 2u_t}{2u}$. The Bayesian approach for this situation is studied in [10] for discrete formulation.

Acknowledgments The research is supported by the Ghent Analysis and PDE Center. The author is grateful to M. Ruzhansky for stimulating discussions.

References

1. S. Agapiou, S. Larsson, A.M. Stuart, Posterior contraction rates for the Bayesian approach to linear ill-posed inverse problems. Stoch. Process. Appl. **123**, 3828–3860 (2013)
2. E. Belitser, On coverage and local radial rates of credible sets. Ann. Stat. **45**, 1124–1151 (2017)
3. E. Belitser, N. Nurushev, Needles and straw in a haystack: robust empirical Bayes confidence for possibly sparse sequences. Bernoulli **26**, 191–225 (2019)
4. L. Cavalier, Nonparametric statistical inverse problems. Inverse Probl. **24**, 1–19 (2008)
5. L. Cavalier, Y. Golubev, Risk hull method and regularization by projections of ill-posed inverse problems. Ann. Stat. **34**, 1653–1677 (2006)
6. H.W. Engl, M. Hanke, A. Neubauer, *Regularization of Inverse Problems*. Mathematics and Its Applications, vol. 375 (Kluwer Academic, Dordrecht, 1996)
7. M. Giordano, R. Nickl, Consistency of Bayesian inference with Gaussian process priors in an elliptic inverse problem. Inverse Probl. **36**, 085001 (2020)
8. Y. Golubev, The principle of penalized empirical risk in severely ill-posed problems. Prob. Theory Rel. Fields **130**, 18–38 (2004)
9. J. Kaipio, E. Somersalo, *Statistical and Computational Inverse Problems*. Applied Mathematical Sciences, vol.160 (Springer, New York, 2005)
10. H. Kekkonen, Consistency of Bayesian inference with Gaussian process priors for a parabolic inverse problem. Inverse Probl. **38**, 035002 (2022)
11. H. Kekkonen, M. Lassas, S. Siltanen, Posterior consistency and convergence rates for Bayesian inversion with hypoelliptic operators. Inverse Probl. **32**, 085005 (2016)
12. B. Knapik, J.-B. Salomond, A general approach to posterior contraction in nonparametric inverse problems. Bernoulli **24**, 2091–2121 (2018)

13. B.T. Knapik, A.W. van der Vaart, J.H. van Zanten, Bayesian inverse problems with Gaussian priors. Ann. Stat. **39**, 2626–2657 (2011)
14. B.T. Knapik, A.W. van der Vaart, J.H. van Zanten, Bayesian recovery of the initial condition for the heat equation. Commun. Stat. Theory Methods **42**, 1294–1313 (2013)
15. B. Mair, F.H. Ruymgaart, Statistical inverse estimation in Hilbert scales. SIAM J. Appl. Math. **56**, 1424–1444 (1996)
16. F. Monard, R. Nickl, G.P. Paternain, Efficient nonparametric Bayesian inference for x-ray transforms. Ann. Stat. **47**, 1113–1147 (2019)
17. F. Monard, R. Nickl, G.P. Paternain, Consistent inversion of noisy non-abelian x-ray transforms. Commun. Pure Appl. Math. **74**, 45–99 (2021)
18. F. O'Sullivan, A statistical perspective on ill-posed inverse problems. Stat. Sci. **1**, 502–527 (1986)
19. K. Ray, Bayesian inverse problems with non-conjugate priors. Electron. J. Stat. **7**, 2516–2549 (2013)
20. A.M. Stuart, Inverse problems: a Bayesian perspective. Acta Numer. **19**, 451–559 (2010)
21. S.J. Vollmer, Posterior consistency for Bayesian inverse problems through stability and regression results. Inverse Probl. **29**, 125011 (2013)

The Interferon Influence on the Infection Wave Propagation

A. Mozokhina and V. Volpert

Abstract Viral infection spreads in the infected tissue or in cell culture as a reaction-diffusion wave due to virus replication in the infected cells and its transport between the cells. The speed of wave propagation correlates with the virus virulence and the severity of the disease, while the total viral load with its infectivity, that is, the rate of transmission from infected to uninfected individuals. The first barrier to the progression of viral infection after its penetration into the organism is provided by the innate immune response mediated mainly by immune cells and interferon. The effectiveness of this response determines whether infection progression in the organism will be stopped at this stage or it will further develop stimulating the adaptive immune response. The interferon mediated immune response down-regulates virus penetration in the host cells and its replication inside the cells. In this work, we consider the interferon influence on the infection spreading in the tissue, and on virus infectivity and virulence.

Keywords Viral infection · Immune response · Interferon · Reaction-diffusion equations · Wave speed · Viral load

Mathematics Subject Classification 92D30, 35Q92, 92D25

1 Introduction

At the end of 2019, a new respiratory disease, which later became known as COVID-19, was detected in the Chinese city of Wuhan. The nonspecific nature of the symptoms, the long incubation period compared to the infectious time,

A. Mozokhina
Peoples' Friendship University of Russia, Moscow, Russia

V. Volpert (✉)
Institut Camille Jordan, UMR 5208 CNRS, University Lyon 1, Villeurbanne, France
e-mail: volpert@math.univ-lyon1.fr

D. Cardona et al. (eds.), *Extended Abstracts 2021/2022*, Research Perspectives Ghent Analysis and PDE Center 3, https://doi.org/10.1007/978-3-031-48579-4_26

the spread mainly by airborne droplets and the high contagiousness of the virus determined its rapid spread around the world, and turned into a pandemic. This disease has not yet been completely eliminated and continues to have an impact on various aspects of the life of the society. There were other widespread epidemics induced by respiratory virus infections, such as Spanish influenza in the first quarter of twentieth century, Middle East respiratory syndrome (MERS) in 2012, severe acute respiratory syndrome (SARS) in the beginning of the twenty-first century, and the annual flu epidemics. Our knowledge about the pathogenesis, epidemical, and immunological aspects of respiratory viral infections is still not comprehensive enough. Mathematical modelling of viral infections can help in their further understanding and characterization.

It has been shown in [3, 4] that infection spreads in cell culture as a reaction-diffusion wave. The main characteristics of infection progression are the virus replication number, the speed of wave propagation, and the total viral load. The virus replication number determines whether the infection spreads in cell culture, or, mathematically, whether the wave solution can exist. The speed of wave propagation corresponds to the severity of the disease as it determines how fast the infection spreads in the tissue or in the cell culture, and correspondingly, how much tissue will be damaged at given time. The total viral load, i.e., the total amount of extracellular virus for respiratory virus infections determines its infectivity through the amount of virus particles in airborne droplets. In [3], some estimates of all these characteristics were obtained. In this work, we evaluate the dependence of these characteristics on the parameters of innate immune response, namely, on the interferon (IFN) system.

After the virus enters the organism and starts releasing its genetic material, cells of the organism recognize a pathogen and activate the interferon (IFN) system. Interferon (IFN) is a collective name for macro-molecules which induce production of proteins suppressing virus replication in different ways. There are three types of IFNs: IFN-I, IFN-II, and IFN-III, which differ from each other by their molecular structure, cells producing them and production rate [5]. Produced IFNs activate IFN-stimulated genes (ISGs) in infected cells and in uninfected cells nearby [7] inducing production of proteins which directly suppress different stages of virus replication and penetration into cells. Interestingly, IFNs do not influence virus replication directly, they do it through the activation of ISGs and subsequent production of proteins. The ISGs induced by IFN-I and IFN-II partially overlap [5], so they lead (partially) to the production of the same proteins, and thus we can ignore the difference between IFNs of different types at least at the first stages of modelling.

ISGs encode a variety of antiviral proteins with diverse modes of action [6–8]. In the infected cells, these proteins lead to reducing/stopping virus replication including down-regulation of viral transcription, translation, assembly, viral RNA synthesis, and cell exit. In the uninfected cells, the ISGs proteins lead to impeded virus entry into the cell reducing the rate of cell infection. In turn, in evolutionary process, most of viruses have developed strategies of countering the IFN system. These strategies include blocking IFN induction/expression, acting on IFN binding receptors, perturbation of the intracellular IFN signaling pathway, directly down-regulating the level of expression of ISGs [6].

In the current work, we investigate the influence of this mutual counteraction of IFN system and virus on the basic characteristics of the infection spreading in the tissue or cell culture with one-dimensional reaction-diffusion models.

2 IFN Action in Infected Cells

In the first model, we address the IFN action only in infected cells. We consider the following system of equations:

$$\frac{\partial U}{\partial t} = -aUV, \; \frac{\partial I}{\partial t} = aUV - \beta I, \tag{1}$$

$$\frac{\partial V}{\partial t} = D_1 \frac{\partial^2 V}{\partial x^2} + \frac{b_1}{1+k_1 C} I_{\tau_1} - \sigma_1 V, \; \frac{\partial C}{\partial t} = D_2 \frac{\partial^2 C}{\partial x^2} + \frac{b_2}{1+k_2 V} I_{\tau_2} - \sigma_2 C. \tag{2}$$

Here U is the concentration of uninfected cells, I is the concentration of infected cells, V is the virus concentration, and Z is the IFN concentration. As a result of virus penetration, uninfected cells become infected, as described by the bilinear term in right-hand sides of Eq. (1). The last term in the right-hand side of the second equation in (1) describes death of infected cells, which is supposed to be significantly faster than for uninfected cells, the latter being ignored. Equation (2) describe the virus and IFN dynamics: both diffuse in the extracellular space, they are produced by infected cells and both are neutralised/degrade with rates σ_1 and σ_2, respectively. Virus production is down-regulated by the IFN and, vice versa, the IFN production is down-regulated by virus. This behaviour is taken into account in the inverse dependence of virus and IFN production rates on IFN and virus respectively in the second terms of right-hand sides of Eq. (2). In (2), time delay is taken into account as $I_{\tau_i}(x,t) = I(x, t-\tau_i)$, where τ_i is the time intervals after which infected cells start to produce virus/IFNs.

There is a family of stationary points of this system (U_0, $I = 0$, $V = 0$, $C = 0$), where $U_0 = U(0)$ is the initial concentration of uninfected cells. The stationary point $(U_0, 0, 0, 0)$ corresponds to the uninfected, or healthy, state: in this state, there is no infection. Stability of this point can be determined by linearization. The linearized kinetic system about $(U_0, 0, 0, 0)$ has the following form

$$\frac{dU}{dt} = -aU_0 V, \; \frac{dI}{dt} = aU_0 V - \beta I, \; \frac{dV}{dt} = b_1 I - \sigma_1 V, \; \frac{dC}{dt} = b_2 I - \sigma_2 C, \tag{3}$$

and the stability condition of the stationary point is given by the inequality:

$$R_0 = \frac{ab_1 U_0}{\beta \sigma_1} < 1. \tag{4}$$

Here R_0 is the virus replication number, it is the same as for the system without IFN considered in [3]. The condition for infection spreading is $R_0 > 1$ (Fig. 1a), as otherwise the healthy state is stable, and infection decays. If $R_0 > 1$, then the healthy state loses its stability, and a travelling wave solution of system (1)–(2) can exist. Its existence is proved in [1] under some additional assumptions.

In the case of wave existence, the lower estimate of the wave speed c for system (1)–(2) is given by the following inequality [2]

$$c \geq c_0 = \sqrt{\min_{\mu > \mu_0} \frac{D_1 \mu^2 (\mu + \beta)}{(\mu + \sigma_1)(\mu + \beta) - a u_0 b_1 e^{-\mu \tau_1}}}, \tag{5}$$

where $\mu_0 > 0$ is the value for which the denominator vanishes. This speed equals the wave speed in the system without IFN considered in [3]. Since the wave speed correlates with the severity of the disease, we conclude that it is not influenced by the local IFN production. However, IFN decreases viral load [2].

In model (1)–(2), the IFN influence only on infected cells is considered. The analysis of the results shows that IFN reduces the virus infectivity, but it does not change the conditions of infection emergence (R_0), neither the severity of the disease (wave speed c).

3 IFN Action in Uninfected Cells

In the next model, we add the IFN action on uninfected cells resulting in the developing of the so-called antiviral state [7]. In the antiviral state, virus penetratation into uninfected cell is reduced, and the rate of cell infection decreases:

$$\frac{\partial U}{\partial t} = -\frac{a}{1 + k_3 C} U V, \quad \frac{\partial I}{\partial t} = \frac{a}{1 + k_3 C} U V - \beta I, \tag{6}$$

$$\frac{\partial V}{\partial t} = D_1 \frac{\partial^2 V}{\partial x^2} + \frac{b_1}{1 + k_1 C} I_{\tau_1} - \sigma_1 V, \quad \frac{\partial C}{\partial t} = D_2 \frac{\partial^2 C}{\partial x^2} + \frac{b_2}{1 + k_2 V} I_{\tau_2} - \sigma_2 C.$$

This behaviour is taken into account in Eq. (6) by means of the inverse dependence of the rate of cell infection on the IFN concentration. Equations for the virus and IFN concentrations remain the same.

As before, this system has a stationary point ($U = U_0, I = 0, V = 0, C = 0$), where $U_0 = U(0)$ is the initial concentration of uninfected cells. The stability condition is also given by inequality (4). Hence, the antiviral state of uninfected cells does not influence the condition of infection emergence. Transition to the wave variable and linearization about the stationary point $(U_0, 0, 0, 0)$ are the same as for system (1)–(2). Therefore, the estimate of the wave speed is given by (5).

We expect that the total viral load in this system is less than in system (1)–(2) since the rate of cell infection is down-regulated by interferon. Hence, the antiviral state of uninfected cells does not influence the conditions of infection emergence and the disease severity, but it reduces the virus infectivity.

We take into account in Eq. (6) the IFN inhibition of cell infection. However, virus can down-regulate the IFN action. In this case, the factor $a/(1 + k_3 C)$ is replaced by the following expression:

$$\frac{a}{1+\frac{k_3}{1+k_4V}C} = \frac{a+k_5V}{1+k_3C+k_4V}, \tag{7}$$

where $k_5 = ak_4$. In this case, the stationary points and the linearized system do not change. Therefore, the virus replication number and the wave speed remain the same as before.

4 IFN Redistribution

Further, we consider the influence of IFN redistribution by blood circulation. After its production, IFN is quickly delivered in the infected tissue by the circulatory system. The characteristic time for this delivery is much less than the characteristic time of the infection progression. Therefore, we can suppose that the IFN concentrations is space-independent:

$$\frac{\partial U}{\partial t} = -aUV, \ \frac{\partial I}{\partial t} = aUV - \beta I, \ \frac{\partial V}{\partial t} = D_1\frac{\partial^2 V}{\partial x^2} + \frac{b_1}{1+k_1Z} I_\tau - \sigma_1 V, \tag{8}$$

$$\frac{dZ}{dt} = b_2 J(I) e^{-k_2 J(V)} - \sigma_2 Z. \tag{9}$$

Here

$$J(I) = \int_{-\infty}^{\infty} I(x,t)dx, \quad J(V) = \int_{-\infty}^{\infty} V(x,t)dx,$$

and $Z = Z(t)$ is the total IFN concentration. Here we assume that the total concentration of IFN in the infection area is directly proportional to the total concentration of infected cells, and it is inversely proportional to the total virus concentration. This system also has the stationary point (U_0, $V = 0$, $I = 0$, $Z = 0$), where U_0 is the initial concentration of uninfected cells. This point corresponds to the absence of the disease.

In this case, we obtain a coupled system of equations for the total viral load $J(V)$ and the minimal wave speed c [2]:

$$J(v) = \frac{cu_0}{\beta\sigma_1}\left(b_1 - \alpha J(v)e^{-k_2 J(v)}\right), \tag{10}$$

$$c_0^2 = \min_{\mu>\mu_0} \frac{D\mu^2(\mu+\beta)}{(\mu+\beta)(\mu+\sigma_1) - ab(J)u_0e^{-\tau\mu}}. \tag{11}$$

In this case, both the viral load and the wave speed decrease with stronger IFN response, i.e., with increasing IFN production or decreasing its degradation (Fig. 1b). Analytical results on the wave speed shown in Fig. 1b are in agreement with the numerical simulations [2]. The dependence of the total viral load on

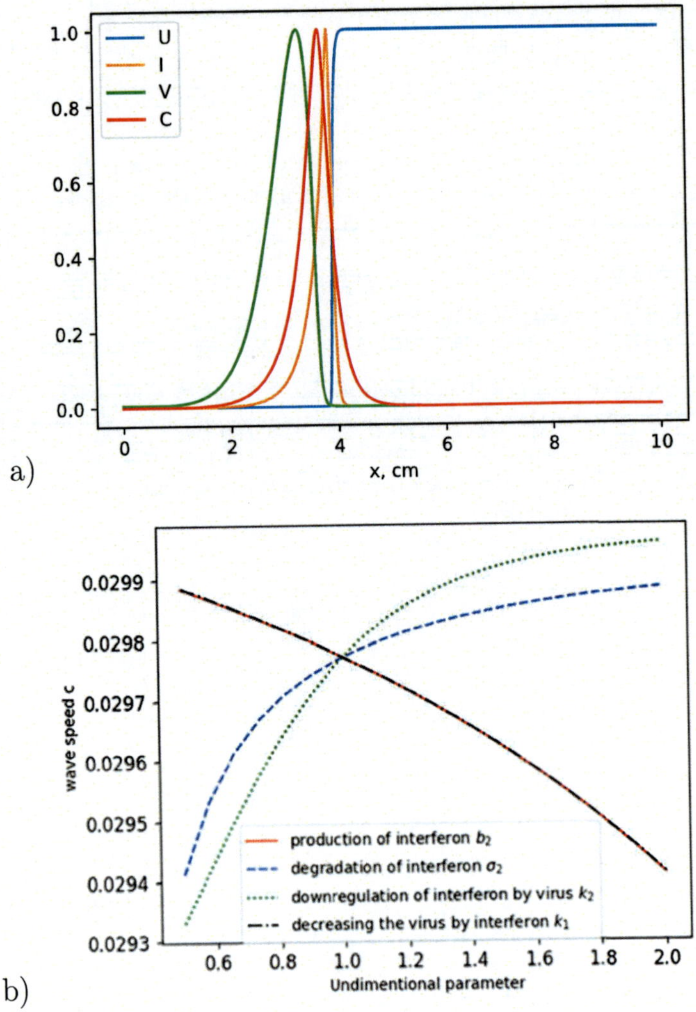

Fig. 1 Calculations of the systems with IFN: (**a**) in system (1)–(2), the concentrations of the uninfected cells U, infected cells I, virus V, and IFN C propagate as a wave from the left to the right with time. All values are normalized by their maximum: $U_{max} = 1$, $I_{max} = 0.76$, $V_{max} = 163{,}729$, $C_{max} = 1.015$. (**b**) In system (8)–(9), the dependence of the wave speed on the parameters of IFN production and its influence on infection. The wave speed is determined by formulas (10) and (11). The values of parameters are as follows: $a = 0.01$, $b_1 = 80{,}000$, $\beta = 0.1$, $\sigma_1 = 0.1$, $D_1 = 0.001$, $\tau = 10$. IFN-related parameters are $b_2 = 13.5$, $\sigma_2 = 3.5$, $k_2 = 10^{-5}$, $k_1 = 1$, $\tau_1 = 10$, $\tau_2 = 5$, in b) the the normalized values of parameters are indicated on the horizontal axis. The initial conditions are $U_0 = 1$, $V(0) = 5000$, $C(0) = I(0) = 0$. Wave speed is measured in cm/h and the total viral load $J(V)$ is measured in $copy/\text{cm}^2$

the parameters of IFN is similar to the dependence of the wave speed on these parameters. These results show that in the case of fast distribution of IFN in the infection site, both the infectivity and the severity of the disease decrease.

5 Conclusions and Perspectives

In this paper, the influence of the IFN production on the basic characteristics of respiratory viral infections are investigated with mathematical models taking into account antiviral influence of IFN on infected and uninfected cells, and also the counteraction of infection on the IFN production. Modelling results show that the IFN influence on infected cells and the antiviral state in uninfected cells do not reduce the severity of the disease induced by infection. However, both of them reduce virus infectivity. The severity of the disease decreases if IFN redistribution by the blood circulation is considered. From these results, we can conclude that the essential role in disease resolving plays not the IFN production itself but rather its efficient distribution in the infection site. Thus, the therapeutic methods of improving the blood circulation and resolving swelling can help in reducing severity of the respiratory viral disease. These modeling results bring new insights on the role of the blood circulatory system in the immune response. They open new perspectives in treatment of viral infections by means of the enhancement of the immune response due to intensification of the circulatory system.

Acknowledgments This work is supported by the Ministry of Science and Higher Education of the Russian Federation (project number FSSF-2023-0016).

References

1. L. Ait Mahiout, B. Kazmierczak, V. Volpert, Viral infection spreading and mutation in cell culture. Mathematics **10**(2), 256 (2022). https://doi.org/10.3390/math10020256
2. L. Ait Mahiout et al., The influence of immune response on spreading of viral infection. Lobachevskii J. Math. **43**(10), 2699–2713 (2022). https://doi.org/10.1134/s1995080222130285
3. L. Ait Mahiout et al., Virus replication and competition in a cell culture: application to the SARS-CoV-2 variants. Appl. Math. Lett. **133**, 108217 (2022). https://doi.org/10.1016/j.aml.2022.108217
4. L. Ait Mahiout et al., Infection spreading in cell culture as a reaction-diffusion wave. ESAIM: Math. Model. Numer. Anal. **56**(3), 791–814 (2022). https://doi.org/10.1051/m2an/2022019
5. V. Fensterl, G.C. Sen, Interferons and viral infections. BioFactors **35**(1), 14–20 (2009). https://doi.org/10.1002/biof.6
6. M.G. Katze, Y. He, M. Gale, Viruses and interferon: a fight for supremacy. Nature Rev. Immunol. **2**(9), 675–687 (2002). https://doi.org/10.1038/nri888
7. J. Shim et al., Influenza virus infection, interferon response, viral counter-response, and apoptosis. Viruses **9**(8), 223 (2017). https://doi.org/10.3390/v9080223
8. Swiss Institute of Bioinformatics, Antiviral state inhibition. Online, Accessed 24 Apr 2023, https://viralzone.expasy.org/1580

Machine Learning-Based Analysis of Human Motions for Parkinson's Disease Diagnostics

S. Nõmm

Abstract Body language is known to be a powerful medium for carrying information about the mental and physical states of a human being. In medicine, visual observations were used by physicians to base the diagnosis on diseases that affect human motor functions. In neurology, different motor tests have been used for more than a century to diagnose cognitive and neurodegenerative disorders such as Parkinson's disease or Alzheimer's disease. Technological advances in motion acquisition techniques have made it possible to register human movement parameters in a way that is invisible to the naked eye. Coupled with advances in machine learning, this has sparked studies that have focused on supporting diagnostics on the basis of human motion analysis. The talk delivered in the colloquium summarises the research of our workgroup in this area.

Keywords Parkinson's disease · Machine learning · Human motion analysis

Mathematics Subject Classification 93B15, 93C55, 93B05

1 Introduction

The talk provides an overview of the activities of the research group aiming to develop a workflow that includes hardware and software to register and analyse human motions observed during the test battery with the goal of supporting the diagnosis of Parkinson's disease. Both the battery and the analysis are chosen with respect to the disease or disorder of interest. There are three main motivating factors to support the digitisation of motor testing procedures. The first is that the machine is able to record the motion parameters that are invisible to the naked eye. The second is that human evaluation inevitably brings about a subjective component.

S. Nõmm (✉)
Department of Software Science, School of Information Technology, Tallinn University of Technology, Tallinn, Estonia
e-mail: sven.nomm@ttu.ee

D. Cardona et al. (eds.), *Extended Abstracts 2021/2022*, Research Perspectives Ghent Analysis and PDE Center 3, https://doi.org/10.1007/978-3-031-48579-4_27

The third is the natural ability of the machine to save the recording for future studies. The classic machine learning workflow combined with the hardware for motion acquisition answers the main goal of the research, whereas individual steps of it have to be tuned for the task. This particularly concerns the steps of feature engineering and result interpretation. The main novel component is the set of integral-like parameters called the *motion mass*. Using the feature sets based on the motion-mass concept, we have received results that in some aspects exceed those previously published in the area. More precisely, the talk summarises the following results. In the area of gross motor analysis [6] where the concept of motion mass has been introduced, [3] describes the clinical application of first-generation prototypes, [4] has demonstrated the applicability of the method for gait analysis, and [12] demonstrated relations between human motions and the level of skin conductivity, which in turn is used to describe the level of emotional arousal. In the area of fine motor movements, [8] has adapted motion mass parameters for the fine motor case, [7] has demonstrated applicability of the approach for the analysis of Peupelreuter's test of overlapping contours, [9] the first step in using deep learning and explainable artificial intelligence approaches for the analysis of Luria's alternating series tests [2], finally [13] has performed targeted feature selection for the analysis of spiral drawing test.

2 Problem Statement

The problem is to provide a binary classifier that answers the question of whether the motor test was performed by a patient with PD or a healthy control subject (HC). Applying the machine learning workflow, one needs to solve two problems specific to the current task. The first is the design and selection of features. The second is the interpretation of the results in terms of the particular medical area.

3 Proposed Solution

This problem is solved for gross motor movements, when limb motions are analysed, or fine motor movements are observed during drawing and writing tests. For both cases, different motion capture technologies may be used. In the gross motor case *stickman*-like body model, depicted in Fig. 1 is usually provided in the form of points that describe the coordinates of the joints in four-dimensional space, where the fourth dimension is time. In the fine motor case, drawing and writing tests are usually performed using a digital table or a tablet PC that allows one to record the coordinates of the stylus tip and pressure applied to the screen. On the basis of this information, numerous features describing the kinematic and pressure parameters are computed.

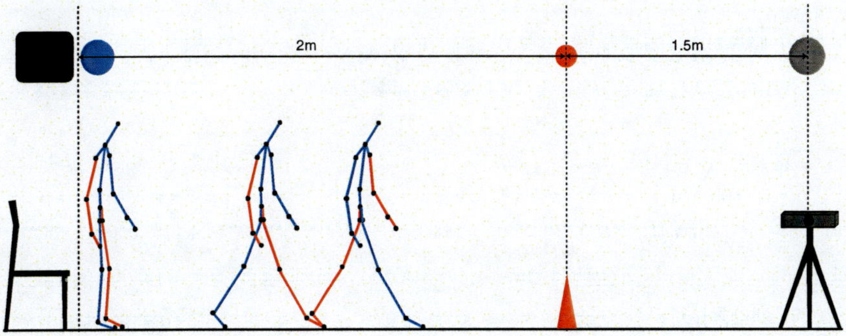

Fig. 1 Experimental setting [4]

3.1 *Motion Mass*

Motion mass parameters were introduced in [6] as measures of amount and smoothness of movement. Subsequently, the concept was extended to [4]. For the set of human body joints $J = \{j_1, \ldots j_n\}$ trajectory mass L_J, the acceleration mass A_J and the combined Euclidean distance E_J are defined as follows:

$$A_J = \sum_{j \in J} \sum_{t=1}^{t_m} a_{j_t}$$

$$L_J = \sum_{j \in J} l_j$$

$$E_J = \sum_{j \in J} s_j$$

where a_{j_t} is the acceleration of the joint j observed at time t, l_j is the trajectory length followed by the joint j during action (motion), and s_j is the length of the straight line segment that connects the positions of the joint j at the beginning and at the end of the experiment. The fourth parameter was the length of the time interval in which the motion took place. The trajectory mass explains the amount of motion, and the acceleration mass describes the smoothness of the motion. Euclidean distance and time were included as references, allowing us to normalise the acceleration and trajectory masses.

Adaptation of the motion-mass parameters for the case of drawing and writing tests has resulted in a larger number of parameters, which in turn was reflected by the number of features to be associated with the tests. The first adaptation was proposed in [8] later extended by Nõmm et al. [7, 9] and was adapted for cases related to tremors in [13]. Furthermore, [13] proposes to extend the set of features by complementing the velocity and acceleration with those based on higher-order

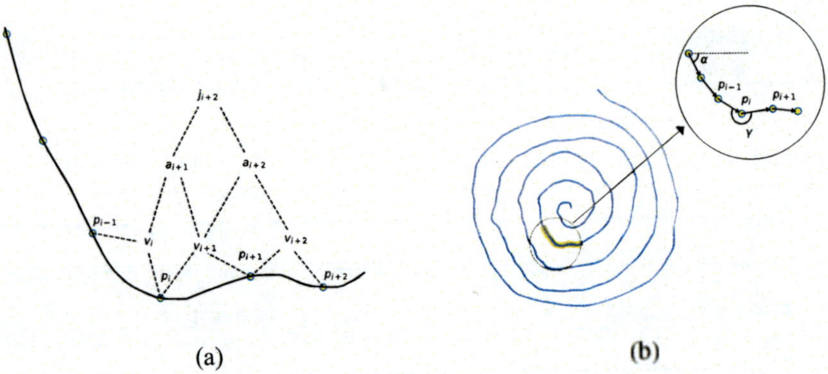

Fig. 2 A visual representation of the differential and angular features [13]. (**a**) demonstrates how kinematic features are computed, and (**b**) demonstrates the meaning of geometric features

derivatives known as jerk, snap, crackle, pop, snatch, and shake. Geometry-based features, such as angular velocities and their derivatives, are also introduced. Figure 2 illustrates the meaning of features based on geometry and kinematics.

3.2 *Machine Learning Workflow*

Computations of motion mass and other parameters that describe kinematic and pressure characteristics of fine motor motions may be considered the feature engineering step. Then the classical machine learning workflow was applied. We used the Fischer score and the XG Boost for feature selection. The nested cross-validation approach was used. The performance of statistical classifiers, such as decision trees, *k*-nearest neighbours, support vector machines, boosted trees, and random forest, was then evaluated. The interested reader is referred to [5, 9, 13] and [4] for a detailed presentation and discussion of the results achieved.

4 Discussion

Proposed motion acquisition techniques coupled with statistical classifiers have demonstrated highly accurate prediction results that allow the application of clinical testing permits. However, the purely numerical approach does not take into account the geometric properties of the drawn contours or the trajectories of the joints. This causes the gap between the *modus operandi* used by medical professionals and the output of diagnostic support systems. On the one hand, applications of deep learning, such as [1] may be seen to mimic the human practitioner, but then kinematic and pressure information can be lost. The results of [11] are based on altering the colour of the drawn contours to encode the pressure or the kinematic parameter. Later, [10] and [14] have extended the procedure, altering not only the colour but also the thickness of the lines to encode two parameters that describe the pressure and kinematic properties of the motions.

5 Conclusions

During the recent time, machine learning-powered analysis of human motions to support the diagnostic process has developed into the strong direction of research in the areas of machine learning and artificial intelligence. There are three main subdirections that need to be investigated. The first is transparency, the second is data augmentation, and the last is semantic analysis of the drawing tests.

Acknowledgments This work in the project "ICT programme" was supported by the European Union through the European Social Fund.

References

1. Z. Galaz, P. Drotar, J. Mekyska, M. Gazda, J. Mucha, V. Zvončák, Z. Smekal, M. Faundez-Zanuy, R. Castrillon, J.R. Orozco-Arroyave et al., Comparison of CNN-learned vs. handcrafted features for detection of Parkinson's disease dysgraphia in a multilingual dataset. Front. Neuroinformatics 35 (2022)
2. J.R. Hodges, *Cognitive Assessment for Clinicians*, 2nd edn. (Oxford Medicine, Oxford, 2007)
3. H. Kähär, P. Taba, S. Nõmm, K. Medijainen, Microsoft kinect-based differences in lower limb kinematics during modified timed up and go test phases between men with and without parkinson's disease. Acta Kinesiologiae Universitatis Tartuensis **23**, 86–97 (2017)
4. A. Krajushkina, S. Nomm, A. Toomela, K. Medijainen, E. Tamm, M. Vaske, D. Uvarov, H. Kahar, M. Nugis, P. Taba, Gait analysis based approach for parkinson's disease modeling with decision tree classifiers, in *2018 IEEE International Conference on Systems, Man, and Cybernetics (SMC)* (IEEE, 2018), pp. 3720–3725
5. A. Netšunajev, S. Nõmm, A. Toomela, K. Medijainen, P. Taba, Parkinson's disease diagnostics based on the analysis of digital sentence writing test. Vietnam J. Comput. Sci. **8**(04), 493–512 (2021)
6. S. Nõmm, A. Toomela, An alternative approach to measure quantity and smoothness of the human limb motions. Estonian J. Eng. **19**(4), 298–308 (2013)
7. S. Nõmm, K. Bardõš, I. Mašarov, J. Kozhenkina, A. Toomela, T. Toomsoo, Recognition and analysis of the contours drawn during the poppelreuter's test, in *2016 15th IEEE International Conference on Machine Learning and Applications (ICMLA)* (2016), pp. 170–175
8. S. Nõmm, A. Toomela, J. Kozhenkina, T. Toomsoo, Quantitative analysis in the digital luria's alternating series tests, in *2016 14th International Conference on Control, Automation, Robotics and Vision (ICARCV)* (2016), pp. 1–6
9. S. Nõmm, K. Bardõš, A. Toomela, K. Medijainen, P. Taba, Detailed analysis of the luria's alternating seriestests for parkinson's disease diagnostics, in *2018 17th IEEE International Conference on Machine Learning and Applications (ICMLA)* (2018), pp. 1347–1352'

10. S. Nõmm, S. Zarembo, K. Medijainen, P. Taba, A. Toomela, Deep CNN based classification of the archimedes spiral drawing tests to support diagnostics of the Parkinson's disease. IFAC-PapersOnLine **53**(5), 260–264 (2020). 3rd IFAC Workshop on Cyber-Physical & Human Systems CPHS 2020
11. C.D. Rios-Urrego, J.C. Vásquez-Correa, J. F. Vargas-Bonilla, E. Nöth, F. Lopera, J.R. Orozco-Arroyave, Analysis and evaluation of handwriting in patients with Parkinson's disease using kinematic, geometrical, and non-linear features. Comput. Methods Programs Biomed. **173**, 43–52 (2019)
12. A. Toomela, S. Nõmm, T. Kõnnussaar, V. Tammik, Why behavioral indicators may fail to reveal mental states: individual differences in arousal-movement pattern relationships. Front. Psychol. **10**, 270 (2019)
13. E. Valla, S. Nõmm, K. Medijainen, P. Taba, A. Toomela, Tremor-related feature engineering for machine learning based parkinson's disease diagnostics. Biomed. Signal Process. Control **75**, 103551 (2022)
14. S. Zarembo, S. Nõmm, K. Medijainen, P. Taba, A. Toomela, CNN based analysis of the Luria's alternating series test for Parkinson's disease diagnostics, in *Asian Conference on Intelligent Information and Database Systems* (Springer, 2021), pp. 3–13